ANCIENT TEXTILES SERIES VOL. 25

ICONIC COSTUMES: SCANDINAVIAN LATE IRON AGE COSTUME ICONOGRAPHY

By

ULLA MANNERING

OXBOW | books
Oxford & Philadelphia

First published in the United Kingdom in 2017. Reprinted in 2019 by
OXBOW BOOKS
The Old Music Hall, 106–108 Cowley Road, Oxford OX4 1JE

and in the United States by
OXBOW BOOKS
1950 Lawrence Road, Havertown, PA 19083

Hardcover Edition: ISBN 978-1-78570-215-0
Digital Edition: ISBN 978-1-78570-216-7 (epub)

A CIP record for this book is available from the British Library and the Library of Congress

Typeset in India by Lapiz Digital Services, Chennai

For a complete list of Oxbow titles, please contact:

UNITED KINGDOM
Oxbow Books
Telephone (01865) 241249
Email: oxbow@oxbowbooks.com
www.oxbowbooks.com

UNITED STATES OF AMERICA
Oxbow Books
Telephone (800) 791-9354, Fax (610) 853-9146
Email: queries@casemateacademic.com
www.casemateacademic.com/oxbow

Oxbow Books is part of the Casemate Group

Front cover: Gold figures © National Museum of Denmark

Contents

List of illustrations

List of tables

Acknowledgements

The publisher and I gratefully acknowledge the permission granted to reproduce the copyright material in this book. Every effort has been made to trace copyright holders and to obtain their permission for the use of copyright material. The publisher apologises for any errors or omissions and would be grateful if notified of any corrections that should be incorporated in future reprints or editions of this book.

I wish to express my gratitude to the Danish Research Council for the Humanities (FKK) that funded my PhD research and Dronning Margrethe II's Arkæologiske Fond for financial support for the translation of my PhD dissertation, *Billeder af dragt: En analyse af påklædte figurer fra yngre jernalder i Skandinavien* (University of Copenhagen, 2006). I would also like to thank Ulla Lund Hansen my PhD supervisor, Bente Magnus and Torsten Capelle† my PhD opponents, and Lise Bender Jørgensen my academic mentor.

My heartfelt thanks to John Peter Wild and Margrethe Watt who kindly read and reviewed the entire manuscript, and Morten Axboe who reviewed the chapter on bracteates, and who all offered invaluable suggestions for improvement. All mistakes are now my own.

I also thank all the museums and universities in Scandinavia who helped provide study material and images, and the following: Anders Andersen for making the drawings of the gold-foil figures from Slöinge, Bjørn Skaarup for making the drawings of some of the gold-foil figures belonging to the National Museum of Denmark; Pernille Foss for helping redraw already published drawings and also making new drawings; Margrethe Watt for kindly providing drawings of the Danish gold-foil figures along with the late Eva Koch, and Peter Andreas Toft for helping with the correspondence analysis. For this publication, I would like to thank Morten Axboe, Jan Peder Lamm, Bente Magnus, Charlotte Rimstad, Marianne Vedeler, Martin Stoltze and John Peter Wild for use of images; and Anna Nørgård, Susanne Lervad and Lise Ræder Knudsen for help in obtaining images. A special thanks goes to Roberto Fortuna at the National Museum of Denmark for the wonderful photographs of the many archaeological textiles.

At the Centre for Textile Research (DNRF64) I would like to thank: my wonderful colleagues Marie-Louise Nosch and Eva Andersson Strand who over the years repeatedly encouraged me to rework the text, for their persistent encouragement and support to publish in English, Cherine Munkholt for translating the original PhD text and her fantastic work editing and coordinating the publication; Sidsel Frisch for her immense help with the picture editing and for insightful suggestions in terms of content editing the revised text, and Louise Ludvigsen, Line Lerke, Stine Bøttern, Ziff Jonker, Philip Dons Madsen, Manon Leroy, Elina Stamatatou, Jane Malcolm-Davies, Mikkel Nørtoft, Egzona Haxha and Camilla Ebert for editorial assistance.

I would also like to thank the Danish National Research Foundation (DNRF64) and the National Museum of Denmark for their support and Farumgaard-Fonden for their financial support for the printing of this book.

Above all, I would like to thank my family for their infinite patience and support.

Introduction

This is a revised edition of my PhD dissertation which was successfully defended at the University of Copenhagen, Denmark in June 2006. Taking its point of departure in an archaeological textile and clothing perspective, this book presents a selection of the rich and varied iconographic material from Scandinavia depicting clothed human figures.

The source material consists of five object categories: gold foils, gold bracteates, helmet plaques, jewellery and textile tapestries, which complement each other with regard to choice of material, content and function. The objects selected for the study are, with a few exceptions, produced and found in Scandinavia, *i.e.* present-day Denmark, Sweden and Norway, and only in a few cases is material found outside Scandinavia included in the analysis. The majority of finds are dated to the Scandinavian Late Iron Age (AD 400–1050) while a few finds are dated to the Scandinavian Middle Ages (AD 1050–1536).

Through a recording system developed for the present analysis of iconographically recorded costumes, the content of the object categories selected for this study is described and compared, and the results are compared with the existing knowledge of archaeological textiles and clothing finds. Here, I have primarily focused on finds from contemporary Scandinavia and the northern part of Germany, which in this period belonged to the Scandinavian cultural area. To include the European continental clothing tradition would be a next step, but it was not feasible within the scope of the PhD project.

In 2005 I began working at the Danish National Research Foundation's Centre for Textile Research (CTR) at the University of Copenhagen, and since 2010 I have been employed at the National Museum of Denmark monitoring the collaborative research programme working with the Danish collections of textiles and skin objects from the Bronze and Early Iron Ages. Knowing that my dissertation, due to its Danish language, was not available to the majority of my overseas colleagues, I have, in this period, published a few articles in English about Scandinavian costume iconography (Mannering 2008; 2013; Mannering and Andersson Strand 2009), but it is a great pleasure to finally be able to present the entire work in English.

To revise a PhD dissertation is not a straightforward task as formalities and the target group are suddenly different, and with time, one's knowledge and the archaeological record, too, are subject to change. Since 2005, many new finds pertinent to this study have appeared, but I soon realised that it would be too complicated a task to incorporate all of them in the comparative material, and instead chose to focus on updating the archaeological textile and clothing chapter, and include many new photos. I hope the readers will appreciate this.

Margrethe Watt, the prime researcher of gold-foil figures, which is the largest object category included, will publish her long-awaited book on the Scandinavian gold-foil figures in 2016 (Watt forthcoming), and no doubt, with its full overview, descriptions and drawings of all Scandinavian gold-foil figures, it will create renewed debate and interpretations of the gold-foil figures themselves, and Scandinavian iconography in general. This may affect the present study, but also make it easier to see and evaluate the material that I did not include here.

Throughout the years, I have heard from many Scandinavian researchers that they have found my methodology useful, and I hope it will now acquire a second life in a wider context. I am also aware that, for instance my gender definitions have been criticised (Danielsson 2007), but my priority has been primarily to disseminate knowledge of prehistoric clothing. As the present work lies in the cross-disciplinary fields of textile and costume history, art history, and traditional archaeological methodology, I welcome debate and hope others will feel inspired to continue the work started. The primary objective of this book is, thus, not to find an interpretation of the depictions themselves, but through them to shed light on the clothing tradition within which they are produced.

The analytical tool presented in the book is developed to ease comparisons of large bodies of material in detail, and, in general, regardless of what they are composed of, their function, chronology, interpretation and context. The tool is used to record and analyse the numerous details of iconographic costumes, and to facilitate a clear and easy description. In most object categories, the clothed figures are characterised by a few, simple and visually clear details. This very conscious and intentional use of explicit costume shapes enables us to understand and interpret the Late Iron Age clothing tradition.

The analysis of the almost 1000 selected clothed human figures employs an internal numbering system referring to the different figures encountered in the various object groups. The numbering consists of an abbreviation of the location as well as a running number. Together with published photos and drawings, my own examination of most Scandinavian gold-foil figures and new drawings created for this purpose form the basis of the documentation and analysis presented. Margrethe Watt and Eva Kock† kindly permitted me to use their unpublished drawings, but only the most necessary drawings are published here. Please note that not all photos are accompanied by a scale.

If there are textile terms in the text that are unfamiliar to some readers, which cannot be satisfied by recourse to the dictionary or Wikipedia, I recommend the glossary presented by Möller-Wiering (2011) or the introduction in Gleba and Mannering (2012).

1

Clothing and iconography

Knowledge of prehistoric clothing in Scandinavia is primarily derived from various archaeological textile sources. However, the number and quality of sources that can be used for clothing studies are highly variable, and based on these textile sources alone, it is difficult to create a clear and detailed picture of the clothing customs in prehistory which can meet the increasing demand both in academic research and among the general public to visualise the past and bring it to life. In order to compensate for the lacuna in the archaeological textile record and to obtain a deeper insight into Scandinavian prehistoric clothing, it is necessary to try new approaches, as for instance combining knowledge from the textile finds with the study of contemporary iconographical sources.

Iconography is not a new source in clothing research, but it is most often used as a single source to prove and underline a known aspect. The study of Scandinavian Late Iron Age costume iconography in a broader perspective and as a source in parallel with the archaeological textiles has not been undertaken until now. Thus, with the present study, the statement made by Scandinavian Late Iron Age costume iconography is brought into this discussion.

This study is chronologically focused on the Scandinavian Late Iron Age (AD 400–1050), which consists of three sub-periods termed the Early Germanic Iron Age (AD 400–520/40), Late Germanic Iron Age (AD 520/40–AD 800) and Viking Age (AD 800–1050) (Lund Hansen *et al.*, 1995, 18 and Jørgensen and Jørgensen 1997a, 38) (Table 1.1), and is primarily based on finds from present-day Denmark, Sweden and Norway (Fig. 1.1). As Late Iron Age Scandinavia is considered a cohesive unit in a cultural historical and textile technological perspective, the aim of this study is to demonstrate how a group of selected depictions of costumed anthropomorphic (humanlike) figures complement already existing sources of archaeological textiles, and to what extent this provides new information on chronological, regional and social differences in clothing customs in this period. In this context, the term *costume* denotes both the various individual garments as well as the outfit in its entirety identified in the iconography, while the term *clothing* is reserved for archaeological textile finds.

Scandinavia is renowned for its wealth of archaeological artefacts, of both organic and inorganic origin that can provide information on textile technology, production and various clothing customs. The earliest depiction of costumed humanlike figures dates to the Mesolithic (Fig. 1.2). The Scandinavian Bronze Age is characterised by a rich and varied iconography depicting anthropomorphic figures (Fig 1.3) and images, and especially petroglyphs (rock carvings) have long been seen as a significant source of information on Bronze Age society. For instance the comprehensive study of ships on bronze objects has demonstrated that systematic research methods can provide new knowledge on prehistoric society, religion and cosmology (Kaul 1998 and 2004; Milstreu and Prøhl 2004). For reasons still unknown, the Scandinavian tradition of depicting anthropomorphic figures disappears with the Bronze Age culture, and it is not before the onset of the Late Iron Age, after a period of almost 1000 years, that humanlike figures and more complex depictions are reintroduced into

Table 1.1 Overview of the Scandinavian chronological periods relevant to this study.

Period	*Date*
Early Stone Age	–4000 BC
Late Stone Age	4000–1700 BC
Early Bronze Age	1700–1100 BC
Late Bronze Age	1100–500 BC
Pre-Roman Iron Age	500–1 BC
Early Roman Iron Age	AD 1–200
Late Roman Iron Age	AD 200–400
Early Germanic Iron Age	AD 400–520/40
Late Germanic Iron Age	AD 520/40–800
Viking Age	AD 800–1050
Middle Ages	AD 1050–1536

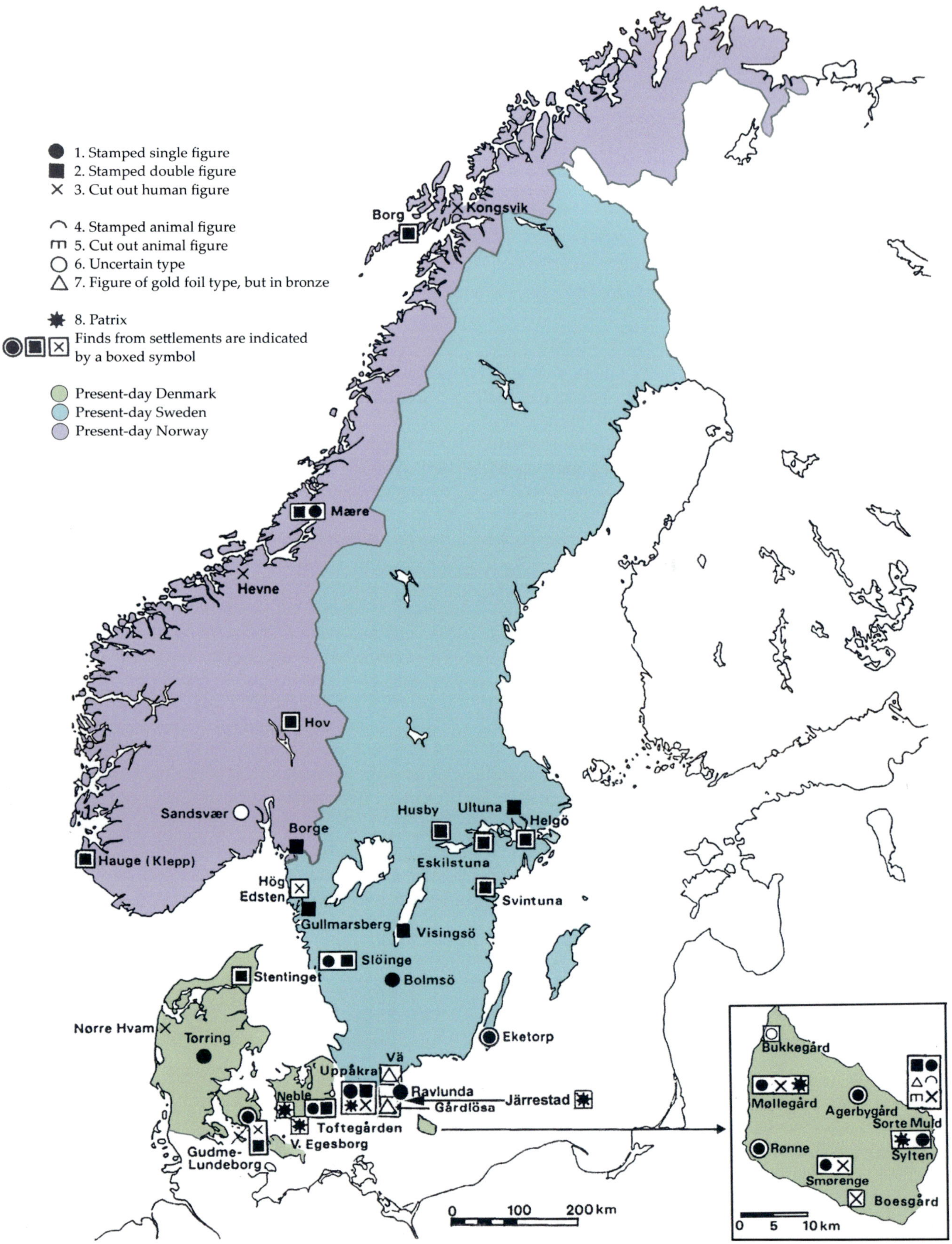

Fig. 1.1 Gold-foil figure find-spots from Scandinavia (Map: Sidsel Frisch, after Watt 1999b, 185).

Fig. 1.2 Auroch bone from Ryemarksgård in Denmark with incision of five clothed human figures. The bone is dated to c. *8000 BC (National Museum of Denmark).*

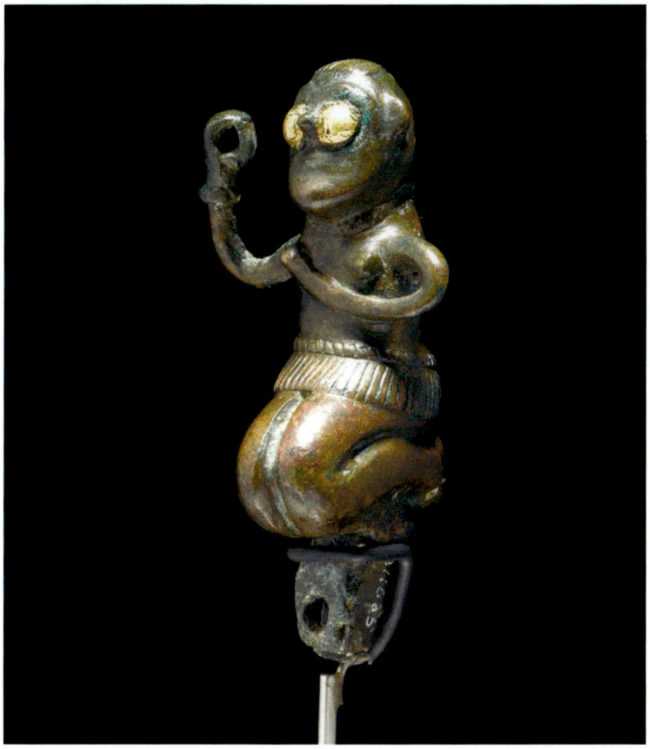

Fig. 1.3 Bronze female figure from Fårdal in Denmark wearing a corded skirt. The figure is dated to the Scandinavian Late Bronze Age (1100–500 BC) (National Museum of Denmark).

the Scandinavian prehistoric material culture, and a local production of more intricate images with anthropomorphic figures manifests itself.

Wooden anthropomorphic idols are certainly an exception, and these are found throughout Northern Europe from the Stone Age to the Middle Ages, although in Scandinavia, they reach their apogee at the turn of the 1st millennium, *i.e.* in the Pre-Roman and Early Roman Iron Ages (Sanden and Capelle 2001, 11). From a clothing perspective, the wooden figures have regrettably not much to offer, as they are either naked or the costume details are difficult to discern (Fig. 1.4). The wooden figures, however, demonstrate a continuity and experience in depicting the human body that cannot be documented in any other object group. It is still unclear as to what caused the stagnation of the image production in the Early Iron Age (500 BC–AD 400), but it is most likely the influence of the Roman Empire that again gave impetus to the production of anthropomorphic figures in Scandinavia. Sources of inspiration could have been, for instance circus glass (Fig. 1.5), metal drinking cups (Fig. 1.6), small bronze statuettes (Fig. 1.7), coins and bejewelled weaponry which occur in quite large numbers in Scandinavia in the first centuries AD (Werner 1966). Scandinavian copies of Roman bronze figures demonstrate that copying took place, but an independent Scandinavian production of anthropomorphic figures had not yet materialised (Kjærum and Olsen 1990). It is first with the production of bracteates that Scandinavian iconography starts flourishing again, and thereby the study of Scandinavian costume from an iconographical perspective becomes possible. This is probably also the reason why iconography in Scandinavian Iron Age research has long played a secondary role, and is primarily used as individual examples to illustrate or underline a specific issue. Another explanation for this situation could be that a large part of the Scandinavian Late Iron Age iconography is found on high-status metal objects, where the economic and aesthetic values rather than the iconographic statement have been in focus. This fact has transformed most iconography into objects of art (see, for example Holmqvist 1955; 1977; Kolstrup 1995) rather than a coded message loaded with social, religious and cultural meanings (for exceptions see, for example Hedeager 1999; Høilund Nielsen 1998; 2002c; Göransson 1999; Pesch 2005; Danielsson 2007; and Helmbrecht 2011).

The object categories selected for the study

Depictions of costumed anthropomorphic figures are found on many object categories, such as bracteates, gold-foil figures (*guldgubber*), jewellery, weaponry, helmet plates, tapestries, picture stones, rune-stones as well as a large number of single objects of various provenance and function. Five of the above-mentioned sources: gold-foil figures, bracteates, decorative bronze helmet plates, jewellery and tapestries, were selected for further analysis, as they supplement one another in terms of the choice of material, composition, function and chronology (Fig. 1.8).

Fig. 1.4 Wooden male figure from Rude Eskilstrup in Denmark. The figure is dated to c. AD 500 (National Museum of Denmark).

Fig. 1.5 Circus glass from Nordrup in Denmark. The glass is dated to the 3rd century AD (National Museum of Denmark).

Fig. 1.6 Silver cups from Hoby in Denmark. The cups are dated to the 1st century AD (National Museum of Denmark).

Gold-foil figures constitute the primary and the largest of the five selected object categories. These are small depictions of anthropomorphic figures, and, in rare cases, also zoomorphic figures. Despite this object category containing the smallest (in physical size) pictorial representations, it is surprising how much information the gold-foil figures contain. Undoubtedly they are the most detail-rich object category in this study. Gold-foil figures contain depictions of clothed, unclothed and clearly naked anthropomorphic figures, of which the first group is the most pertinent here. Geographically, gold-foil figures are found throughout most of Scandinavia, primarily in settlements or central places with settlement continuity. For the same reason, gold-foil figures also have a wide chronological frame ranging from the middle of the Early Germanic Iron Age to the Viking Age (Fig. 1.8).

Bracteates are primarily found as stray finds or as votive deposits, and, to a lesser extent, as burial finds. Within bracteate research, a great deal of effort has been devoted to developing an accurate chronological classification, which is, however, hampered by the often large differences in production and deposition dates (Fig. 1.8). Although bracteates constitute the second largest group of selected objects, only a small part of the material can be used for costume studies. However, they are included in the analysis, as they constitute the earliest independent Scandinavian iconography in the period under study. Bracteates are generally dated to a production period from AD 450 to 540, and are primarily found within the Scandinavian area.

Fig. 1.7 Various bronze figures of Roman or Roman provincial origin (left) and local Danish origin (right). The figures are dated to the first half of the first millennium AD (National Museum of Denmark).

A considerably smaller, but no less informative object category is the decorative bronze plates from helmets; an object category which can be dated to the period AD 500–680, *i.e.* the end of the Early Germanic Iron Age to the middle of the Late Germanic Iron Age. The bronze helmet plates are all found in Sweden, and are primarily derived from well-dated graves. In addition, an English find with close iconographical connections to the Swedish bronze helmet plates is included in the analysis (Fig. 1.8).

The jewellery category constitutes a heterogeneous group of fibulae and pendants. Finds were collected from the entire period in question, although finds from the latest part of the period predominate. These objects are primarily found in Sweden and Denmark, while Norway is only represented by a single find. The context within which these objects are found is quite varied, and they occur in burials, settlements, in hoards and as stray finds. The selected jewellery corpus is considered representative, but not fully comprehensive, as this is an object group where new finds constantly appear, and only clothed figures are included (Fig. 1.8).

The final category to be analysed is tapestries. This small, but impressive, find group of textiles is dated to the Viking Age, with a few to the Scandinavian Middle Ages, and thus have a slightly different chronological distribution than the other categories. Tapestries are found in Sweden and Norway, and include both grave finds and stray finds[1] (Fig. 1.8).

Altogether, the selected iconographical material represents figures that may look quite different and represent different grades of detail: anthropomorphic figures in clearly identifiable costumes and those without an easily identifiable costume or no clothing at all. Although the clothed figures are the focus of this analysis, in some cases, it is difficult to make a clear distinction between partly clothed or unclothed figures, which is why some of these are included in the analysis in a few instances. In the course of the analysis, each object category is dealt with separately.

Another way of grouping the selected material is according to the function of the depictions, and whether the figures constitute the complete object or part of a larger decorative scheme on an object (Table 1.2). Gold-foil figures and pendants from the jewellery category constitute the first kind. In these cases, the figure is, more or less, synonymous with the object. Bracteates, decorative bronze plates on helmets, tapestries and some of the objects within the jewellery category constitute the other type. Here, the figures are part of the decoration and do not interfere with the function of the object. The images may also display one

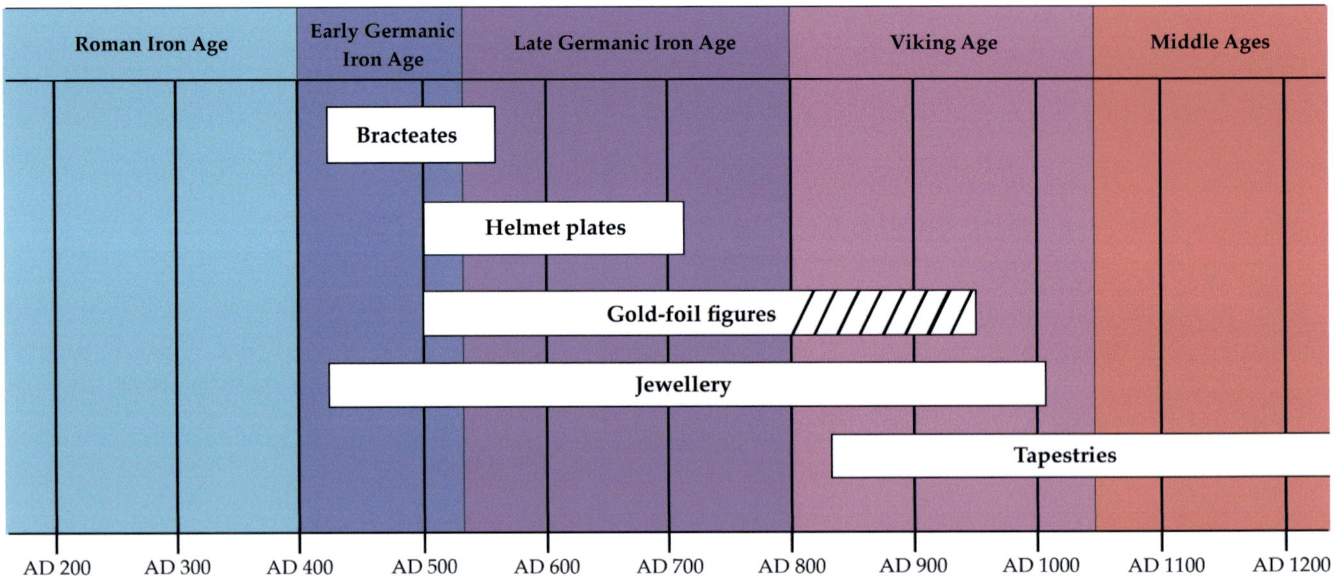

Fig. 1.8 Chronological overview of the five object categories: bracteates, gold-foil figures, helmet plates, jewellery and tapestries selected for this study.

Table 1.2 Different iconographic and material characteristics of the five object categories selected for this study.

Depictions comprising the entire object	*Depictions considered as decorations on the object*
Gold-foil figures, part of the jewellery category	Bracteates, helmet plates, part of the jewellery category, tapestries
1–2 figures	Many figures
Monochrome depictions	Polychrome and monochrome depictions
No animals	Animals occurring
Stationary figures	Figures in action or narrative sequences

or two figures, or sometimes more intricate narrative scenes. The gold-foil figures and pendants constitute the simplest depictions, with one to two figures, while bracteates, decorative bronze plates, tapestries and some of the fibulae from the jewellery category display more intricate images with many figures. It is primarily in the latter categories that humans appear together with animals, something which does not occur among the gold-foil figures. The majority of the selected study material constitutes objects crafted in metal, such as gold, silver and bronze or a combination of these. The metalwork is characterised by a fascinating sense for detail, and there is no doubt that the choice of material was significant and influenced the construction of the images. Tapestries, however, are textiles made of organic material, such as wool or plant fibres, and belong to a complete different production mode. Nevertheless, the tapestries, too, are highly detailed and supplement with aspects that cannot be found in the metal group. For instance depictions made of metal are monochrome, whereas textiles are polychrome. As this is the only category that contains colours, the polychrome aspect is not included in the analysis, although, in many cases, this has helped the identification of the figures. Thus, the selected object categories supplement each other from an iconographical and chronological perspective.

Method

The method of combining iconography with knowledge of archaeological textiles and clothing items is not new. For instance, Iron Age clothing from Danish and North German bogs has been interpreted, to a large extent, in the light of the depictions of Germanic peoples on Roman monuments or descriptions of them in Roman texts (Hald 1950; 1980; Schlabow 1976; Bender Jørgensen 1986). Provincial Roman clothing, too, has been analysed with a similar comparison between a rich iconographical collection of material and a few intact textiles (Wild 1968; 1985; Garbsch 1965; 1985; Böhme 1985). Similarly, the well-known corpus of Pharaonic art has been combined with the comprehensive assemblage of complete or partly intact clothing items from graves (Vogelsang-Eastwood 1992; 1993), which is also the case for research into Anglo-Saxon (Owen-Crocker 1986; 2004) or early medieval clothing on the Continent and in Norway (Müller 2003; Vedeler 2007a). However, only very few of the above-mentioned examples treat the iconographic costumes in a systematic quantitative manner, which makes it difficult to use these works to qualify observations as frequent or rare, or to qualify social differences. A further objective of this study is thus to create a methodical framework for costume studies, and to bring this focal point into textile research.

The strength of iconographical material lies in its ability to demonstrate the relationship between the body and clothing in a given period. For instance, the ancient Greek concept of being clothed or naked was quite different to what the Romans understood as being correctly attired (Bonfante 1989; 2000). Despite the fact that the Romans admired and emulated Greek culture, and especially Greek art, their concrete attitude to clothing, nakedness and the body was very different. To a Roman citizen, clothing equalled ethnic, cultural and social status, an attitude that in many aspects continues to the present. However, to utilise iconography as a source of enquiry in clothing research requires a systematic and in-depth analysis of the source, so as to avoid circular conclusions or placing incorrect emphasis on the available information. Therefore, an analytical tool to compile a typology of the iconographical costumes was developed for this study. With this method, different costume items can be incorporated in the systematic study of heterogeneous materials, including parameters of geographic and chronological importance as well as information on interpretation and context. Altogether, this enables us to interpret the potential value and content of the iconographical costumes alone and in combination with archaeological textile finds. A fundamental element in this study is, thus, the compilation of a registration and analysis method with which the different parts of the costumes can be easily described and compared.

Although the selected object categories encompass various types of image compositions and details, the method is designed to facilitate the comparison of the different categories. In developing this method, a balance was sought between obtaining a satisfactory description of the desired characteristics on the one hand, and to exclude unnecessary details on the other. Thus, the method had to strike a balance between norm and variation without exhausting the material or recording too many details, *i.e.* to identify an ideal type that allows some latitude or deviation. At the same time, the method was compiled in a manner that would make it simple to add new types, as well as facilitate identification and analysis.

The first step in the costume analysis was an evaluation of the available iconographical material to ascertain its suitability, *i.e.* if the costumes are adequately clear and detailed for analytical purposes. The borderline between what is identifiable and usable and what is not is difficult to define sharply, and depends, to a large extent, on personal experience. The figures selected for analysis were then assigned a unique number, which followed them during the analysis. During this process, the gender of the individual figures was also determined, and where it could not be determined, the figure was then omitted from the analysis. When encountering an anthropomorphic figure, determining its gender may, at first glance, seem to be easy and straightforward, but gender definitions are known to be based on culture as well as physiology, and depend on the viewer's preconceptions of what constitutes masculinity, femininity or the non-gendered. When identifying a naked figure, biology comes to one's aid, but for clothed figures, factors such as our knowledge of the clothing tradition and its cultural background, play an important role in the identification. In our modern understanding of clothing and gender, dresses and skirts, especially long items, denote femininity, while trousers are generally used to characterise males. It is through this modern lens that most gender identifications of prehistoric iconographical material are made. However, it cannot be automatically deduced that these gender connotations are applicable to a prehistoric context. Nevertheless, from the archaeological clothing finds, it is known that in Late Iron Age society, too, clothing was gender specific to a large extent, and that specific clothing items existed for men and women. The later Old Norse written sources abound with tales of gods that disguise themselves or change gender identity simply by donning a garment belonging to the opposite sex that altogether could be based on the clothing traditions examined here (Saxtorph 1992, 44, 217–218). However, it is not clear if these differences were primarily based on the use of different clothing items, or whether this could also refer to production characteristics or the way that different clothing items were worn together. From a clothing perspective, it is, therefore, an advantage if the figure's gender can be identified before it is included in the costume analysis.

The iconographical material selected from the five object groups shows both a difference in gender distribution as well as how the details are presented. Therefore, it is necessary to operate with flexible principles for what constitutes the masculine and feminine. As the figures on the double gold-foil figures, in most cases, are depicted in quite different looking costumes and with quite different hair-styles, it is possible to see this as an intention to denote male and female genders. Based on the double gold-foil figures from Slöinge in Sweden, the different elements constituting the figures were subjected to a combination analysis in an earlier study, demonstrating that, in many cases, it is possible to interpret the gender of the gold-foil figures through the costume (Mannering 1998a). Through this preliminary test, it was evident that, in most cases, gold-foil figures with long costume items depict women, and figures with short costume items denote men – a guideline, which is supported by characteristic hair-styles and jewellery (Mannering 2004, 69–72). Collating this finding with the knowledge of gender specific characteristics in the archaeological sources provided the following characteristics, which are used in this enquiry to identify the gender of a figure:

- Women generally wear more costume items than men;
- Women generally wear longer costume items than men;
- Women most often wear dresses;

- Women often have long hair, and longer hair than men;
- Women can wear their hair in a knot at the nape of the neck;
- Women usually wear jewellery, such as fibulae and necklaces;
- Men usually wear trousers or have clearly delineated legs;
- Men often have short hair;
- When two figures are depicted, they most often display different characteristics that can be linked to the male and female gender (rather than two figures of the same gender).

For instance, a gold-foil figure with two people depicted in the same kind of costume has only been recorded once in the material selected for this study, although this occurs more commonly in the jewellery category. According to the gender guidelines, the depiction is interpreted as two men (see Borg 4, Fig. 2.6), whereas the excavator maintains that it is an image of a man and a woman. The disagreement in interpretation is primarily based on the different starting point for the interpretation – the meaning of the depiction, compared to that of the costume (Stamsø Munch *et al.*, 2003, 249). When only a single figure is depicted, it can be more difficult to determine the gender, as there is no basis for direct comparison, but also in these cases, the above-mentioned guidelines can be used. Nevertheless, a small number of clothed gold-foil figures from Uppåkra in Sweden and Sorte Muld on the Danish island of Bornholm, are especially difficult to identify by gender in accordance with the aforementioned guidelines (Watt 2004, 192–194). For instance, Uppåkra 32 was identified as a female in this analysis due to its long costume and long ponytail, whereas similar gold-foil figures like Sorte Muld 240 and 266 (see Figs. 2.14 and 2.15) were recorded as males due to their short hair. Margrethe Watt, however, defines these figures as being gender-neutral (Watt 2004, 189). In this analysis, completely gender-neutral figures were excluded, while in the cases where only some of the gender guidelines were applicable, the identification was based on a personal estimate. Scholars understandably disagree on the methods of identifying single figures of uncertain gender. Yet, as this category is very small compared to the number of clothed figures, it cannot disrupt the overall results of the costume analysis. Figures with no costumes, as are seen, for example on the bracteates and some of the gold-foil figures, demonstrate that, even within the same object category, there can be different ways of depicting the body.

The recording method was primarily developed during the registration of the gold-foil figures, and was a two-step process. In the first step, the gold-foil figures were separated into different elements which were registered according to the "first incident" method, *i.e.* each time a new type was recorded, it was allocated a new number. The various costume elements were determined by their shape and location on the figures (Mannering 1999 and 2004). Characteristics, such as the shape of the feet, were recorded, but not included in the analysis, as this feature turned out not to contain specific differences. The recording of the costume items was then adjusted to fit the figures in the other object categories selected for this study, but in fact it was only necessary to add a very few supplementary types, and after including the new costume items and patterns, the recording method was systematised into a general tool. For the costume analysis, a combination of primary sources, photographs and sketches was used in the analytical phase. As far as the gold-foil figures are concerned, the Scandinavian material was analysed through primary studies. Only a small proportion of the gold-foil figures, as well as some objects in the other categories, were analysed on the basis of published documentary material, such as sketches and photos.

Definition of costumes

The recording of the costumes is based on a system constituting 63 different costume types (Fig. 1.9). These are determined by their shape, labelled by a letter and a number indicating the type of patterning. The latter reflects the maximum number of variations, which may be seen in the material under analysis. Further, the types are grouped according to where they are placed on the body, *i.e.* if they represent outerwear, innerwear or leg wear. Outerwear can occur without innerwear and vice versa. This system is further combined with a terminology that facilitates both the description of the various costume types and the differentiation of the various elements from one another. The terms used are theoretical, and are not based on concrete finds or clothing items.

Outerwear is divided into six different groups (OG1-OG6) according to the shape of the costumes (Fig.1.10). The first three groups comprise cloaks that may be defined as simple outerwear of varying shapes, length and method of wear. They are worn over the shoulder and may either be closed, or open in front or at the side, and may sometimes be closed with a fibula. The cloak is worn by both genders, but its design and the manner in which it was worn vary according to gender and chronology. There are 21 types in this category.

Outerwear Group 1 (OG1) comprises cloaks that are closed in front and have three sub-types. OG1a is a closed cloak with a curved front edge. This includes types A1–A2, A5–A8, A15–A17, which are all female types. OG1b is a bell-shaped closed cloak with the female types A9–A10. OG1c is a teardrop-shaped closed cloak. This includes types A19–A21, of which the first is a short male type and the latter two longer female types. Outerwear Group 2 (OG2) comprises cloaks that are open in front with the female types A3–A4. Outerwear Group 3 (OG3) constitutes cloaks with two pointed corners.

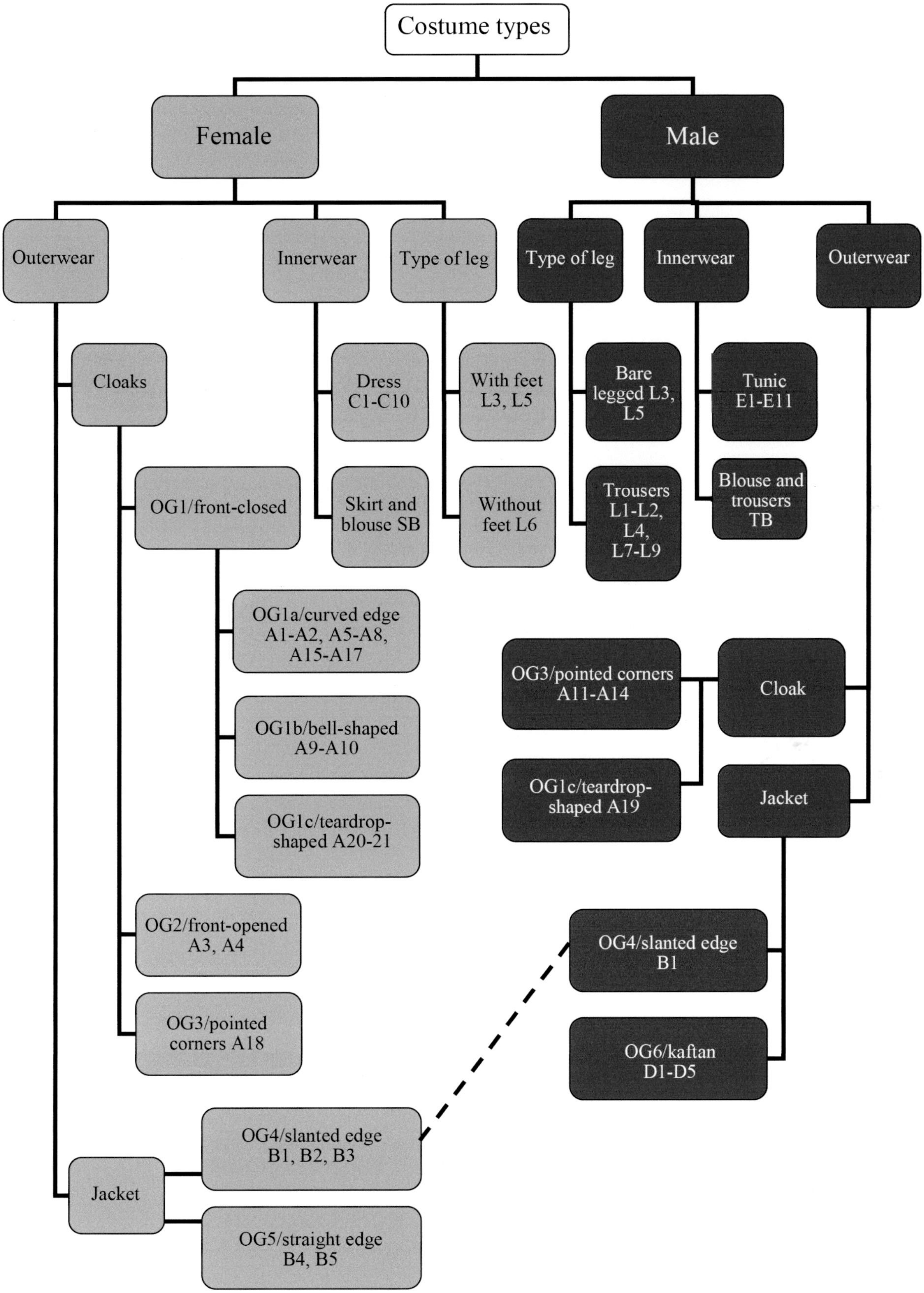

Fig. 1.9 Overview of the different outerwear groups and innerwear types for males and females.

Cloak types: Outerwear Group 1-3

♀ **A1**	Closed, plain (without patterning) female cloak, with a curved front edge, held together in front.		♀ **A2**	Plain, open female cloak, with a curved front edge, held together in the front, with a border at the hem.	
♀ **A3**	Plain, open female cloak, with a curved front edge resting on the shoulder.		♀ **A4**	Vertical-striped, open female cloak with a curved front edge resting on the shoulder.	
♀ **A5**	Vertical-striped, closed female cloak with a curved front edge, held together in front.		♀ **A6**	Diamond or lozenge-patterned, closed female cloak with a curved edge, held together in front.	
♀ **A7**	Diamond or lozenge-patterned, closed female cloak with a border and curved edge, held together in front.		♀ **A8**	Checked, closed female cloak with a curved front edge, held together in front.	
♀ **A9**	Plain, closed, bell-shaped female cloak with a curved edge, ending with a border in front.		♀ **A10**	Diamond or lozenge-patterned, closed, bell-shaped female cloak with a straight edge.	
♂ **A11**	Plain, closed male cloak with two pointed corners, and open on one side.		♂ **A12**	Vertical-striped, closed male cloak with the hem ending in two pointed corners, open on one side, with a dotted border along the edge (combination of A13 and A14).	
♂ **A13**	Closed male cloak with the hem ending in two pointed corners, open on one side with a dotted border along the hem.		♂ **A14**	Vertical-striped, closed, male cloak with two pointed corners, open on one side.	

Outerwear Groups 1-3: Cloak types

Type	Sex	Description	Illustration
A15	♀	Scale-shape patterned, closed female cloak with a curved front-edge held together in front.	
A16	♀	Dotted, closed female cloak with a curved front-edge, held together in front.	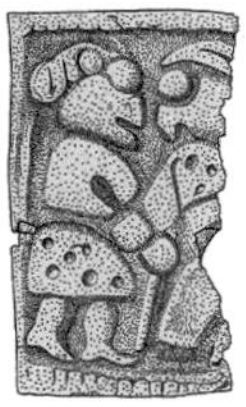
A17	♀	Closed female cloak in a combined vertical-striped and dotted pattern, with a curved front-edge, held together in front.	
A18	♀	Plain, closed female cloak with two pointed corners, and open at the side.	
A19	♂	Closed, teardrop-shaped, short male cloak with a hem below that ends in a point at the centre of the figure. The patterning was not registered.	
A20	♀	Closed, teardrop-shaped, long female cloak with a hem below that ends in a point in the middle of the figure. The pattern was not registered.	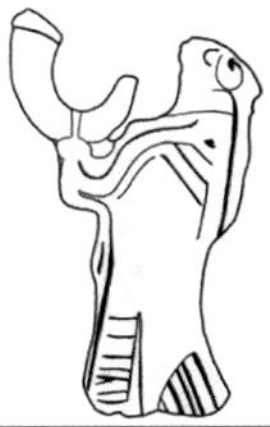
A21	♀	Closed, dotted, tear-drop-shaped, long female cloak with the hem below ending in a point in the middle of the figure.	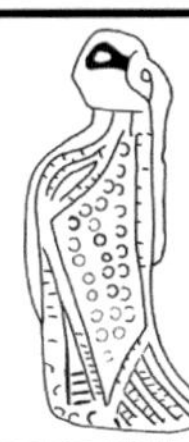

Fig. 1.10 Outerwear and innerwear recorded during the costume analysis of the selected object categories. For a full overview, see Fig. 2.44.

Outerwear Groups 4-5: Jacket types					
♀♂ **B1**	Plain unisex jacket with curved front-edge, and fringing along the edge.		♀ **B2**	Plain female jacket with a curved front-edge.	
♀ **B3**	Diamond or lozenge-patterned female jacket with a curved front-edge.		♀ **B4**	Plain female jacket with a straight lower edge.	
♀ **B5**	Diamond or lozenge-patterned female jacket with a straight lower hem.				

Outerwear Group 6: Kaftan types					
♂ **D1**	Male jacket with a slanted front-edge and a straight lower hem with a zigzag border. Also termed a kaftan.		♂ **D2**	Male jacket with a slanted front-edge and a straight lower hem with a horizontal-striped border. Also termed a kaftan.	
♂ **D3**	Male jacket with a slanted front-edge and a straight lower hem with a dotted border. Also termed a kaftan.		♂ **D4**	Male jacket with a slanted front-edge and a straight lower hem with a simple border. Also termed a kaftan.	
♂ **D5**	Diamond or lozenge-patterned male jacket with a slanted front-edge and a straight lower hem. Also termed a kaftan.				

Innerwear: Dress types

Type	Description	Illustration	Type	Description	Illustration
♀ **C1**	Plain female dress.		♀ **C2**	Plain female dress with a border at the hem.	
♀ **C3**	Plain female dress with a border running along the lower hem and up along the front edge.		♀ **C4**	Vertical-striped female dress.	
♀ **C5**	Vertical-striped female dress with a border at the hem.		♀ **C6**	Horizontal-striped female dress.	
♀ **C7**	Horizontal-striped female dress with a border at the hem.		♀ **C8**	Diamond or lozenge-patterned female dress.	
♀ **C9**	Dotted female dress.		♀ **C10**	Checked female dress.	
♀ **SB**	Skirt and blouse combination. Exclusively female.				

Fig 1.10 cont.

Innerwear: Tunic types							
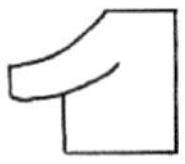 ♂	**E1**	Plain male tunic.		 ♂	**E2**	Plain male tunic with a border at the hem.	
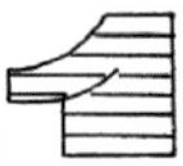 ♂	**E3**	Horizontal-striped male tunic.		 ♂	**E4**	Diamond or lozenge-patterned male tunic.	
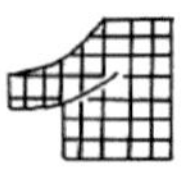 ♂	**E5**	Checked male tunic.		 ♂	**E6**	Scale-shape patterned male tunic.	
 ♂	**E7**	Dotted male tunic.		 ♂	**E8**	Vertical-striped male tunic.	
 ♂	**E9**	Male tunic with a border along the edge and centre marking.		 ♂	**E10**	Braided-patterned male tunic.	
 ♂	**E11**	Vertical-striped male tunic with a border at the hem.		 ♂	**TB**	Trousers and blouse combination. Exclusively male.	

Innerwear: Leg types	
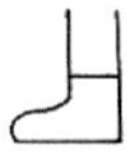 ♂ **L1** Narrow ankle-length trouser legs. Exclusively male.	 ♂ **L2** Wide ankle-length trouser legs. Exclusively male.
 ♀♂ **L3** Smooth lower legs with no sign of any (costume) separation at the ankle. Primarily male.	 ♂ **L4** Narrow breeches. Exclusively male.
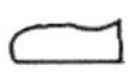 ♀♂ **L5** Feet only. Primarily female.	♀ **L6** No feet. Exclusively female.
 ♂ **L7** Wide breeches. Exclusively male.	 ♂ **L8** Trousers with clearly delineated crotches. Exclusively male.
 ♂ **L9** Trousers with clearly delineated crotches or lower legs with patterning. Exclusively male.	

Fig 1.10 cont.

This includes types A11–A14 and A18, of which the former are male types and the latter a female type.

The jacket is yet another outer garment, but differs from the cloak by being cut into shape, and having clearly defined sleeves. It is found in various shapes, with various gender attributes, but no associated fibulae. There are ten types of jacket, which may be divided into three basic groups. Outerwear Group 4 (OG4) constitutes jackets with a curved edge. The shape is reminiscent of the cloak, but has a hint of a sleeve, and does not reach below the knee. This includes types B1–B3, of which the first is common to both genders, while the other two are exclusively female types. Outerwear Group 5 (OG5) constitutes jackets with a straight edge, which do not reach below the knee. This includes types B4 and B5, which are both female types. Outerwear Group 6 (OG6) constitutes jackets with a curved front edge and a straight lower hem. They can be both knee-length and ankle-length, and often have wide borders. This item of costume is also termed a *kaftan*. It includes types D1–D5 which are exclusively male types.

Innerwear is recorded as gender-specific, and this is done to ensure that the costume items are easily distinguished from one another, although the depictions contain very few specific costume details. The main difference between male and female innerwear is seen in the length of the costume items and the type of leg coverings with which they are combined. Leg coverings are grouped according to their design and amount of leg that is left visible.

The term *dress* is used for an ankle- or floor-length female garment. As the upper body in female depictions is frequently concealed by outerwear, most details are hidden, although long sleeves seem to be a general characteristic. The dress is always closed in front and has no fibulae attached. The dress can be longer at the back than in front, or have a marked line giving the impression of terminating in a train. The dress encompasses ten types (C1–C10).

The female blouse is close-fitting, possibly long-sleeved and ends at the waist. The blouse is closed in front, but may have a fibula at the neck, indicating the presence of a slit. The female blouse is always combined with a skirt, which is an ankle- or floor-length garment beginning at the waist. The skirt may be longer at the back than in front, or have a marked line giving the impression that it terminates in a train. This combined type is referred to as SB (skirt/blouse).

The term *tunic* is used for a thigh- or knee-length male garment. It has a straight hem. It may be either short or long-sleeved. The tunic is closed in front and has no attached fibulae. It may be used with or without a belt. It encompasses 11 types (E1–E11).

Trousers can be of varying length and design. Depending on the type of outer and innerwear they are combined with, different parts become visible. When trousers are combined with a tunic, only the lowest parts of the trouser legs are visible (L1–L2, L4, and L7–L9), while trousers with a clearly evident crotch are often combined with a blouse (TB), or with a blouse and belt (TBB) or a bare chest. In the case of the females depicted, the lower part of the figure is divided into a long leg, a foot or no feet. In the case of males, where no specific trouser types are indicated, the length of leg is given as either a long or short lower leg. If there is no marking between the leg and foot, it is not possible to determine if the figure wears shoes or not. The shape of the feet can either be narrow and pointed or short and clumpy. There is plenty of evidence to suggest that the figures wear shoes or boots, often pointed or long-toed. The legs and feet encompass three types (L3, L5 and L6).

In the following chapters, where a table includes a type with an "X" added, this refers to a special shape or supplementary characteristic, such as a belt or a train. Where an F or M is added, it is to distinguish between types which occur among both genders. The analysis and numbering reflect the registration process. Thus, there is no correlation, for instance between the numbering of the patterning on female dresses and male tunic types.

Note

1 The Bayeux Tapestry found in France, dated to the AD 1070s, is not included in the analysis, as it represents a different geographical and cultural historical context (Rud 1988).

2

Gold-foil figures

Gold-foil figures are tiny pieces of gold sheet metal depicting males, females and animals. They are found exclusively in Scandinavia and have no direct parallels outside this region. Approximately 2900 gold-foil figures from, at least, 37 different sites in Denmark, Sweden and Norway have been found (see Fig. 1.1) (Table 2.1).[1] A precise estimate is difficult to make as not only do new finds continuously appear, but also many of the finds made before 1900 were registered somewhat inadequately.

Gold-foil figures can be produced in several ways. The commonest method is to stamp the image onto a strip of gold sheet metal from below with a patrix or die stamp (Watt 1987). Thus, the stamped gold-foil figures are often uniformly square or oblong in shape, usually with an accentuated framing of the motif. Several, more or less, complete die stamps have been found in both Denmark and Sweden confirming this method of production (Watt 1999b; 2004). All preserved die stamps (patrices) are made of bronze. The majority are similar in shape to the gold-foil figures, but two specimens also have a completely or partially preserved handle (Ulriksen 2000; Hellerström *et al.*, 2000). Only in three instances are gold-foil figures from a preserved die stamp known (Watt 1999b, 181–183; 2004, 175; Lamm 2004, 105–107), and it is striking that die stamps and the corresponding gold-foil figures produced by them have never been discovered together on the same site. The die stamps are, thus, an important supplement to the images of the gold-foil figures and are also included in the costume analysis. A stamped gold-foil figure, here termed *an impress*, may thus appear in many identical versions. However, generally only one or two of each specimen have been recovered. In addition, impresses that are very closely related, although not entirely identical, are also found. This phenomenon arises when a (worn) die stamp is copied or reused. These groups of stamps are referred to as die families and die series respectively (Watt 1999b; 2004, 214).

Another method of production is to scratch the image directly onto the gold sheet, before the figures are engraved or cut out (Watt 1991a). Figures may also be directly cut out of the gold sheet without preliminary sketching, resulting in the simplest types. These figures are thus unique, despite closely resembling one another. A further difference between impressed and engraved or cut-out gold-foil figures is that the former are paper-thin, whereas the latter are often cut from somewhat thicker gold sheet (Watt 1992, 204). A number of cut-out animals were found in Sorte Muld, but no gold-foil figures representing animals and humans together, as seen in some of the other selected object categories have been found (Watt 2002b) (Fig. 2.1). As the cut-out gold-foil figures seldom hold information on costume, they are not included in the costume analysis.

Gold-foil figures may be divided into images of one or two figures, also termed *single or double figures* (Tables 2.2–2.6). The group of single gold-foil figures may either be produced as stamped or engraved and cut-out figures, whereas the double figures without exception are stamped. Gold-foil figures are often approximately 1 cm^2 large, and taking their size into account, the amount of detail these minute images contain is remarkable.

Interpretation

Like other objects of precious metal, gold-foil figures have always been the focus of much interest. The first work emerged in 1725, where the earliest gold-foil figures found were published as a special variant of bracteates or coins (Melle 1725/1997; Vedel 1886). However, gradually it came to be accepted that gold-foil figures were an object category of their own, encompassing different style and symbolic content, and in 1791 they were given their name: *guldgubbar* or gold-foil figures (Sjöborg 1791; Lamm 2004). Only a few scholars have studied this object category in depth, and many finds are still only partially published, while others await full publication. The largest collection of gold-foil figures from Denmark has been found at Sorte Muld on the island of Bornholm (Fig. 2.2) and Lundeborg (Watt forthcoming; Thomsen *et al.*, 1993) (Fig. 2.3). Older Danish finds are kept at the National Museum of Denmark in Copenhagen, and

Fig. 2.1 A selection of gold-foil animals in the shape of pigs or bears from Sorte Muld, Bornholm, Denmark (photo: Martin Stoltze).

Table 2.1 Gold-foil figure finds in Denmark, Sweden and Norway, Early Germanic Iron Age (EGIA), Late Germanic Iron Age (LGIA) and Viking Age (VIK).

Location: Denmark	*Region*	*Year found*	*Dating*	*Context*
National Museum	unknown	unknown	–	unknown
National Museum (NM1-NM16)	perhaps Norway	unknown	–	unknown
National Museum (NM51)	unknown	unknown	–	unknown
Bornholm	unknown	unknown	–	unknown
Rønne	Bornholm	1710	–	–
Smørenge	Bornholm	1828/1996	–	–

(Continued)

Table 2.1 Gold-foil figure finds in Denmark, Sweden and Norway, Early Germanic Iron Age (EGIA), Late Germanic Iron Age (LGIA) and Viking Age (VIK). (Continued)

Location: Denmark	*Region*	*Year found*	*Dating*	*Context*
Sylten	Bornholm	1831	–	–
Sorte Muld	Bornholm		–	central place
Sorte Muld	Bornholm	1985–1987	6th–7th century AD	central place
Møllegård	Bornholm	1992	–	settlement
Bukkegård	Bornholm	1989	–	stray find
Agerbygård	Bornholm	1992	–	settlement
Boelsgård	Bornholm	1997	–	?
Toftegård	Zealand	1993	LGIA–VIK	trading place
Gudme	Funen	1984/1992	6th–7th century AD	central place
Lundeborg	Funen	1988–1992	AD 600–650	landing place /trading place
Stentinget	Jutland	1990	LGIA	settlement/trading place
Tørring	Jutland	1967	–	stray find
Nørre Hvam	Jutland	1824	–	deposit/treasure
Ravlunda 1a (Grönsö)	Scania	19th century	–	stray find
Ravlunda 1b	Scania	18th century	–	stray find
Uppåkra	Scania	1996–2003	LGIA	cult house of a central place complex
Eketorp	Öland	1966	EGIA–LGIA 6th century AD	deposit/treasure hoard at ring fort
Bolmsö	Småland	1911	AD 650–700	grave
Brahekyrkan	Småland	1999	LGIA-VIK	stray find in connection with a settlement layer
Slöinge	Halland	1992–1995	AD 710–720	hall/settlement
Hög-Edsten	Bohuslän	1864	EGIA–LGIA	treasure
Gullmarsberg	Bohuslän	1912	–	stray find, perhaps in connection with a central place
Svintuna	Östergötland	1936	7th century AD	stray find in connection with a settlement
Ulltuna	Uppland	1900	7th–8th centuries AD	stray find in connection with a grave
Helgö	Uppland	1954–1964	7th–9th centuries AD	trading place
Husby	Närke	1997–1998	6th–7th centuries AD	stray find, perhaps in connection with a hall
Eskiltuna	Södermanland	1961–1962	8th century AD	stray find, perhaps in connection with a cult place
Kongsvik	Nordland	1740s	–	unknown
Kirkeseter	Sør-Trøndelag	1810	–	unknown
Sandsvær	Østfold	1825	–	unknown
Borge	Østfold	1825	–	unknown
Hauge	Rogaland	1897	VIK	unknown
Mære	Nord-Trøndelag	1966–1967	VIK	hall/central place?
Borg	Nordland	1983/1987/1988	LGIA/VIK	hall/central place
Hov	Oppland	1993/2004–2005	7th century AD	house

This information is based on sources known and published up to and including 2005 only.

Table 2.2 Locations in Denmark of gold-foil figures, their total number, and the actual number used in the costume analysis.

Location	*Amount*	*Double gold-foil figures*	*Single gold-foil figures*	*Total number of impresses*	*Impresses used in the costume analysis*
Agerbygård	1		1m	1	0
Boesgård	1		1m	1	0
Bornholm	3		1f, 1m, 1?	3	1m
Bukkegård	1		1m?	1	0
Gudme	2	1	1f	2	1df,1f
Lundeborg	103	102	1m	70	46d
Møllegård	2		2m	2	1m
National Museum	45	16	3f, 2f?, 20m, 4m?	39	8d, 1f, 7m
Nørre Hvam	1		1m	1	0
Rønne *	20 (16)		5f, 10m	16	4f, 3m
Smørenge	2		1f, 1m?	2	1f
Sorte Muld	>2350	13	25f, >134m	>430	11d, 19f, 133m
Stentinget	1	1		1	1d
Sylten	1		1f	1	1f
Toftegård	8	6	2m?	7	5d
Tørring	1		1f	1	1f
Total	>2542	139	>220	>578	245

Registration and processing were completed in the beginning of 2005, double gold-foil figures (d), female represented singly (f), male represented singly (m). * According to Maregrethe Watt, these fold-foil figures should now be ascribed to a location on Bornholm in Denmark and be termed "von Melle".

Table 2.3 Locations in Sweden of gold-foil figures, their total number, and the actual number used in the costume analysis.

Location	*Amount*	*Double gold-foil figures*	*Single gold-foil figures*	*Total number of impresses*	*Impresses used in the costume analysis*
Bolmsö	2		2m	1	0
Brahekyrkan	1	1		1	1d
Eketorp	15		9m, 6f	3	2m, 1f
Eskiltuna	2	2		2	2d
Gullmarsberg	1	1		1	1d
Helgö	26	26		>18	18d
Husby	1	1		1	1d
Hög-Edsten	3		3m	1	0
Ravlunda	2		2m	1	0
Slöinge	63	43	2f, 5m, 2?	28	20d, 2f, 5m
Svintuna	1	1		1	1d
Ulltuna	1	1		1	1d
Uppåkra	122	6	116	58	5d, 10f, 31m
Total	240	82	147	>117	101

Registration and processing were completed in the beginning of 2005, double gold-foil figures (d), female represented singly (f), male represented singly (m).

Table 2.4 Locations in Norway of gold-foil figures, their total number, and the actual number used in the costume analysis.

Location	*Amount*	*Double gold-foil figures*	*Single gold-foil figures*	*Total number of impresses*	*Impresses used in the costume analysis*
Borge	1	1		–	0
Kongsvik	10?			–	0
Kirkeseter	2		2m	–	0
Sandsvær	1	1		–	0
Borg	5	5		4	4
Hauge	16	16		8	8
Hov	20	?		?	0
Mære	19	19		17	12
Total	74	>42	2	>29	24

Registration and processing were completed in the beginning of 2005, double gold-foil figures (d), female represented singly (f), male represented singly (m).

Table 2.5 Scandinavian gold-foil figure die stamps, their total number and the actual number used in the costume analysis.

Location	*Region*	*Year found*	*Number*	*Double figures*	*Single figures*	*Impresses used in the costume analysis*
Vester Egesborg/PAT1	Zealand	2000	1	1		1d
Neble/PAT2-PAT3	Zealand	1985, 1994	2		2f	2f
Møllegård/PAT4	Bornholm	1990	1		1m	1m
Sylten/PAT5	Bornholm	1997	1		1m	1m
Smørenge/PAT6	Bornholm	2002	1		1m?	1m?
Uppåkra/PAT7-PAT10	Scania	1997–1998/ 2001–2004	5		2m, 2f	2f, 2m
Järrestad/PAT11	Scania	1999	1	1		1d
Total			11	2	9	11

Registration and processing were completed in the beginning of 2005, double gold-foil figures (d), female represented singly (f), male represented singly (m).

Table 2.6 Scandinavian bronze-foil figures, their total number and the actual number used in the costume analysis.

Location	*Region*	*Year found*	*Number*	*Single bronze-foil figures*	*Impresses used in the costume analysis*
Sorte Muld/BZG1	Bornholm	1986–1987	2	1f	1f
Vä/BZG2	Scania	1945	1	1f	1f
Gårdlösa/BZG3	Scania	1981–1983	1	1f	1f
Total			4	3	3

Registration and processing were completed in the beginning of 2005, double gold-foil figures (d), female represented singly (f), male represented singly (m).

only a few of these have been published, and primarily as comparative material in other connections (Mackeprang 1943; Axboe 1983; Holmqvist 1960). Norwegian gold-foil figures are primarily available through a few poorly illustrated articles (Gustafson 1900; Lidén 1969; Haraldsen 1995) (Figs 2.4 and 2.5). An exception is the publication of the finds from Borg in Lofoten in Norway (Stamsø Munch *et al.*, 2003) (Fig. 2.6). The Swedish material is the easiest accessible material with the largest part collected and published in three articles (Lamm 2004; Watt 1999b and 2004) (Figs 2.7–2.9).

A fascinating aspect of the gold-foil figures is that, as yet, no fully convincing explanation for their presence and use has been presented. Gold-foil figures have no apparent practical function unlike most archaeological objects originating

Fig. 2.2 A selection of gold-foil figures from Sorte Muld, Bornholm, Denmark (scale c. 1:1) (Bornholms Museum; photo: Lennart Larsen).

Fig. 2.3 Gold-foil figures from Lundeborg, Funen, Denmark (Svendborg Museum; photo: Lennart Larsen).

from prehistoric times. Moreover, their find contexts rarely provide any supplementary information on their use. Thus, interpretations have oscillated between a contextual and iconographical approach, and today, these two approaches are almost inseperable. The earliest attempt, at the beginning of the 18th century, to interpret a paired gold-foil figure shows that the figures therein were at first perceived as two men: a bishop (the figure with a long garment) and a warrior (the figure with a short garment) (HA1, Fig. 2.4). This interpretation implied a Christian connotation, which due to their Scandinavian pre-Christian dating can be excluded (Lamm 2004, 115–125). In 1725, Jacob von Melle postulated that, the known single figures were reproductions of Nordic male and female household gods (Melle 1725/1997). In 1899, Gabriel Gustafson launched a new interpretation of the paired figures as a courtship scene (Gustafson 1900). He identified the figures as a man and a woman, but it was not until 1909 that gold-foil figures were linked with the mythological representation of the Old Norse sources (*Skírnismál*, the Lay of Skírnir), symbolised by the Old Norse god Freyr's wooing of the giantess Gerdr (Olsen 1909). Due to the apparently random scattering of the gold-foil figures in the soil, they were perceived as part of an agrarian fertility cult, in which the gold-foil figures functioned as sacrifices to obtain a rich harvest. Gro Steinsland later utilised this interpretation of the motif of the paired gold-foil figures to posit that the marriage myth, *hieros gamos* (sacred matrimony) between Frejr and Gerdr functioned as the legitimisation of Nordic aristocratic

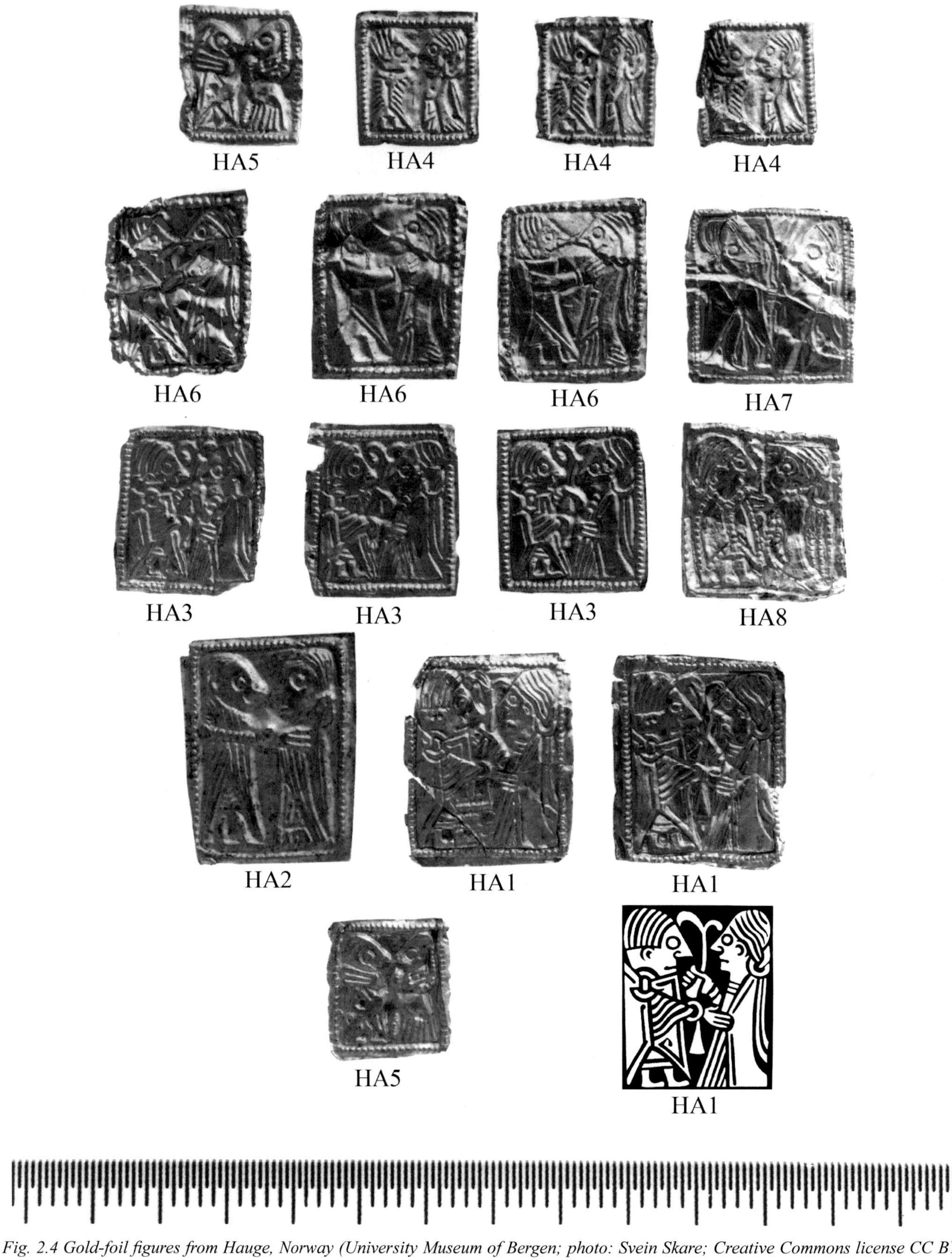

Fig. 2.4 Gold-foil figures from Hauge, Norway (University Museum of Bergen; photo: Svein Skare; Creative Commons license CC BY-NC-ND 3.0, https://creativecommons.org/licenses/by-nc-nd/3.0/deed.en; HA1: drawing: Egil Bakka).

lineages: a marriage, which according to legend was the foundation of the Swedish *Yngling* dynasty (Steinsland 1989, 286–288, 293–295). Rudolf Simek has since questioned this interpretation. He sees the narrative not as an actual marriage, but rather as bride-stealing that, in his opinion, scarcely deserves an exemplary model. According to Simek, the only role models for legitimate marriages can be those between Thor and Sif or Odin and Frigg (Simek 2002, 107). As the above-mentioned hypotheses primarily apply to the paired or double gold-foil figures prevalent in Norway and Sweden, an alternative explanation has yet to be sought for the numerous single gold-foil figures that characterise the majority of the southern Scandinavian finds (Watt 2004, 217; Simek 2002, 108).

Concrete explanations for how the image functioned in practice, too, are lacking. However, as more sites with gold-foil figures are known today, and a clearer picture of the contexts in which they are found emerges, it is evident that the gold-foil figures were not buried or scattered in open fields. Finds from Borg and Mære in Norway, as well as Slöinge and Uppåkra in Sweden demonstrate that, in many instances, the gold-foil figures were deposited or found in connection with so-called cult houses or large residential halls (Lundqvist 1996a; Stamsø Munch *et al.*, 2003; Watt 2004).

Furthermore, on the basis of the finds in Uppåkra, it has been suggested that the gold-foil figures could have been fixed to the inner roof-bearing upright posts in the cult houses (Larsson and Lenntorp 2004, 24), whereas the finds from Sorte Muld have been perceived as a type of temple currency used in connection with cult activities. This could account for the presence of the numerous gold-foil figures found scattered around in the cultural layer in Sorte Muld. New surveys of the Sorte Muld settlement have also indicated traces of housing constructions, suggesting the possibility of find contexts comparable to Uppåkra (Watt 2004, 215; Larsson and Lenntorp 2004). Taking the contents of the images into account, the depictions may represent a secular elite with religious power, gods or other mythological figures (Watt 2003, 26; Lamm 2004). Karl Hauck who has largely been engaged in the interpretations of the iconography of bracteates has suggested, on several occasions, that the gold-foil figures primarily represent divine figures that can be attributed to Old Norse mythology, and that the images can be seen as depictions of Odin, Thor or Freye (Hauch 1994; Lamm 2004, 121–125; Watt 2004, 217).

Fig. 2.5 Gold-foil figures from Mære, Norway (NTNU University Museum).

Fig. 2.6 Gold-foil figures from Borg, Lofoten, Norway (Tromsø Museum-Universitetsmuseet; photo: K. Kristensen).

Thus, scholarly research in recent decades has emphasised the link between gold-foil figures and Scandinavian religion, ideology and social organisation (*e.g.* Steinland 1989; 1990; 1997; Watt 1992; and Hauck 1993). To summarise, there is as yet no generally accepted and unequivocal explanation for the function of the gold-foil figures. Although they are considered a homogeneous artefact group, the question may be asked whether they should be considered as a comprehensive unit at all. Even if similar in terms of production, it cannot be excluded that double gold-foil figures, single gold-foil figures and engraved animals were utilised in different ways and thus should not be interpreted in the same manner (Watt 2002b; Simek 2002).

Fig. 2.7 Gold-foil figures from Bolmsö, Småland, Sweden (Statens Historiska Museum; http://creativecommons.org/licenses/by/4.0/ photo: Gunnel Jansson).

Fig. 2.8 Gold-foil figures from Eketorp, Öland, Sweden (Statens Historiska Museum; http://creativecommons.org/licenses/by/4.0/ photo: Gunnel Jansson).

Fig. 2.9 Gold-foil figures from Uppåkra, Skåne, Sweden (Lund University Historical Museum).

Context

Gold-foil figures have been found in various, more or less, well-documented contexts: as stray finds, stray finds with a possible connection to a dwelling, grave finds, in hoards and as settlement finds. In describing the function of these places, various terms, such as cult place, central place, settlement or trading place are used, yet their significance, or the difference between these terms, is seldom defined more precisely (Watt 1991b–c; Fabech and Ringtved 1995; Helgesson 1998). At the same time, the presence of gold-foil figures is a defining element in itself. Essentially, it cannot be denied that there is a link between gold-foil figures and large residential halls with representative rooms or proper cult houses as well as activities in central or trading places (Helgesson 1998, 43). Producing and distributing gold-foil figures constituted part of a common Scandinavian tradition, and links between die stamps found in various parts of Scandinavia demonstrate a widespread exchange of motifs and ideas of images across a vast distance. Whether the activities in which the gold-foil figures featured were also homogenous is still not possible to determine.

Margrethe Watt has convincingly demonstrated that, there was a close relationship between the finds from Uppåkra in Sweden, and Sorte Muld, Lundeborg and Toftegård in Denmark respectively (Watt 2004, 214). For instance, several die-linked items were found in the above-mentioned settlement complexes, suggesting that the die stamp belonged to a peripatetic artisan who produced the gold-foil figures (Lamm 2004, 47). It further implies that the production of gold-foil figures in the above-mentioned places must have occurred during the same periods. Thus, there are close ties between the settlement complexes in the southern Scandinavian region that evidently functioned in parallel, and, possibly, separately from the production of gold-foil figures in the remaining Swedish and Norwegian areas.

Dating

Absolute dating of gold-foil figures is a difficult task as the bulk of the Scandinavian gold-foil figures have come to light in many different ways ranging from uncertain or unknown find contexts to localities that functioned continuously through several centuries, and where only an upper and lower time limit can be given (Table 2.1). As for the image interpretations, the dating is partly based on context and partly on style. The context-based framework for the dating of the gold-foil figures falls approximately within the period from the onset of the 6th to the 10th centuries AD, depending on the locality.

Gold-foil figures from Uppåkra are considered to be among the earliest finds, being deposited in the early part of the Germanic Iron Age. Thorough excavations on this site indicate, among others, at least, seven housing phases, but as the gold-foil figures were primarily located in the topsoil, a more precise dating and link between the gold-foil figures cannot be established (Larsson and Lenntorp 2004).

The Sorte Muld gold-foil figures are considered to be more or less contemporary, as they come from a cultural layer that may be linked to the demolition of a building in the course of the Late Germanic Iron Age (Watt 2004, 214–216). Those from Lundeborg are dated to a period from the late 6th century AD to the mid-7th century AD, most likely around the year AD 600 (Thomsen *et al.*, 1993, 87–90).

The gold-foil figures from Slöinge are linked to a dendrochronological dating of the uprights or posts that still remained in the hole where the bulk of the gold-foil figures were recovered. The find context thus indicates that the gold-foil figures from Slöinge were deposited around AD 710–720 when the building was constructed (Lundqvist 1996a, 15, 45–47). On the basis of the iconographical resemblance to the gold-foil figures from Slöinge, the gold-foil figures from Helgö in Sweden have similarly been dated to the beginning of the 8th century AD (Lamm 2004, 132). The treasure hoard with the gold-foil figures from Eketorp on the Swedish island of Öland was dated to the transition from the Early to Late Germanic Iron Age on the basis of stratigraphic observation (Stenberger 1973; Lamm 2004, 65). The only gold-foil figures to be found in a burial context are those from Bolmsö in Sweden. The grave is dated to AD 650–700, but the gold-foil figures are considered to belong to the 6th century AD on the basis of their iconographical resemblance to the finds from Eketorp and Hög Edsten (Kville) in Sweden (Lamm 2004, 62–64; Arrhenius 1971, 138). The Norwegian gold-foil figures constitute the youngest finds in terms of dating. Thus, the finds from Hauge in Norway are dated to the Viking Age on the basis of the iconographical content of the costumes and jewellery (Gustafson 1900, 94; Watt 2004, 216), while those from Borg are dated, on the basis of House I: 1A, to AD 700–1000, *i.e.* the latter part of the Late Germanic Iron Age and the Viking Age (Stamsø Munch 1991, 329–331; Stamsø Munch *et al.*, 2003, 13, 252). The gold-foil figures from Mære are also assumed to belong to the Viking Age (Lidén 1969, 16), but finds and structures in both Borg and Mære demonstrate continuity back in time to the Germanic Iron Age. In view of the find situation in Uppåkra, these dates require revision and Watt suggests that the Norwegian finds reflect a later deposition, rather than a later production (Watt 2004, 216).

At least two different dating scenarios may be postulated, based on the information gleaned from the various localities presented and their contexts: first, that the gold-foil figures were produced within a relatively short period in a few central places and trading places, and were possibly employed over a longer period in different regions, and secondly, that the production of gold-foil figures in southern Scandinavia reached its peak in the early part of the Late Iron Age, while it continued in the northern parts into the Viking Age. In comparison with the shifts in shape and style of other contemporary artefacts in the archaeological record, gold-foil figures have a surprisingly long functional period, which possibly could be narrowed down further when the study of costumes and jewellery is included in the interpretations. For instance, a comprehensive evaluation of datable objects visible on the gold-foil figures demonstrates that, these primarily constitute Late Germanic Iron Age object types like disc-on-bow fibulae, disc fibulae and plate fibulae, or the more general types appearing throughout the entire Iron Age, such as penannular brooches (Tables 2.7–2.12). Using these objects as the primary dating element is indeed risky as they could represent archaic elements, but considered together with the archaeological clothing elements, they definitely provide invaluable information.

Distribution

In Denmark, gold-foil figures have been found in 15 known localities, while the rest are, more or less, of unknown provenance (Table 2.2 and Fig. 1.1).[2] It has been discussed whether some of the 16 double gold-foil figures (National Museum 1–16, Fig. 2.10) belonging to the National Museum of Denmark could originate from the Kongsvik find in northern Norway, which from an early archive record dated to the 1740s is known to have included ten double gold-foil figures. Yet, as the gold-foil figures in the National Museum of Denmark already became part of the museum collection at the end of the 17th century, this hypothesis is not valid. However, as several of these double gold-foil figures belonging to the National Museum of Denmark are die linked to two of the gold-foil figures from Hauge (Figs 2.10 and 2.4), it is likely that some of these double gold-foil figures originate from Norway, and today they are registered as being possibly of Norwegian provenance (Axboe 1983; Stamsø Munch *et al.*, 2003, 249).

Compared to Sweden and Norway, Denmark possesses the largest number of recorded gold-foil figures. However, apart from a few very rich sites, finds from the rest of the country are modest in number, often only single finds of a gold-foil figure. The largest find from before 1985 is from Bornholm which was discovered in 1710 (Melle 1725/1997).[3] Twenty gold-foil figures were recorded on that occasion, of which 16 are preserved today (Fig. 2.11). Most of the Danish gold-foil figures originate from Bornholm, while there are only a few finds from Zealand (Fig. 2.12) and Jutland. Funen is represented by two sites, Gudme and Lundeborg, the latter having the second largest number of gold-foil figures.

Table 2.7 Costume analysis of the gold-foil figures selected for this study from the National Museum of Denmark and the rest of Denmark, except for Lundeborg and Sorte Muld, male (M), female (F), not dressed (ND).

Find ID	*Museum number*	*Gender*	*Outerwear*	*Dress*	*Tunic*	*Leg type*	*Belt*	*Other objects*
Bornholm (NM20)	NM 4714	M	D1			L5		staff
Gudme 1	Dnf. 57/83	F	A5	C3		L5		disc-on-bow fibula
Gudme 2	6220/x547	M/F		C1	ND	L8/L6		
Møllegård 2	Dnf. 4/92	M	D2			L5		staff?
National Museum 1	NM 8681	M/F	A11/A3	C4	E3	L3/L6		penannular brooch/ branch necklace
National Museum 3	NM 8681	M/F	-/A3	C1	E1	L3/L5		
National Museum 4	NM 8679	M/F	A14/A3	C1	E1	L2/L6		penannular brooch/–
National Museum 6	NM 8681	M/F	A11/A4	C6-1F?	E1	L3/L6		
National Museum 9	NM 8681	M/F	A11/A1	C4		L3/L6		–/branch
National Museum 12	NM 8681	M/F	A11/A1	C4		L3/L6		
National Museum 13	NM 8681	M/F	A11/-	C1		L3/L6		
National Museum 14	NM 8681	M/F	A14/A3	C4F?	E1	L3/L6		penannular brooch?/–
National Museum 17	NM C1056	M	D4			L3		pointed staff
National Museum 21	No no.	F	A15	C5		L5		
National Museum 23	No no.	M	D4			L3		staff
National Museum 25	No no.	M			E6	L3		staff
National Museum 26	NM C6655d	M			E2	L3		staff
National Museum 29	No no.	M	D4			L3		staff?
National Museum 30/31	NM C6655ab	M			E10	L3		
National Museum 32	NM C6655c	M	D1			L5		staff
Rønne* 1 (NM19)	NM 121(8684)	F	A5	C2		L5		beaker
Rønne* 2 (NM18)	NM 121(8684)	F	A17	C2		L5		disc fibula, beaker
Rønne* 3 (NM22)	NM 121(8684)	F	A15	C8		L5		
Rønne* 4 (NM50)	NM 121(8684)	F		SB		L5		
Rønne* 16 (NM45)	NM 121(8684)	M?			E2	L5		
Rønne* 18 (NM52)	NM 121(8684)	M			E2	L3		
Rønne* 19 (NM 55)	NM 121(8684)	M?			E2	L?		
Smørenge 1 (NM34)	MDCCCXLVI	F	A15	C5		L5		
Stentinget	Dnf. 8/91	M/F	A11/A1	C6		L3/L6		
Sylten (NM37)	MMCCXXXIII	F	A15	C5		L5		
Toftegård 1	Dnf. 9/96	M/F	-/A16	C1	E1	L3/L?		
Toftegård 2	Dnf. 12/96	M/F	-/A16	C1	E1	L?/L6		
Toftegård 3	Dnf. 13/96	M/F	-/A3?	C1?	E2	L2/L?		
Toftegård 4	Dnf. 10/96	M/F	-/A16	C5	E1	L3/L5		
Toftegård 6	Dnf. 11/96	F/M		C1	E1	L6/L3		
Tørring	3174	F	A15	C2		L5		disc-on-bow fibula, necklace

* According to Margrethe Watt, these gold-foil figures should now be ascribed to a location on Bornholm in Denmark and now be termed "von Melle".

Table 2.8 Costume analysis of the gold-foil figures selected for this study from Lundeborg in Denmark, male (M), female (F), not dressed (ND).

Find ID	Museum number	Gender	Outerwear	Dress	Tunic	Leg type	Belt	Other objects
Lundeborg 1	Dnf.1/93	M/F		C4?	E1?	L3/L5		
Lundeborg 3	Dnf.3/93	F/M		C1–C3	ND	L5/L3?		
Lundeborg 4	Dnf.4/93	F/M	A3/–	C1	E8	L6/L2		–pendant?
Lundeborg 5	Dnf.5/93	M/F	–/A6	C4F	E2	L3/L?		–/disc-on-bow fibula
Lundeborg 6	Dnf.40/92	?/F		C9		L?/L6		
Lundeborg 7	Dnf.37/92	M/F		C1	E1	L2/L6	X/–	
Lundeborg 13	Dnf.13/92	M/F		C1	E3	L3/L6		
Lundeborg 16	Dnf.16/92	M/F	–/A5		ND	L8/L3		
Lundeborg 18	Dnf.18/92	M/F	–/A5?		ND	L8/L5		
Lundeborg 19	Dnf.19/92	M/F	–/A5	C1	E1	L5/L5		
Lundeborg 25/30	Dnf.25/92, 30/92	M/F	–/A16	C7	E7	L3/L5		
Lundeborg 26	Dnf.26/92	M/F		C1	E1	L3/L5?		
Lundeborg 27	Dnf.27/92	F/M		C4	E8	L5/L3		necklace?/-
Lundeborg 32	Dnf.32/92	M/F	–/A16	C8F	E7?	L3/L6		
Lundeborg 36	Dnf.36/92	M/F		C3?	E1	L3/L6		necklace
Lundeborg 37	Dnf.37/90	F/M		C4	E9	L5/L5		
Lundeborg 38	Dnf.38/90	M/F	–/A5		ND?	L8/L3		
Lundeborg 39	Dnf.39/90	M/F	A11/A3?	C1–C4		L3/L6		
Lundeborg 47	Dnf.47/90	M/F	–/A5	C1	E1	L5/L5		
Lundeborg 50	Dnf.50/90	M/F		C4	E1	L3/L5		
Lundeborg 52	Dnf.52/90	M/F	–/A5	C4	E8	L3/L5		
Lundeborg 53	Dnf.53/90	F?/M?		C4	E8	L3/L3		
Lundeborg 55	Dnf.55/90	M/F	A14/–	C4		L3/L6		
Lundeborg 57	Dnf.57/90	M/F		C5–C8	E1	L5/L5		
Lundeborg 58	Dnf.58/90	M/F	–/A5		E1	L3/L5	X?/ –	
Lundeborg 60	Dnf.60/90	M/F	A12/–	C4		L3/L6		
Lundeborg 64	Dnf.64/90	M/F		C4	E8	L3/L5		
Lundeborg 66	Dnf.66/90	M/F	–/A3		ND	L8/L6		
Lundeborg 70	Dnf.70/90	M/F	–/A1?	C1	E8	L3/L6		
Lundeborg 74	Dnf.74/90	?/?		C1?	E1?	L3/L3		
Lundeborg 75	Dnf.75/90	F/M		C1	E1	L5/L3		
Lundeborg 79	Dnf.79/90	M/F		C4	E8	L3/L6		branch or staff
Lundeborg 80	Dnf.80/90	M/F	–/A16	C1F?	E7	L3/L6		
Lundeborg 81	Dnf.81/90	M/F	A14/A13	C4		L3/L6		
Lundeborg 83	Dnf.83/90	M/F	–/A1	C2	E1	L2/L6?		
Lundeborg 84	Dnf.84/90	M/F	A14/A5?	C4		L3/L6		
Lundeborg 87	Dnf.87/90	M/F	A5/A1	C4		L3/L6		
Lundeborg 90	Dnf.90/90	F?/M?		C1	E1	L5/L3		
Lundeborg 91	Dnf.91/90	M/F	–/A5	C4	E3	L3/L6		
Lundeborg 95	Dnf.95/90	M/F		C9	E1	L3/L5		
Lundeborg 96	Dnf.96/90	M/F		C1F?	ND	L3/L6		
Lundeborg 97	Dnf.97/90	F/M		C4	E1	L?/L?		

(Continued)

Table 2.8 Costume analysis of the gold-foil figures selected for this study from Lundeborg in Denmark, male (M), female (F), not dressed (ND). (Continued)

Find ID	*Museum number*	*Gender*	*Outerwear*	*Dress*	*Tunic*	*Leg type*	*Belt*	*Other objects*
Lundeborg 98	Dnf.98/90	M/F	–/A16	C7	E7	L3/L5		
Lundeborg 99	Dnf.99/90	M/F	–/A5		ND	L?/L5		
Lundeborg 100	Dnf.100/90	M/F	–/A5	C4	E1	L3/L5		
Lundeborg 101	Dnf.101/90	F/M		C1	E1	L5/L3		

Table 2.9 Costume analysis of the gold-foil figures selected for this study from Sorte Muld in Denmark, male (M), female (F), not dressed (ND).

Find ID	*Gender*	*Outerwear*	*Dress*	*Tunic*	*Leg type*	*Belt*	*Other objects*
Sorte Muld (NM36)	F/M		C5	E8	L5/L5		
Sorte Muld 1	M	D1			L3		staff
Sorte Muld 2	M	D4			L5		club
Sorte Muld 3	M	D1			L5		staff
Sorte Muld 4	M	D1			L3	X	staff
Sorte Muld 5	M	D1			L3		staff
Sorte Muld 6	M	D1			L5		staff
Sorte Muld 7	M	D1			L3		staff
Sorte Muld 9	M	D4			L5		staff
Sorte Muld 10	M	D4			L3		
Sorte Muld 11	M	D4			L3		
Sorte Muld 12	M			E1	L3		staff
Sorte Muld 13	F	A6x	C1		L5		fibula
Sorte Muld 14	M	D4			L5		staff
Sorte Muld 15	M			E4	L5		staff
Sorte Muld 16	M	D1			L5		
Sorte Muld 17	M			E8	L3		staff
Sorte Muld 18	M	D5			L3		staff?
Sorte Muld 20	M?	D4			L5		staff?
Sorte Muld 21	M			E4–E8	L3		
Sorte Muld 22	M			E10	L3		
Sorte Muld 25	M			E1	L3	X	staff
Sorte Muld 27	M			E10	L3		staff
Sorte Muld 28	M			E1	L3	X?	staff
Sorte Muld 29	M	A14		E3	L3		
Sorte Muld 30	M			E10	L3		
Sorte Muld 31	M			E2	L3		staff
Sorte Muld 32	M			E2	L3		staff
Sorte Muld 33	M			E8	L1		staff
Sorte Muld 34	M			E8	L3		
Sorte Muld 35	M			E8	L3		staff
Sorte Muld 36	M			E6	L3		staff

(Continued)

Table 2.9 Costume analysis of the gold-foil figures selected for this study from Sorte Muld in Denmark, male (M), female (F), not dressed (ND). (Continued)

Find ID	*Gender*	*Outerwear*	*Dress*	*Tunic*	*Leg type*	*Belt*	*Other objects*
Sorte Muld 38	M			E8	L3		staff
Sorte Muld 39	M			E2	L3		staff
Sorte Muld 40	M	D4x			L1		
Sorte Muld 43	M			E1	L3		
Sorte Muld 44	M			E8	L3	X	
Sorte Muld 45	M			E1	L3		staff
Sorte Muld 46	M			E1	L3		staff
Sorte Muld 47	M			E2	L3		staff
Sorte Muld 48	M			E2	L3		staff
Sorte Muld 49	M			E2	L3		staff
Sorte Muld 50	M			E1	L5		staff
Sorte Muld 54	M			E8	L1		staff
Sorte Muld 55	M			E8	L3		staff
Sorte Muld 56	M			E8	L3		staff
Sorte Muld 57	M			E8	L3		staff
Sorte Muld 58	M			E5	L3		staff
Sorte Muld 59	M			E8	L3		staff
Sorte Muld 60	M			E8	L3		
Sorte Muld 64	M			E4	L3		staff
Sorte Muld 65	M			E4	L5		staff
Sorte Muld 66	M			E4	L3		staff
Sorte Muld 67	M			E4	L3		staff
Sorte Muld 68	M			E8–E6	L3		staff

Table 2.10 Costume analysis of the gold-foil figures selected for this study from Sweden, male (M), female (F), not dressed (ND).

Find ID	*Gender*	*Outerwear*	*Dress*	*Tunic*	*Leg type*	*Belt*	*Other objects*
Brahekyrkan	M/F	–/A1	C1?	E2	L3/L5		
Eketorp A	M			E6	L3	X?	
Eketorp B	M			E6	L3	?	
Eketorp C	F	A5	C7		L5		necklace
Eskiltuna 1	M/F	B1/B1	C4	E1	L1/L5		
Eskiltuna 2	M/F	–/A1	C4	E1	L1/L5		disc fibula
Gullmarsberg	M/F		C1	E1	L5/L5		
Helgö 2	F/M	A7/	C3	E5	L5/L3		disc fibula, necklace
Helgö 3	M/F	–/A10	C1F	E1	L3/L5		
Helgö 4	M/F	–/B3	C1	BB (checked)	L9/L5		
Helgö 5	M/F	–/A6	C8	E1	L3/L5		
Helgö 6	M/F	–/A5	C6	E4	L1/L5		disc-on-bow fibula
Helgö 7	M/F	–/A2	C1	E1	L3/L5		disc-on-bow fibula or necklace

(Continued)

Table 2.10 Costume analysis of the gold-foil figures selected for this study from Sweden, male (M), female (F), not dressed (ND). (Continued)

Find ID	*Gender*	*Outerwear*	*Dress*	*Tunic*	*Leg type*	*Belt*	*Other objects*
Helgö 8	M/F	–/A1	C1F	E1	L3/L5	X/–	
Helgö 9	M/F		C9	E7	L? /L?		
Helgö 10	M/F		C9	BB	L2/L5	X/–	
Helgö 12	M/F	–/B5	C1	E2	L2/L5		
Helgö 13	M/F		C1	E?	L? /L?		beaker?
Helgö 14	F/M	B2/–	C1	E1	L5/L?		disc-on-bow fibula
Helgö 15	F/M	B2/–	C5	E1	L5/L1		
Helgö 16	F/M		C1	E5	L5/L3		disc fibula
Helgö 20	M/F		C9	BB	L2/L5	X?/ –	
Helgö 22	M/F	–/B4	C1	E1	L1/L5		
Helgö 23	M/F		C1?	E1	L3/L5		
Helgö 26	M/K	–/A9	C1F	E1	L3/L5		
Husby	M/F	–/A10?	C1F?	E1–E7	L3/L5		
Slöinge 1a	M/F	B1/B1	C4	E1	L1/L5		
Slöinge 1b	M/F	B1/B1	C4	E1	L1/L5		
Slöinge 1c	M/F	B1/B1	C4	E1	L3/L5		
Slöinge 1d	M/F	B1/B1	C4	E1	L3/L5		
Slöinge 2	M/F	–/A1	C4	E1	L3/L5		disc-on-bow fibula
Slöinge 3	M/F	–/A1	C4	E1	L3/L5		disc fibula
Slöinge 4	M/F	–/A?	C1	E?	L?/L?		
Slöinge 5	M/F		SB	E1	L1/L5		disc-on-bow fibula
Slöinge 6	F/M	A1/–	C4	E1	L5/L1		disc-on-bow fibula
Slöinge 7	F/M	A1/–	SB	E2	L5/L1		
Slöinge 8	F/M	A1/–	C1	E4	L5/L1		fibula
Slöinge 9	F/M	A1/–	SB	E4	L5/L1		disc-on-bow fibula
Slöinge 10	F/M	?	C4?	E?	L5/l?		
Slöinge 11a	M/F	–/A1	C1	E5	L3/L5		disc fibula
Slöinge 11b	M/F	–/A1	C1	E5	L3/L5		disc fibula
Slöinge 11c	M/F	–/A1	C1	E5	L3/L5?		disc fibula?
Slöinge 12	M/F	–/A1	C1	E2–E3	L1/L5	X/–	disc-on-bow fibula
Slöinge 13	F/M		C4	E5	L5/L3		
Slöinge 14	M/F	–/A5	C5	E4?	L3/L5	X?/ –	
Slöinge 15	F	A5	C1		L5		
Slöinge 16	F	A5	C5		L5		disc fibula, dress pin, necklace
Slöinge 17	M			E10	L3	X	
Slöinge 18	M			E8	L3	X?	
Slöinge 19	M			E8	L3		
Slöinge 20	M?			E1–E8	L3		
Slöinge 21	M?			E5	L?		
Slöinge 22	F/M		C1	ND	L5/L3		
Svintuna	F/M	A8/–	C4	E5	L5/L3		
Ulltuna	M/F	–/A1	C4	E4	L3/L5		

Table 2.11 Costume analysis of the gold-foil figures selected for this study from Norway, male (M), female (F), not dressed (ND).

Find ID	*Gender*	*Outerwear*	*Dress*	*Tunic*	*Leg type*	*Belt*	*Other objects*
Borg 1	F/M		C?	E1	L5?/L3?	–/X	
Borg 2	M/F	–/A6	C4	E4	L3/L5		
Borg 3	M/F		C4	E1	L3/L5		
Borg 4	M?/M?			E1/E1	L3/L3		
Hauge 1	M/F	A11/A3	C4	E3	L3/L6		penannular brooch/ necklace, branch, appendage
Hauge 2	M/F	A14/A5	C5	E3	L3/L6		
Hauge 3	M/F	A11/A3	C4	E3	L3/L6		penannular brooch/ necklace, branch, appendage
Hauge 4	M/F	A14/A3	C4F	E1	L5/L6		penannular brooch/-
Hauge 5	M/F	A11/A3x	C4		L3/L6		
Hauge 6	M/F	A11/A3	C4F	E1	L3/L6		
Hauge 7	M/F	A11/A3	C1		L3/L6		-/branch, appendage
Hauge 8	M/F	A13/A3	C4	E1	L3/L6		-/necklace, branch, appendage
Mære 1	M/F	–/A1	C1F?	E1	L1/L6		plate fibula
Mære 2	M/F	–/A3	C1	E1	L3/L6		
Mære 3	M/F		C4	E1	L3/L5		
Mære 4	M/F	A11/A1	C1		L3/L6	?/-	-/branch, necklace
Mære 5	M/F	A11/A3	C1F?		L3/L6		-/branch
Mære 6	M/F	A14/A11	C4		L3/L6		-/fibula
Mære 7	F/M		C1	E1	L5/L3	-/X	
Mære 8	F/M		C4	E1	L5/L3		
Mære 9	F/M	A3/–	C9?	E1	L6/L3		
Mære 10	M/?			E1	L1		
Mære 11	M/?			E1	L1	X	
Mære 12	M/F	–/A1	C1	E1	L3/L6		

Table 2.12 Costume analysis of the die stamps and bronze-foil figures selected for this study from Scandinavia, male (M), female (F), not dressed (ND).

Find ID	*Gender*	*Outerwear*	*Dress*	*Tunic*	*Leg type*	*Belt*	*Other objects*
PAT1	M/F		C4	E8	L5/L5		
PAT2	F	A?	C2F		L5		
PAT3	F	A15?	C1		L6		
PAT4	M	D1			L3		staff
PAT5	M			ND	L8		staff
PAT6	M	D2			L5		staff, beaker
PAT7 s	F	A5	C5		L5		drinking horn, branch, necklace
PAT8 s	F	A5	C2		L5		drinking horn, disc-on-bow fibula
PAT9	M	D2			L5		beaker, club, ball
PAT10	M	D4			L5		beaker
PAT11	M/F		C9	TB (dotted)	L9/L5		
BZG1	F	A5	C5		L5		
BZG2	F	A15?	C?		L5		
BZG3	F	A4	C6		L?		disc-on-bow fibula?

Fig. 2.10 Drawings of gold-foil figures from the National Museum of Denmark (drawings: Bjørn Skaarup and Margrethe Watt).

Most Danish finds are dominated by single gold-foil figures, which are especially prevalent on Bornholm while the bulk of the finds from Zealand and Funen consist of double gold-foil figures. Furthermore, single figures are predominantly males, while singly represented female figures only comprise around 1% of the total number. Out of over 430 impresses from Sorte Muld used in this survey, approximately 6% of the gold-foil figures are singly represented female figures and 3% are double figures. The rest are singly represented male figures (Figs 2.13–2.15).

Fig. 2.11 Drawings of gold-foil figures from Gudme, Tørring, Rønne and Lundeborg in Denmark (drawing: Eva Koch; Tørring: after Fischer 1974, 29). *According to Margrethe Watt, these gold-foil figures should now be ascribed to a location on Bornholm, Denmark and be termed "von Melle".*

In Sweden, gold-foil figures have been discovered in 13 different localities (Lamm 2004) (Table 2.3).[4] Furthermore, a gold-foil figure identical to that from Ravlunda was identified in a private collection in Grönsö Castle. A third identical specimen of unknown provenance belongs to the National Museum of Denmark, and all are assumed to originate from the same find (Watt 1989) (Fig. 2.16). The Swedish locality where the highest number of gold-foil figures has been found is Uppåkra, where 122 gold-foil figures have been recorded (Watt 2004), while

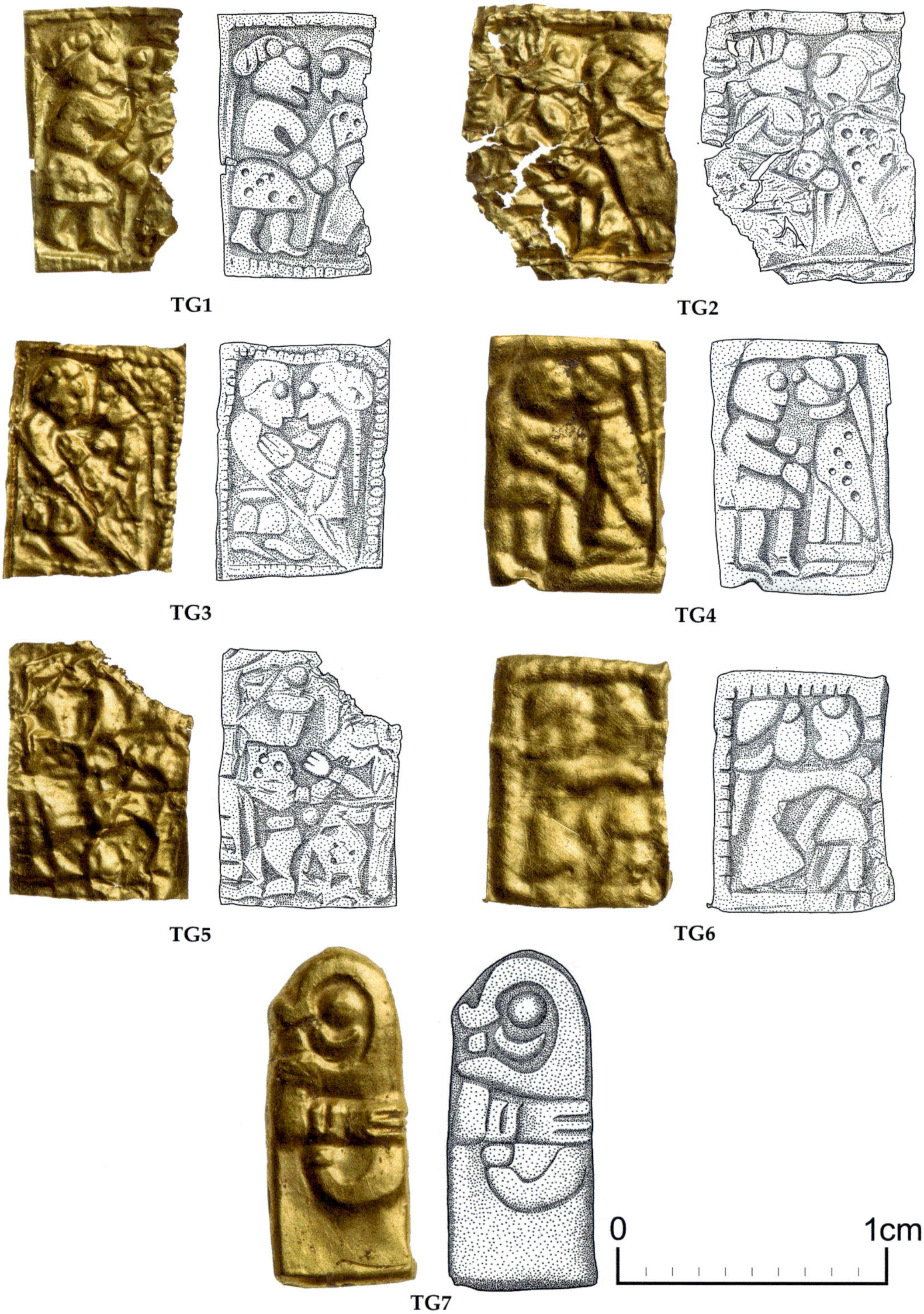

Fig. 2.12 Photos and drawings of gold-foil figures from Toftegård, Zealand, Denmark (Museum Sydøstdanmark; photo: Jens Olsen).

Slöinge contains the second largest number of gold-foil figures in Sweden (Lundqvist 1996a) (Figs 2.17–2.19). In general, the Swedish finds are dominated by single figures that primarily come from Uppåkra, while the material from Slöinge and Helgö is dominated by double figures (Figs 2.20–2.21). Most Swedish gold-foil figures are stamped, and only very few are individually engraved or cut-out figures (Fig. 2.22).

Fig. 2.13 Drawings of gold-foil figures from Sorte Muld, Bornholm, Denmark (drawing: Eva Koch).

Gold-foil figures have been discovered in, at least, eight different locations in Norway, although several of the older finds are no longer preserved today (Stamsø Munch *et al.*, 2003, 249). The 42 preserved Norwegian gold-foil figures are all double figures produced from at least 29 different impresses (Table 2.4). However, it is possible that the finds from Kongsvik and Kirkeseter (Hemne) may also have contained individually engraved

Fig. 2.14 Drawings of gold-foil figures from Sorte Muld, Bornholm, Denmark (drawing: Eva Koch).

or cut out figures (Stamsø Munch *et al.*, 2003, 249; Axboe 1983, 101, 105).

In all of Scandinavia, 11 gold-foil figure die stamps have been found, six in Denmark and five in Sweden (Fig. 2.23), and none in Norway (Table 2.5).[5] Two of the 11 die stamps constitute double figures, while the rest are single figures. Both in terms of style and geographical distribution, these finds are linked to the

Fig. 2.15 Drawings of gold-foil figures from Sorte Muld, Bornholm, Denmark (drawing: Eva Koch).

southern Scandinavian region. Only the die stamps that until now do not have any corresponding gold-foil figures are incorporated in the costume analysis.

Furthermore, four bronze-foil figures from two localities in Sweden and one in Denmark (Table 2.6), produced in a similar manner to the gold-foil figures, are also included in the costume analysis. These bronze-foil figures are all single female figures (Stjernquist 1951, 113–115; Stjernquist 1993a, 27; Lamm 2004, 60, 62–65; Watt 1999b, 178). It is possible that the bronze-foil figures from Sorte Muld and Gårdlösa in Sweden, were used as pendants or amulets (Stjernquist 1993a, 23–24; Lamm 2004, 60). A mounting for attachment may also be observed on rare occasions on the back of single gold-foil figures, such as those from Hög Edsten, Bolmsö and some from Sorte Muld (Lamm 2004, 62–65; Watt 1992, 204–206). Such use of the gold-foil figures was undoubtedly a secondary adaptation.

The costume analysis

The costume analysis of the Scandinavian gold-foil figures was undertaken according to the method explained in Chapter 1. A total of 384 costumed impresses (Tables 2.2–2.6), which is approximately half of the total number of stamps in Scandinavia, and only gold-foil figures with complete images in terms of surface and size, and recognisable individual parts were employed in the analysis. In the cases of double figures where only one of the figures is clothed, both were recorded, although only the costumed figure was utilised in the analysis. Thus, the purpose of this study was

Fig. 2.16 Drawings of gold-foil figures from Husby, Ravlunda, Eskiltuna, Bolmsö, Brahekyrkan and Eketorp, Sweden (after Lamm 2004).

Fig. 2.17 Drawings of gold-foil figures from Slöinge, Sweden (after Mannering 1998a, 20; drawing: Anders Andersson).

SL10 **SL11a** **SL11b** **SL11c**

SL12 **SL13** **SL14** **SL22**

Fig. 2.18 Photo and drawings of gold-foil figures from Slöinge, Sweden (after Mannering 1998a, 21; drawing: Anders Andersson).

not to present all the Scandinavian gold-foil figures in their entirety, but to employ as many images as possible in the costume analysis. This also explains the discrepancy between the number of recorded gold-foil figures and recorded costume items shown in the different tables.

Of the selected gold-foil figures from Denmark, 245 impresses were deemed suitable for costume analysis. These comprise 72 double gold-foil figures, 145 single male figures and 28 single female figures (Table 2.2), the finds from Sorte Muld and Lundeborg constituting the majority. Most of the images utilised are stamped, while only a tiny proportion is engraved and thus unique. Of the latter, three belong to the von Melle find (Rønne 16, 18 (Fig. 2.11) and 19) and three are from Sorte Muld (Sorte Muld P13, Sorte Muld Q19 and Sorte Muld ris (BMR1191x737R)) (Fig. 2.15). Together with the Danish die stamps and bronze-foil figures a total of 314 figures were recorded (Tables 2.7–2.9). The objects that accompany the figures were also recorded. In the case of female figures, the objects are primarily jewellery, such as necklaces and various fibulae. Neck, arm and foot rings were also recorded, but not included as a parameter, as it was difficult to ascertain if this was part of the costume or the marking of anatomical details. Less than 30% of the singly represented females on the gold-foil figures wear a disc-on-bow, a plate or a disc fibula at the neck. Only two females on double gold-foil figures (3%) wear a disc-on-bow fibula. Necklaces are also rare, with 11% among the single female figures and 6% among females on the double gold-foil figures. Furthermore, drinking horns are only observed on two single female figures (7%), while beakers are more common among single male and female figures. In three instances, beakers are recorded with female figures (11%) and in 13 instances, with males (9%). The latter occur primarily among the kaftan-clad males. Objects that accompany males are primarily staffs, clubs and balls, and in two instances, also weapons. Staffs and clubs are frequent in the groups of kaftan-clad males (70%), but also occur among the tunic-clad males (42%). Penannular brooches are observed on three gold-foil figures from the National Museum of Denmark of which two are die-linked to the finds from Hauge (Figs 2.10 and 2.4).

Of the selected gold-foil figures from Sweden, 101 were deemed suitable for the costume analysis. These comprise 50 double gold-foil figures, 38 single male figures and 13 single female figures (Table 2.3). The Uppåkra finds constitute the largest part of the singly represented figures,

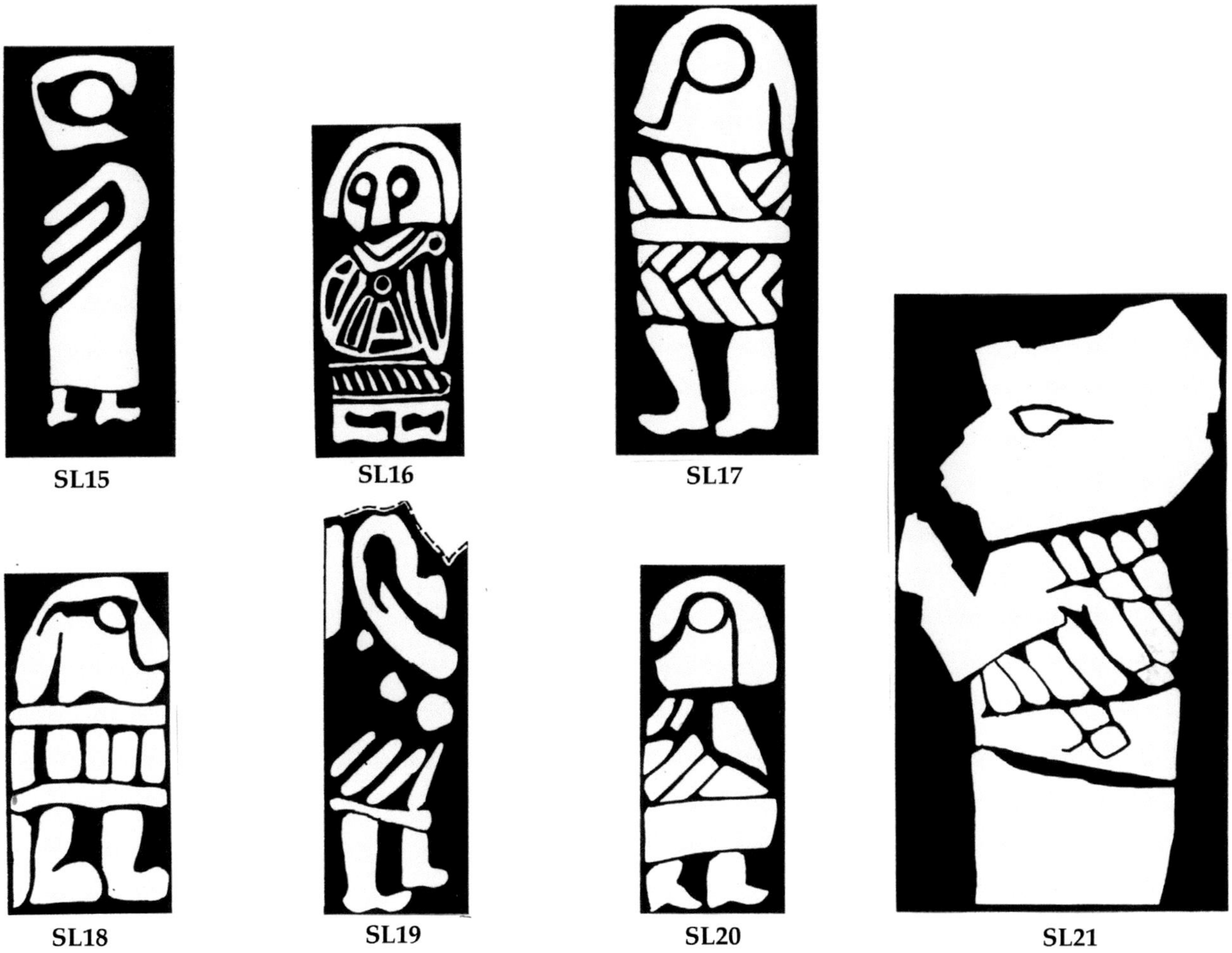

Fig. 2.19 Drawings of gold-foil figures from Slöinge, Sweden (after Mannering 1998a, 21; drawing: Anders Andersson).

while the double figures are primarily found in other Swedish locations. Together with the Swedish die stamps and bronze-foil figures, a total of 158 costume items were recorded (Table 2.10). The objects accompanying the figures are, more or less, similar to those recorded in the Danish area, albeit with a slightly different distribution. Disc fibulae and disc-on-bow fibulae predominate among the females on the double gold-foil figures with 32% of the females on the double figures in comparison to 15% on the single figures. Necklaces occur with females on double and singly represented figures. Among the males a third of those clad in tunics and a little over half of those clad in kaftans carry a staff or club. Only one male holds a weapon, *i.e.* a single-edged sword (Uppåkra 1).

Of the selected gold-foil figures from Norway, 24 double figures were deemed suitable for costume analysis (Table 2.4). A total of 46 figures, 25 male and 21 female, were used in the analysis (Table 2.11). Only a few of the figures from Norway have accompanying objects. Thus, a female has a plate fibula at the neck, while another has an unidentifiable butterfly-shaped object at the neck. This may possibly be a kind of fibula. Moreover, four females may possibly wear necklaces. Among the males, four bear a large penannular brooch at the neck. Other objects observed on several images are branches that most often are held by females.

Altogether, the 11 die stamps and three bronze-foil figures from Denmark and Sweden contain 16 figures, of which seven are males and nine females (Tables 2.5 and 2.6). Of these, two die stamps of single male figures were excluded, as they are already incorporated among the gold-foil figures (PAT4 and PAT9) (Table 2.12). Half the figures on the die stamps and bronze-foil figures have accompanying objects similar to the types seen on the gold-foil figures.

Female outerwear and innerwear

In the Scandinavian female gold-foil figure material selected for the costume analysis, 132 items of outerwear were recorded. They are divided on the basis of their shape into two groups that contain cloaks and jackets.

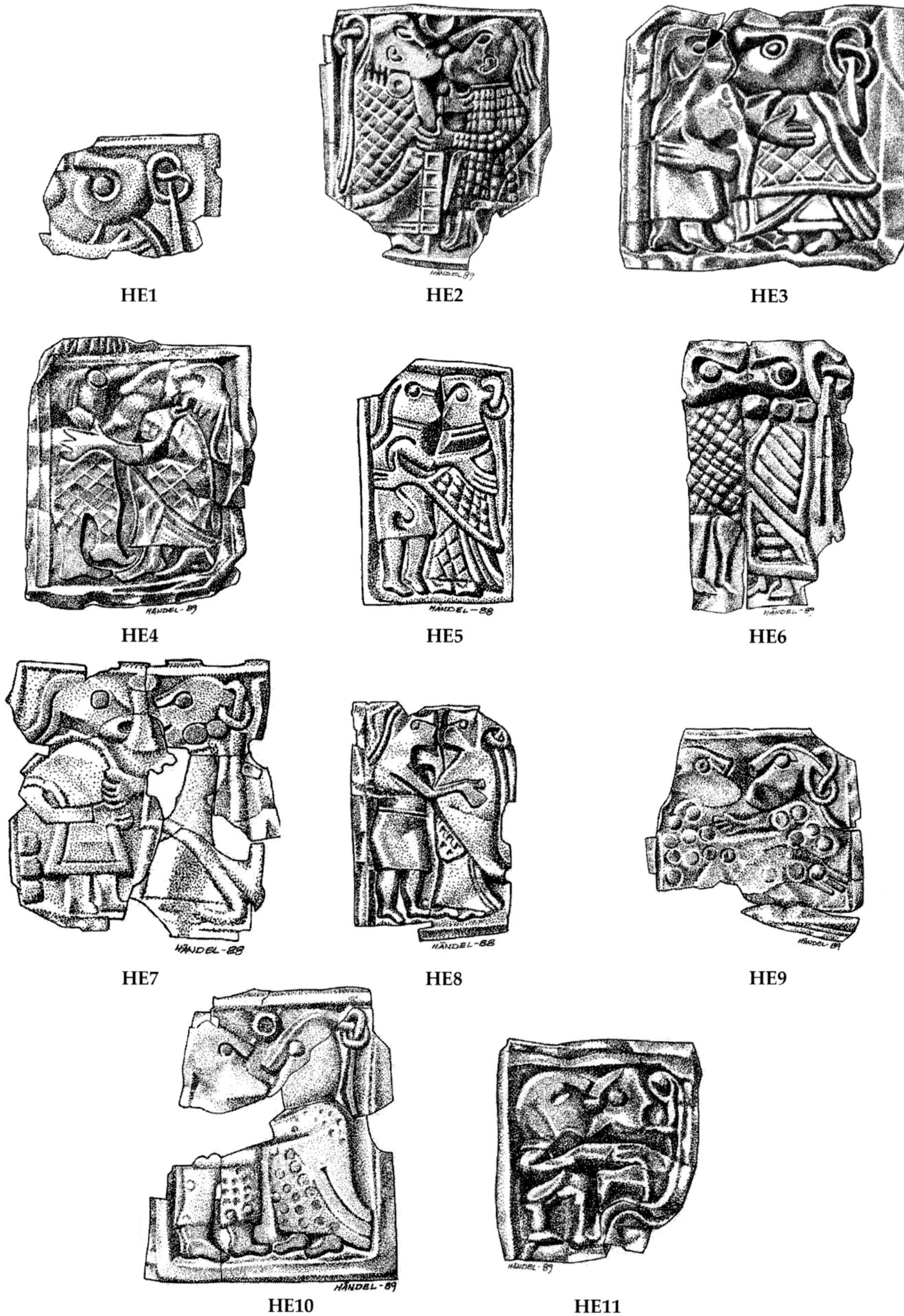

Fig. 2.20 Drawings of gold-foil figures from Helgö, Sweden (Statens Historiska Museum; after Lamm 2004).

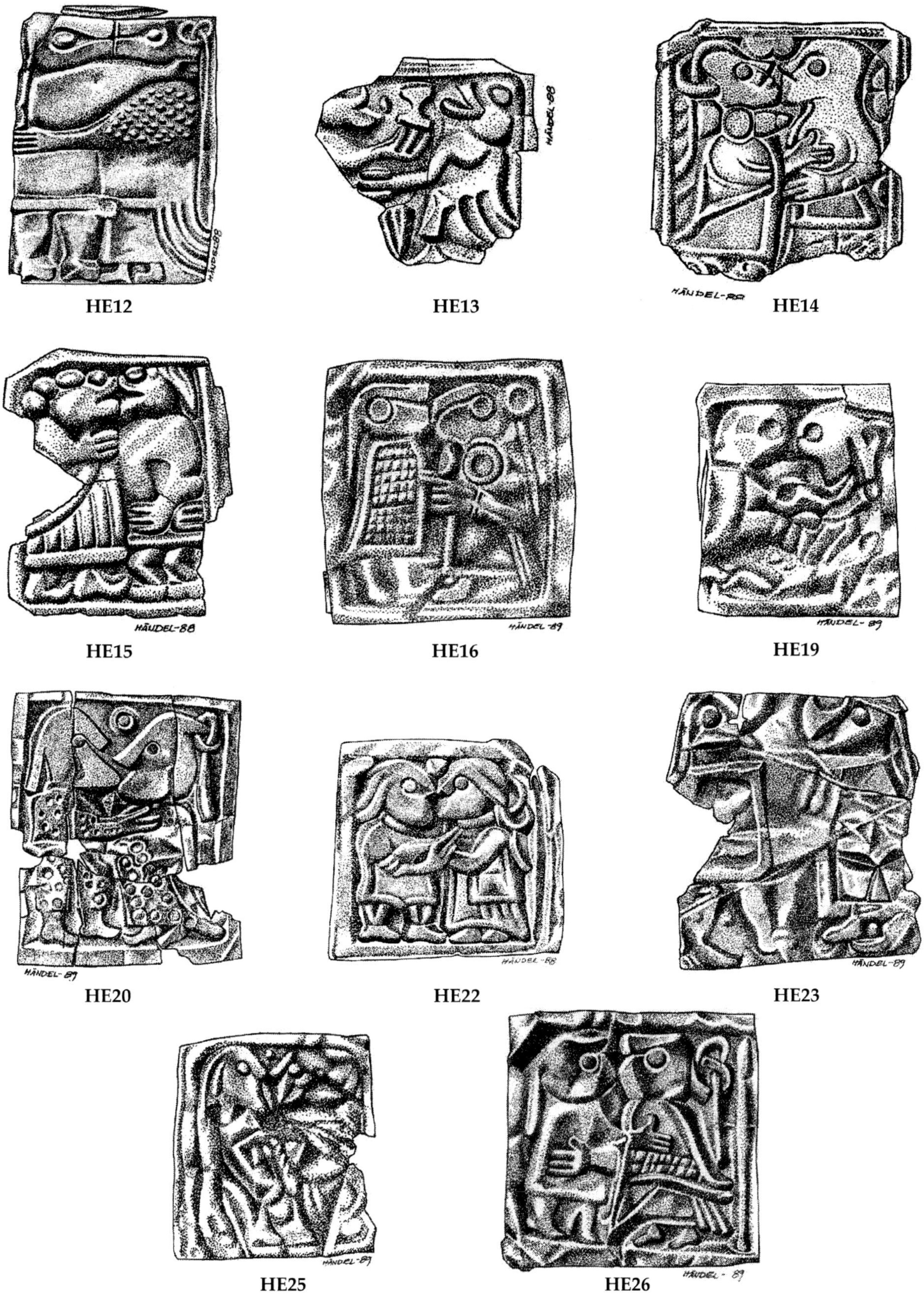

Fig. 2.21 Drawings of gold-foil figures from Helgö, Sweden (Statens Historiska Museum; after Lamm 2004).

UL1 SV1 GL1 HØ1

Fig. 2.22 Drawings of gold-foil figures from Ulltuna, Svintuna, Gullmarsberg and Hög-Edsten, Sweden (Statens Historiska Museum; after Lamm 2004).

Outerwear comprises cloaks that are closed in front at the breast (OG1), which may either have a curved lower edge (OG1a) or be bell-shaped (OG1b), and those that are open in front at the breast and which only rest on the shoulder (OG2); jackets with clearly marked sleeves and a curved lower edge (OG4) or jackets with clearly marked sleeves and a straight lower edge (OG5). Among the females is also a registration of cloak type A11, which is primarily an item of male outerwear (OG3). The cloak is the commonest group of outerwear among females, while jackets occur in one tenth of the material (Fig. 2.24). Here, OG1a is clearly predominant, while the other sub-groups are relatively rare. Geographically, OG1a is, more or less, evenly distributed among Danish and Swedish finds while it is rare in the Norwegian material. Conversely, OG2 cloaks predominate in the Norwegian material, while the Danish examples of this type primarily occur on double figures in the National Museum of Denmark, which may have a connection to Norway. Finally, OG1b cloaks and OG4 and OG5 jackets are dominant in the Swedish finds. The sub-groupings of the female outerwear (OG1–2 and OG4–5) reflect the various modes of costume decoration. The striped, flowing A5 type cloak is the most widespread, followed by the plain (non-patterned) A1 and A3 type cloaks (Figs 1.10 and 2.25). The other cloak types are represented by less than ten of each, and most of them by less than five of each. In the case of jackets, there is a slight predominance of the plain, but rare fringed B1 type, but the other types occur equally rarely. Generally, within the outerwear there is a slight predominance of patterned cloaks in comparison with plain types, while the jacket group consists primarily of plain types. Patterned outerwear constitutes types A4–A8, A10, A15–A17, B3 and B5. Plain outerwear consists of types A1, A3, A11, B1–B2 and B4, while plain items with borders constitute types A2 and A9.

Among the female figures, 187 items of innerwear were recorded. These may be primarily divided into two groups: dresses, and a combination of a skirt and blouse. The dress group can be further divided into ten subtypes according to their patterning or decoration. Generally, the dress completely dominates the female gold-foil figure costumes, while only four female skirt and blouse combinations were recorded. In many cases, the specific shape of the dress is difficult to discern, partly due to the top part of the body being covered by outerwear, and partly due to the type rarely being supplied with details. In cases where the arms of the figures are visible, it appears to be a sleeved costume item. Furthermore, it seems that the dress of the female gold-foil figures only consists of one layer. Both the dress and skirt are long and often only allow the figure's feet to be visible. Of the dress types, the plain or non-patterned C1 type is the most widespread, followed by the vertically-striped C4 type. Then follow C2, the plain dress with a border at the lower hem and C5, the corresponding patterned dress type with a border at the lower hem. The other types which are mainly patterned are represented by less than ten of each type (Fig. 2.26).

When examining the geographical distribution of dress patterns, it is evident that, while the Danish and Swedish finds are, more or less, similarly distributed with a wide variety of types, the Norwegian dress types are significantly less varied with types differently distributed. Thus, the most predominant dress type in the Norwegian material is the vertically-striped, flowing C4 type followed by the plain C1 type. A few examples of C5, the vertical type with a border at the lower hem and the dotted C9 type also occur. The overall distribution of the innerwear patterning demonstrates a clear preponderance of the patterned types in comparison to the plain types and plain types with borders. The skirt and blouse combination is primarily plain. The group of

Fig. 2.23 Photos and drawings of Scandinavian gold-foil figure die stamps and bronze-foil figures (PAT1: after Ulriksen 2000, 6; PAT2-PAT4: after Watt 1999, 179, 182; PAT5: after Nielsen 1999, 80; PAT6: after Watt 2004, 175; PAT7: after Watt 1999, 180; PAT8: after Watt 2004, 188; PAT9: after Watt 1999, 180; PAT10: after Watt 2004, 189; PAT11: after Lamm 2004; BZG1: after Watt 1986, 70; BZG2-BZG3: after Stjernquist 1951; 1993).

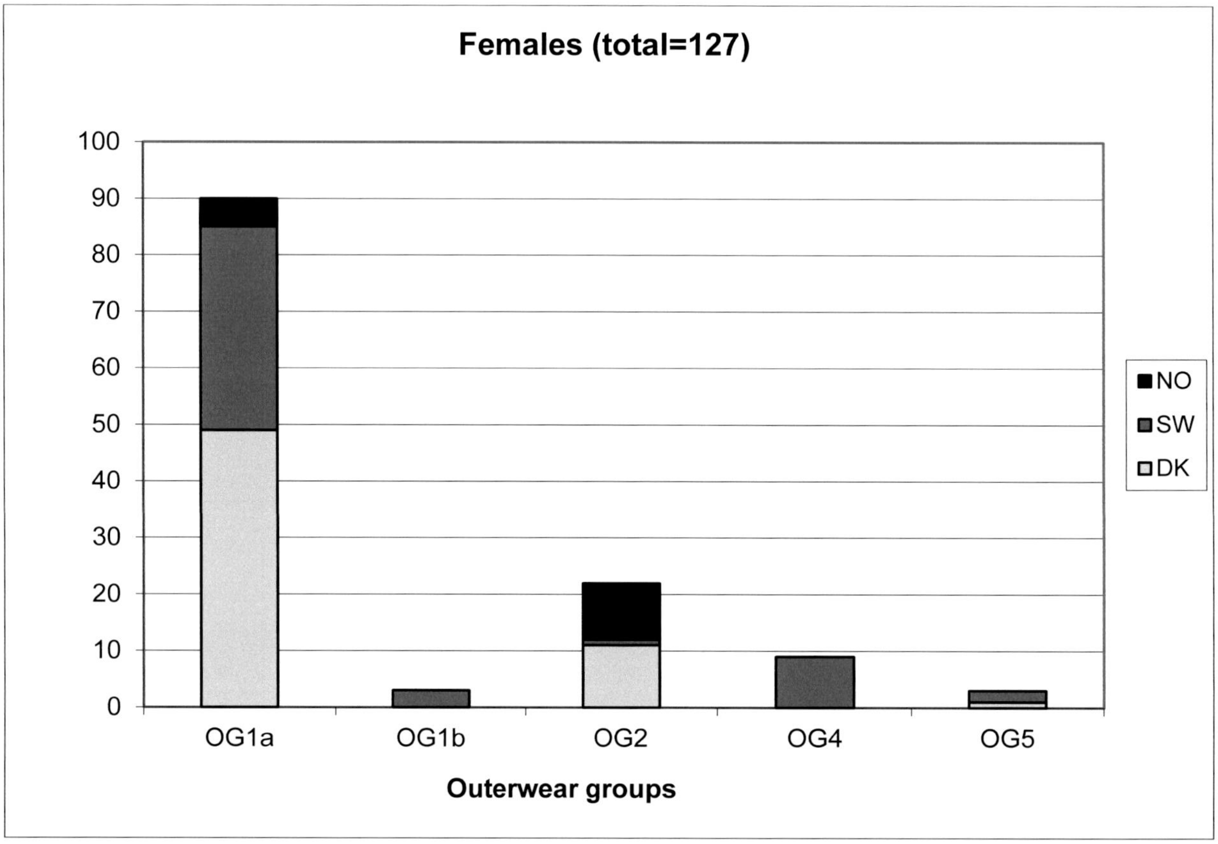

Fig. 2.24 Distribution of the female gold-foil figure outerwear groups in Denmark (DK), Sweden (SW) and Norway (NO).

patterned dresses comprises types C4–C10, the plain type C1 and the plain with borders C2 and C3 (Fig. 2.27). Trains occur on most dress types, with the exception of C2 and C7 (Fig. 2.28). As the dress types C1 and C4 are the most frequently occurring types, they also have the highest number of recorded trains. However, in percentage terms the train is most frequently found in dress type C9 and the skirt and blouse combination, with dress types C1 and C4 following with 49% and 54% respectively. In total, 41% of all dresses have a train.

Female costume combinations

Not all innerwear is combined with outerwear or vice versa. Thus, there are many instances where either cloak type A3 or A5 is the only costume discernible on the gold-foil figures. In total, 68% of all dress types are combined with an item of outerwear (Fig. 2.29).

The dress types that are most often combined with outerwear are C9, C8, C1 and C4. Moreover, the most common combination of cloak and jacket types with the dress or skirt and blouse types comprises plain cloaks with plain dresses (A1/C1 or A3/C1) or plain cloaks with vertically-striped dresses (A1/C4 or A3/C4) (Figs 2.30 and 2.31). These are followed by striped cloaks with vertically-striped dresses (A5/C4 or A5/C5) and striped cloaks with plain dresses (A5/C1 or A5/C2). The rest of the combinations are, more or less, single phenomena (Fig. 2.30). In two cases, a skirt and blouse combination occurs with an A1 cloak. The other two skirt and blouse combinations have no accompanying outerwear.

It is evident from the examination of jackets and dress combinations that a plain jacket with fringes along with a vertically-striped dress (B1/C4) is the most common combination, while the rest are represented by a single case each (Fig 2.31). However, in view of both outer and innerwear having a distribution pattern of a few frequently used types and a smaller selection of more unusual types, the number of variations of outer and innerwear combinations that are portrayed on the images is striking. Presumably, this is a reflection of a desire to make the figures as varied and individual as possible within the framework of contemporary costume conventions. Moreover, seen from the perspective of geographical distribution, there are clear variations in the way the various patterned costume items are combined in the different areas (Figs 2.30 and 2.31). There are only a few combinations that are found in all three areas, thus reflecting an entirely general tendency. The similarity between the Danish and Norwegian finds that may be observed is primarily due to the presence of the double

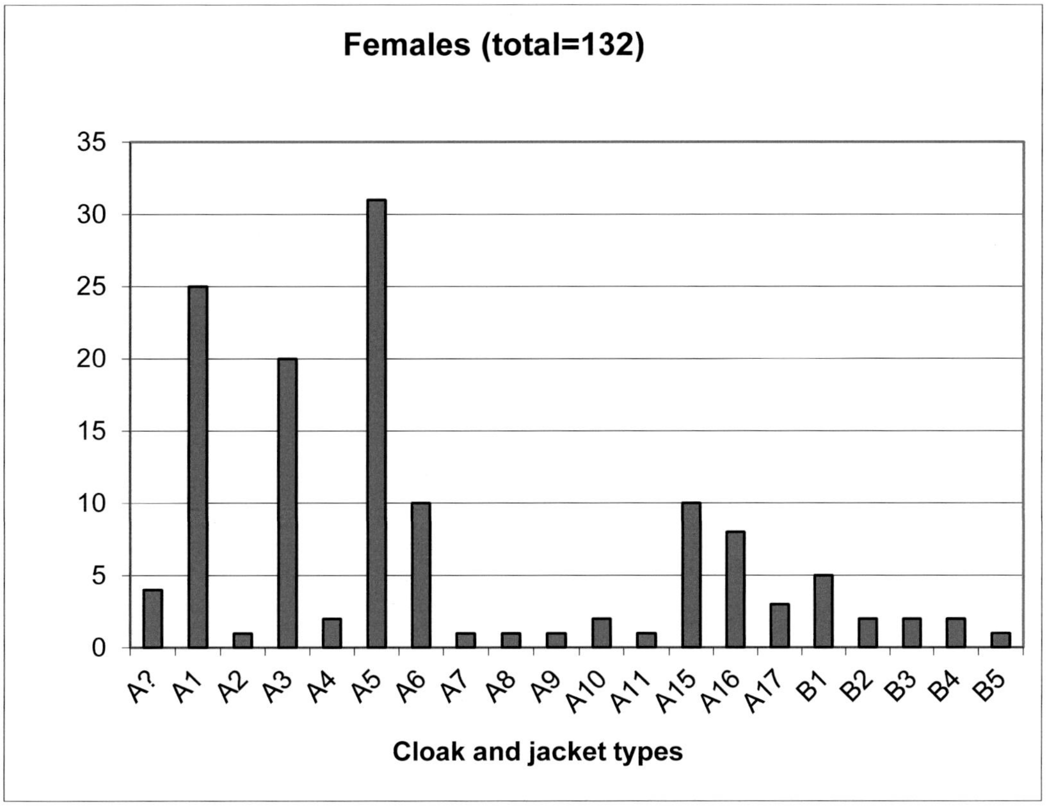

Fig. 2.25 Distribution of the female gold-foil figure cloak and jacket types.

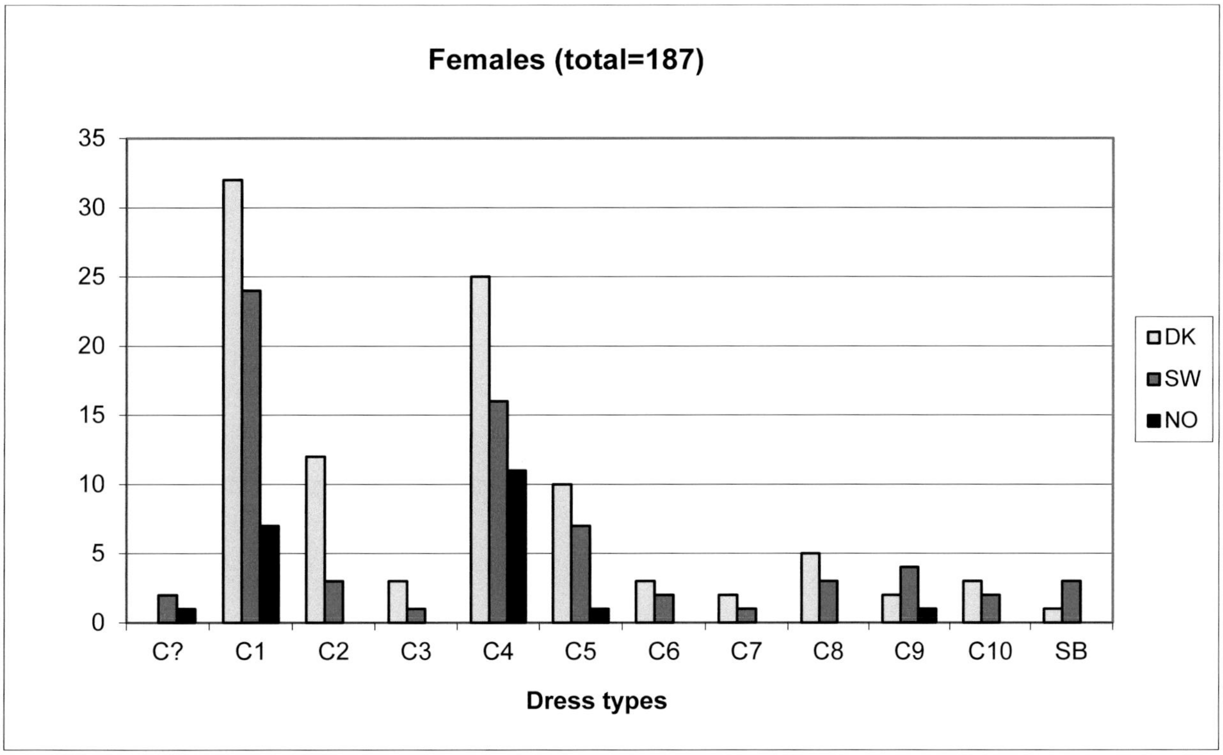

Fig. 2.26 Distribution of the female gold-foil figure dress types in Denmark (DK), Sweden (SW) and Norway (NO).

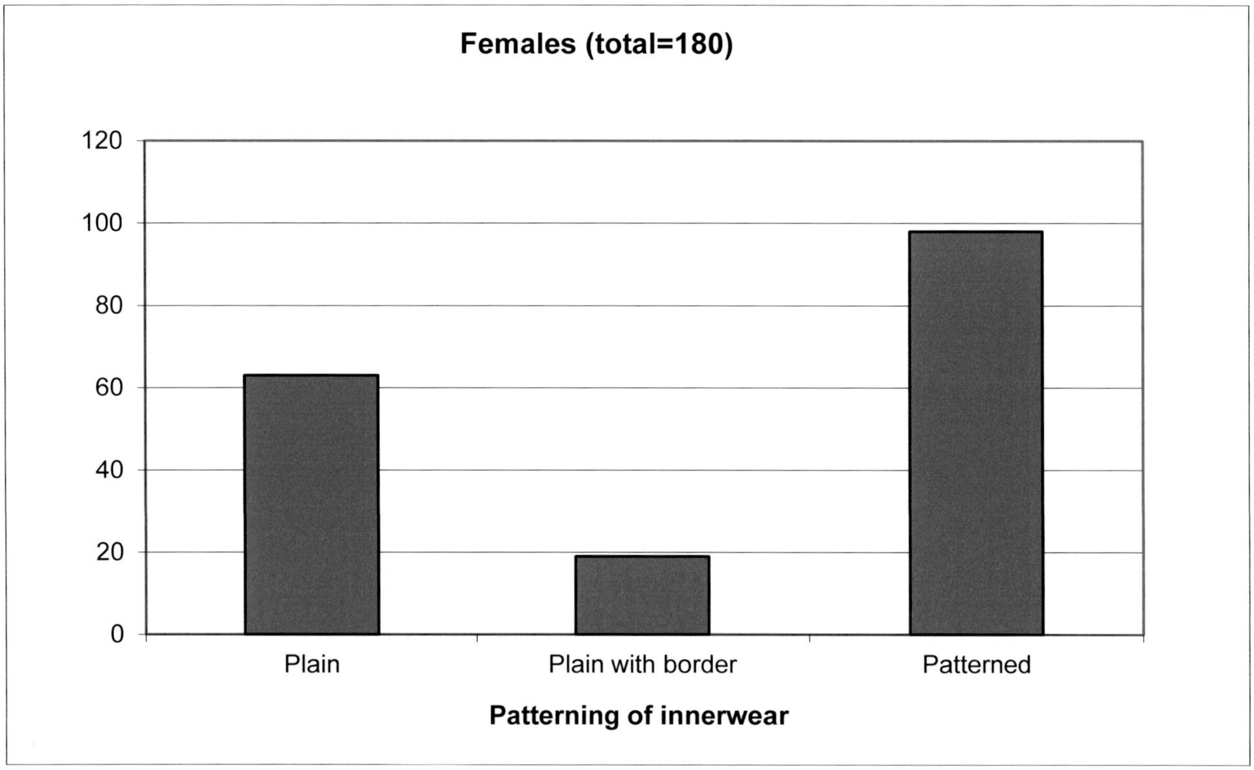

Fig. 2.27 Distribution of the patterning of female gold-foil figure innerwear.

Fig. 2.28 Percentage distribution of the female gold-foil figure dress types with or without trains.

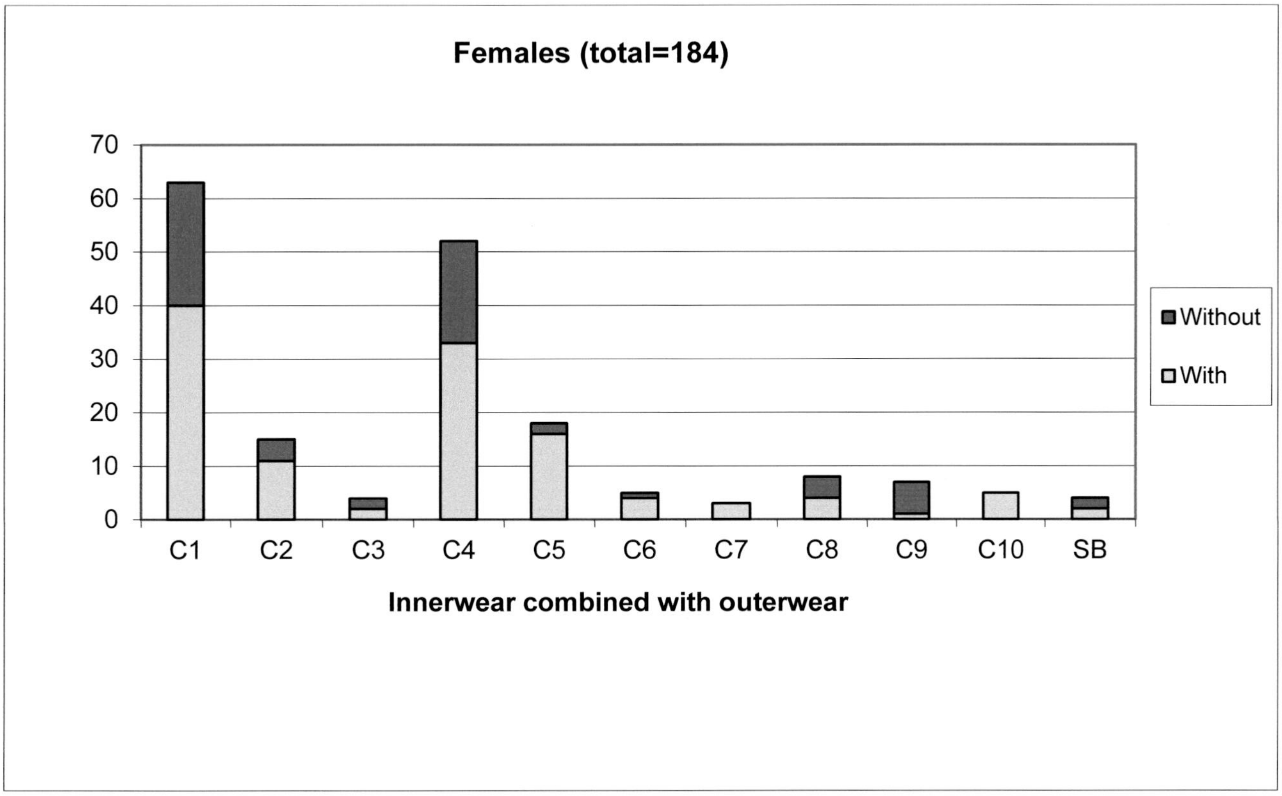

Fig. 2.29 Distribution of the female gold-foil figure innerwear combined with outerwear.

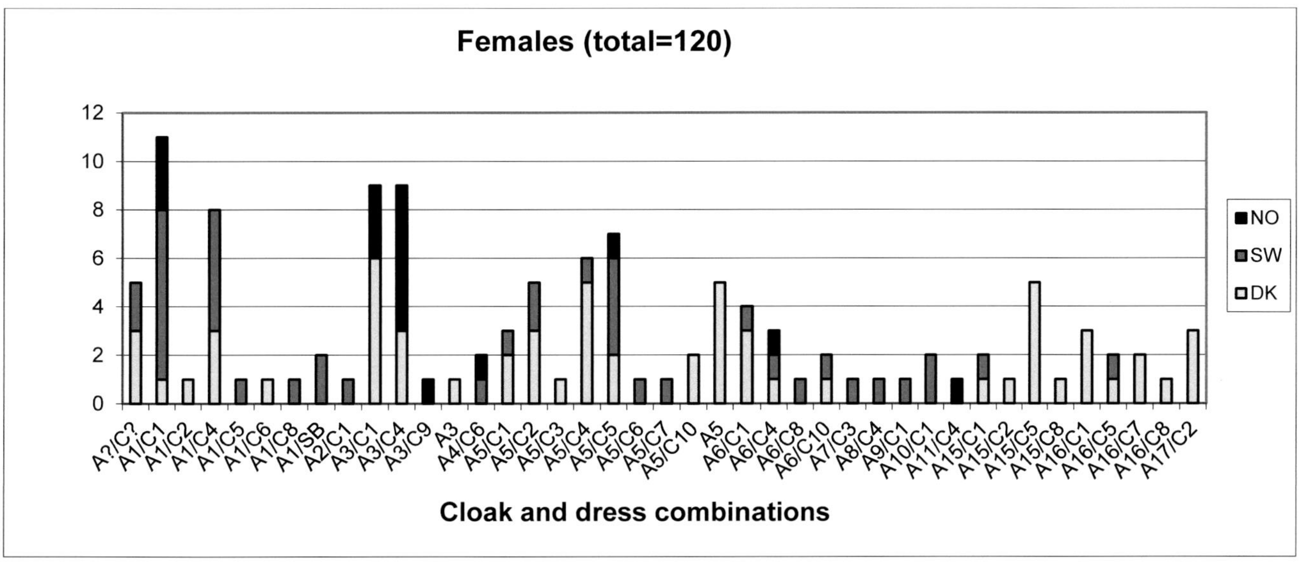

Fig. 2.30 Distribution of the female gold-foil figure cloak and dress combinations in Denmark (DK), Sweden (SW) and Norway (NO).

gold-foil figures in the National Museum of Denmark which have a possible Norwegian provenance. Conversely, there are several similarities between the Danish and Swedish finds, although their distribution is different.

Another element of the female apparel which has been examined more closely is the way the various costume combinations are distributed in relation to whether or not the costume length allows the legs or feet of the figures to be visible. For the sake of clarity, this is shown in relation to the outerwear (OG1–OG2 and OG4–OG5) in combination with innerwear (C or SB). As demonstrated in Figure 2.11, figures with visible feet occur most frequently (Fig. 2.32).

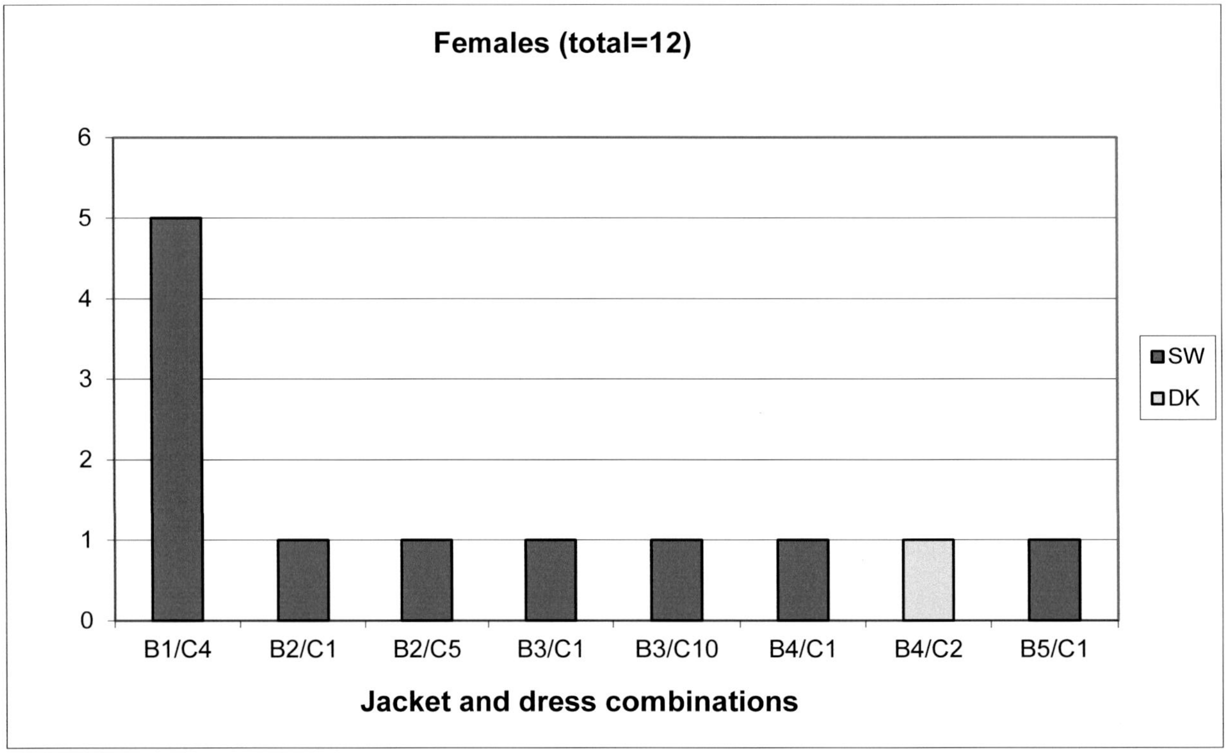

Fig. 2.31 Distribution of the female gold-foil figure jacket and dress combinations in Denmark (DK) and Sweden (SW).

Figures in full-length dress (without feet in evidence) are most conspicuous when Outerwear Group 2 is combined with dresses (OG2/C). Where the Outerwear Group 1 is combined with dresses (OG1/C) and in the cases where the figures are only attired in a dress (C), figures in full-length dress amount to about a quarter of the total. In most cases, the feet of the figures are well delineated, nevertheless, the depictions generally provide no hint of whether or not the figures wear shoes.

A more detailed picture of the female gold-foil figure costume items may be obtained by grouping the costume combinations into some more general categories. These are primarily based on combinations of outer and innerwear, with no allowances made for the patterning variations or differences in leg types (Table 2.13). On the whole, it is evident that the cloak and dress combination, possibly supplemented with an apron, is the most frequently occurring female costume among single and double figures. This combination constitutes 79% of all singly represented females and 51% of all those on the double figures. The same picture is observed when looking at the Danish find material, alone (Table 2.14). Two special costumes may be further added to the costume combinations containing outerwear. The first constitutes females wearing a cloak without visible innerwear and is only found among the gold-foil figures from Lundeborg. The second is a jacket and dress combination primarily found on the double figures from Sweden and a single female figure from Sorte Muld (Sorte Muld 259, Fig. 2.14). Even though there is a general predominance of female figures with outerwear among single and double figures, an almost equal distribution of costume combinations with and without outerwear may be observed in the Danish find material.

On a number of depictions in the cloak and dress and dress only combinations, a special feature worn on top of the dress is observed. This is interpreted as an extra costume item, termed *an apron*. Aprons are recorded on 17 figures, most often on double figures wearing a cloak. While it is difficult to estimate the function of the apron, its presence is so evident that it must be taken into consideration as a separate costume item. Furthermore, the analysis of the female gold-foil figure costumes demonstrates that they consist of several different combinations of outerwear and innerwear, or only outerwear or innerwear. The relationship between the various costume categories in Denmark, Sweden and Norway, varies greatly in the case of the double gold-foil figures, whereas there are more similarities in the costume selection among the single figures from Denmark and Sweden. The costume combinations of the single gold-foil figures from Denmark, however, have clear parallels to the double gold-foil figures from the Swedish area. By scrutinising the selection of basic costume types and their patterning, similarities and differences between the individual areas can be outlined. Thus, the Norwegian

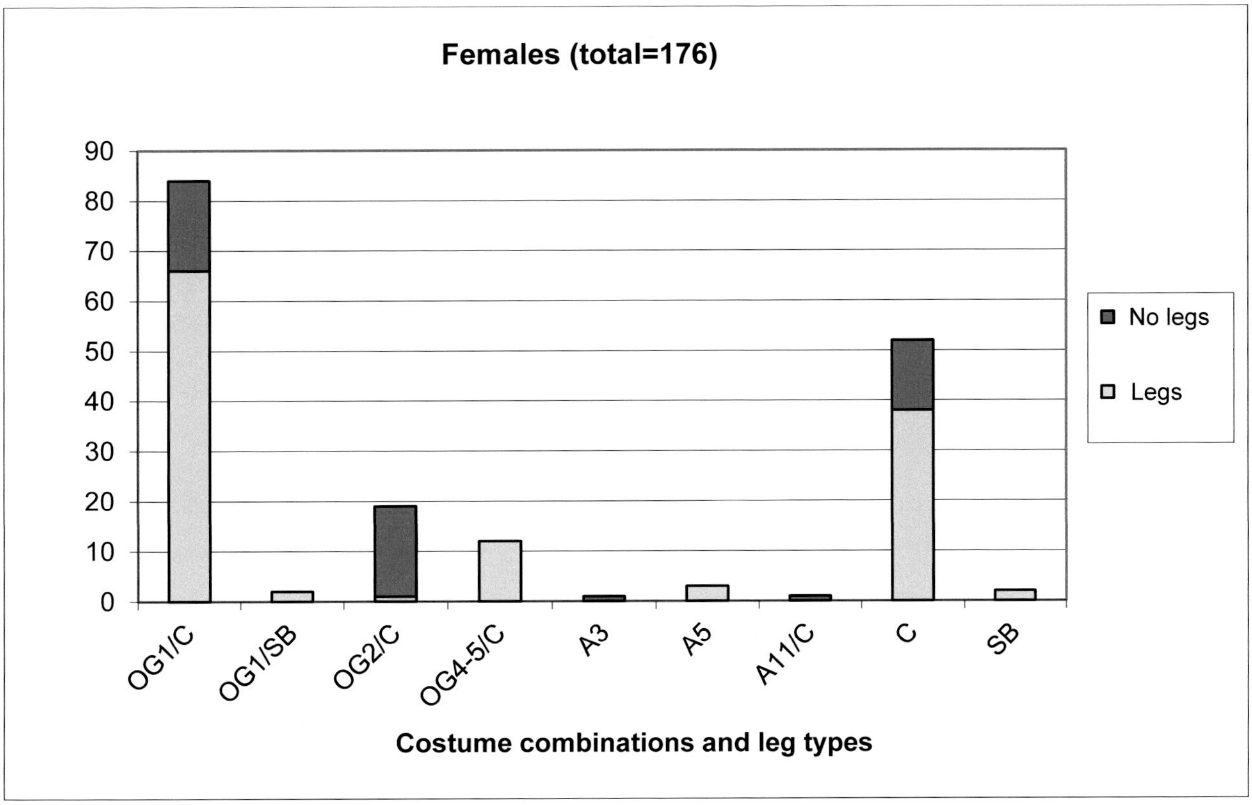

Fig. 2.32 Distribution of the female gold-foil figure costume combinations and leg types.

Table 2.13 Female costume combinations on the gold-foil figures, die stamps and bronze-foil figures from Denmark (DK), Sweden (S) and Norway (N) in absolute and percentage terms.

Costume combination	*Females on single gold-foil figures*			*Females on double gold-foil figures*			*Females on single gold-foil figures (%)*			*Females on double gold-foil figures (%)*			*Total (%)*
Country	S	N	DK	S	N	DK	S	N	DK	S	N	DK	
Cloak						6						9%	3%
Cloak and dress	14		21	21	12	28	82%		68%	41%	57%	38%	50%
Cloak, dress and apron			3	4	4	5			10%	8%	19%	7%	8%
Cloak, skirt and blouse				2						4%			1%
Jacket and dress			1	11					3%	22%			6%
Dress	3		5	12	5	33	18%		16%	23%	24%	45%	30%
Dress and apron						1						1%	1%
Skirt and blouse			1	1					3%	2%			1%
Total	17		31	51	21	73	100%		100%	100%	100%	100%	100%
Costumes in total		48			145								193

female figures wear costume items different to the other areas, both in terms of the selected basic outerwear types and in the frequent use of plain outerwear. A more or less identical picture emerges when considering innerwear, where the Norwegian finds are both less varied and contain other costume items. The greatest similarities are encountered between the Danish and Swedish finds. The Danish finds, possibly due to their greater numbers, are characterised by having the greatest variation in both outer and innerwear, as well as patterning. The Swedish material is distinguished by having special types of outer and innerwear, such as jackets and skirt and blouse combinations.

Table 2.14 Female costume combinations on the gold-foil figures, die stamps and bronze-foil figures from Denmark in absolute and percentage terms with a subdivision of the finds from Sorte Muld (SM), Lundeborg (LU) and the rest of Denmark (DK).

Costume combination	*Females on single gold-foil figures*			*Females on double gold-foil figures*			*Females on single gold-foil figures (%)*			*Females on double gold-foil figures (%)*			*Total (%)*
Country	DK	SM	LU	DK	SM	LU	DK	SM	LU	DK	SM	LU	
Cloak						6						13%	6%
Cloak and dress	9	12		10	4	14	82%	60%		63%	36%	30%	46%
Cloak, dress and apron	1	2		2		3	9%	10%		12%		7%	8%
Jacket and dress		1						5%					1%
Dress		5		4	7	22		25%		25%	64%	48%	37%
Dress and apron						1						2%	1%
Skirt and blouse	1						9%						1%
Total	11	20		16	11	46	100%	100%		100%	100%	100%	100%
Costumes in total		31			73								104

Male outerwear and innerwear

In the Scandinavian male gold-foil figure material selected for the costume analysis, 90 items of outerwear were recorded. They are divided on the basis of their shapes into two groups containing cloaks or jackets. Outerwear comprise cloaks that end in two pointed corners (OG3), jackets with clearly marked sleeves and a curved lower edge (OG4) and jackets with clearly marked sleeves, a straight lower edge and a slanted front edge (OG6), also termed *a kaftan*. In addition, an A5 type cloak that by definition belongs to female outerwear OG1 was observed (Figs 2.33).

The kaftan (OG6) is the item of male outerwear that occurs most frequently. It is represented by 58 specimens divided into five types (D1–D4) (Fig. 2.34). Most of the kaftans are plain bodied, while the characteristic bands may be patterned in several ways. Type D5 has a patterned surface. The most common kaftan type, D4, has a simple border, while the second most common type has zigzag borders. The remaining types are represented by less than ten examples each.

All male cloaks belong to the OG3 group, with the exception of the previously mentioned A5 cloak type, which has pointed ends and closes at the shoulder. The plain A11 type is the most common, followed by the striped, flowing A14 type. The remaining types are only represented by a few examples each. In total, 26 figures wearing this type of cloak are recorded. Most unusual is the jacket type OG4, which is the only jacket type used by both males and females, while OG3 and OG6 are items only worn by males, primarily in Uppåkra (Figs 2.33). Generally, cloaks with pointed ends (OG3) are predominant in the Norwegian material, jackets and kaftans (OG4 and OG6) in the Swedish, and cloaks with pointed ends (OG3) and kaftans (OG6) in the Danish material primarily from Sorte Muld. The Swedish images depicting kaftans are exclusively found in Uppåkra and no kaftan-clad males are depicted among the Norwegian gold-foil figures.

Considering the use of patterns on outerwear, a clear preponderance of plain types over plain types with borders, and least of all of patterned types (Fig. 2.35) is evident. The group of plain outerwear comprises types A11, B1 and D4, while types A13, D1–D3 comprise plain outerwear with borders, and A5, A12, A14 and D5 constitute the patterned outerwear.

Among the male figures, 242 items of innerwear were recorded. These may be divided into two groups: tunics, and a combination of a blouse and trousers. The tunic group may be further divided into ten subtypes according to their patterning or decoration. The specific shape of the tunic is difficult to discern in most cases as the type is rarely provided with details. In the cases where the arms of the figures are visible, the tunics seem to have long sleeves. Tunics can be either thigh-length, knee-length or in rare cases ankle-length, and it seems as if they consist of only one layer. The blouse and trouser combination refers to a costume which bears an iconographic resemblance to the figures not wearing any recognisable costume items, but where, for example a patterning or marking at the wrists, waist or feet indicates that the figures are not naked. In these cases, the combination of a blouse and trousers is defined as a close-fitting costume with a clearly delineated crotch. The most common tunic type is the plain E1, followed by the vertically-striped, flowing E8 type and thereafter E2, the plain tunic with a border along the lower edge

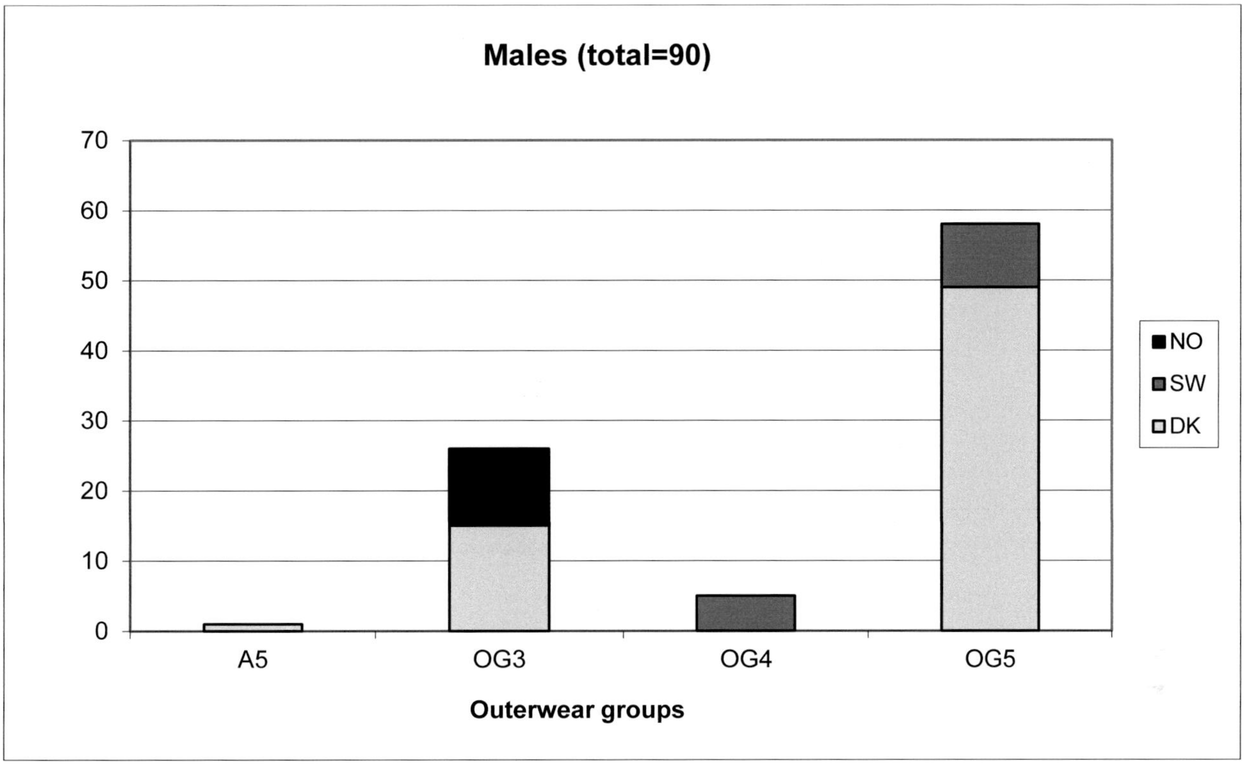

Fig. 2.33 Distribution of the male gold-foil figure outerwear groups in Denmark (DK), Sweden (SW) and Norway (NO).

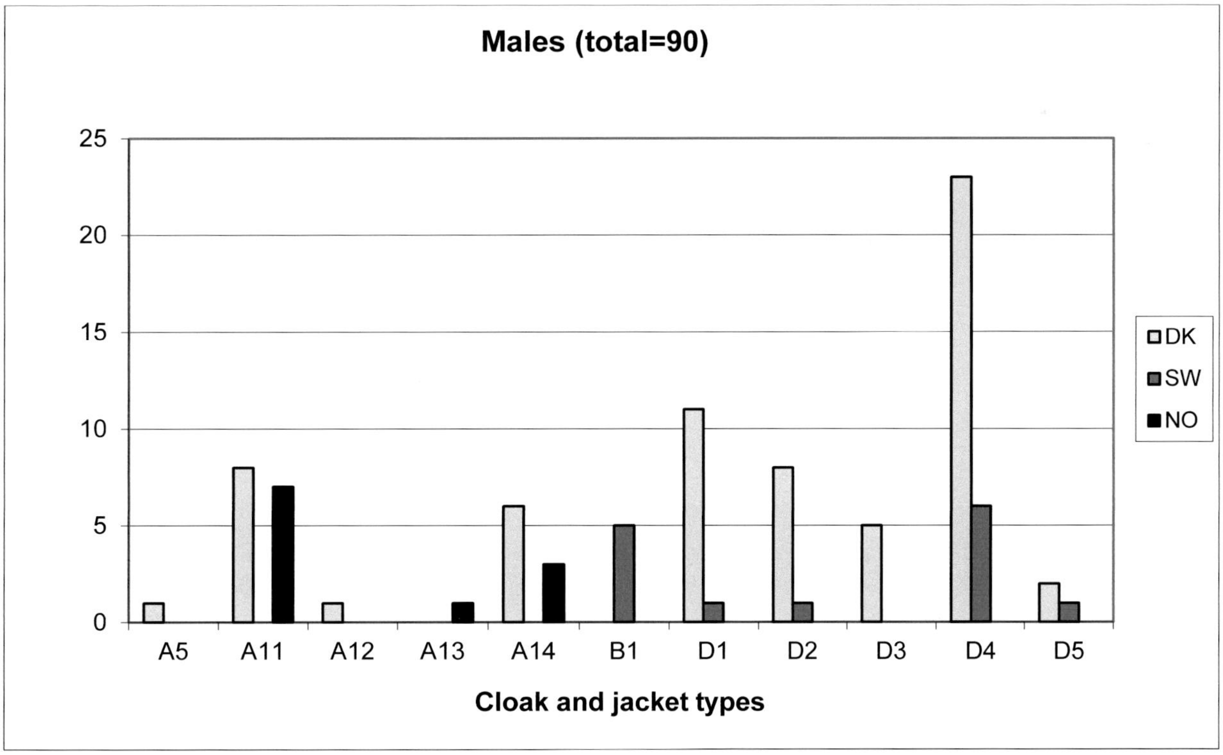

Fig. 2.34 Distribution of the male gold-foil figure cloak and jacket types in Denmark (DK), Sweden (SW) and Norway (NO).

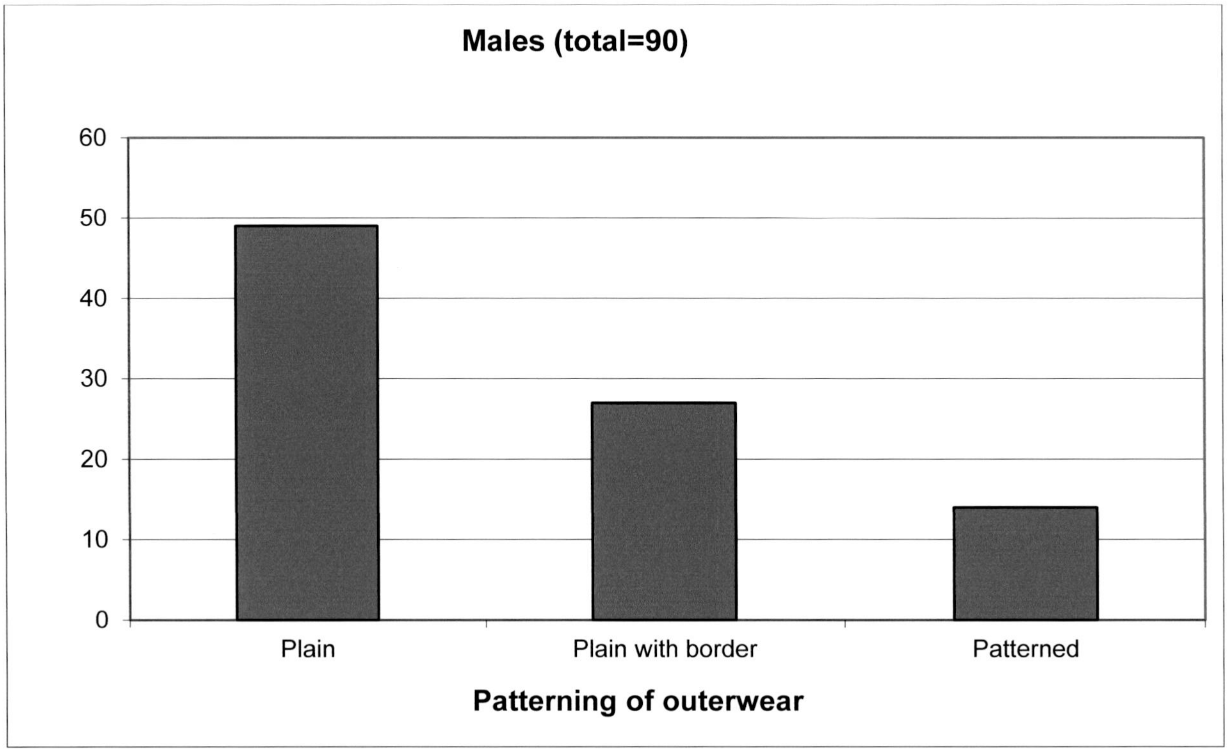

Fig. 2.35 Distribution of the patterning of male gold-foil figure outerwear.

(Figs 2.36). Thus, a preference for the plain tunic type is evident. The diamond or lozenge-patterned types, E4 and E5, are represented by 30 examples, while the remainder are represented by less than ten of each. Generally, there are, more or less, equal numbers of plain and patterned tunic types (Fig. 2.37).

In terms of geographical distribution, the Danish male gold-foil figures display the most varied tunic types, while the Norwegian are the most homogeneous. The distribution of the Swedish tunic types closely resembles the Danish finds, but also includes several of the less distinct trouser and blouse combinations. Six different leg and trouser types were recorded, of which the most frequently occurring type was the simple leg type, with no clearly marked trousers or any noticeable demarcation between leg and foot (L3 and L8). The plain leg type with a marking between leg and foot (L1) is to be found in all areas, whereas the wide or loose-fitting trousers (L2) and the narrow breeches (L4) are only evident in a few examples from Denmark and Sweden, and never occur in the material from Norway. The special patterned trousers with a clearly delineated crotch (L9) are only encountered in two cases from Sweden (Fig. 2.38). Even though the feet are clearly outlined on many figures, in most cases, it is difficult to ascertain whether the figures are wearing shoes or not. Instances where a clear demarcation of the transition from leg to foot is seen are taken as an indication of the presence of footwear.

Male costume combinations

Similar to the female costume, not all male innerwear is combined with outerwear, or vice versa. In total, only 27% of all male gold-foil figures are clad in outerwear, and only 7% wear this over a tunic (Fig. 2.39). Furthermore, only two tunic types (E1 and E3), occur in these combinations.

Belts often occur together with tunics, but, in a few cases, they are also to be seen together with the blouse and trouser combination and with kaftans. As most kaftan-clad males have their arms bent at the waist, the belt may well have been a more commonly used item than is immediately apparent in the depictions.

An examination of the way that the outer and/or innerwear is combined with leg coverings shows that an ensemble of a tunic and plain trousers (L3) occurs most frequently (Fig. 2.40). However, the combinations are geographically distributed differently, and depend on the outer and innerwear types. Thus, in the Danish area, it is primarily costumes, such as kaftan, tunic and/or cloak, which are combined with plain trousers (L3 or L5) or trouser types L1 and L2, whereas in the Swedish area, the recorded

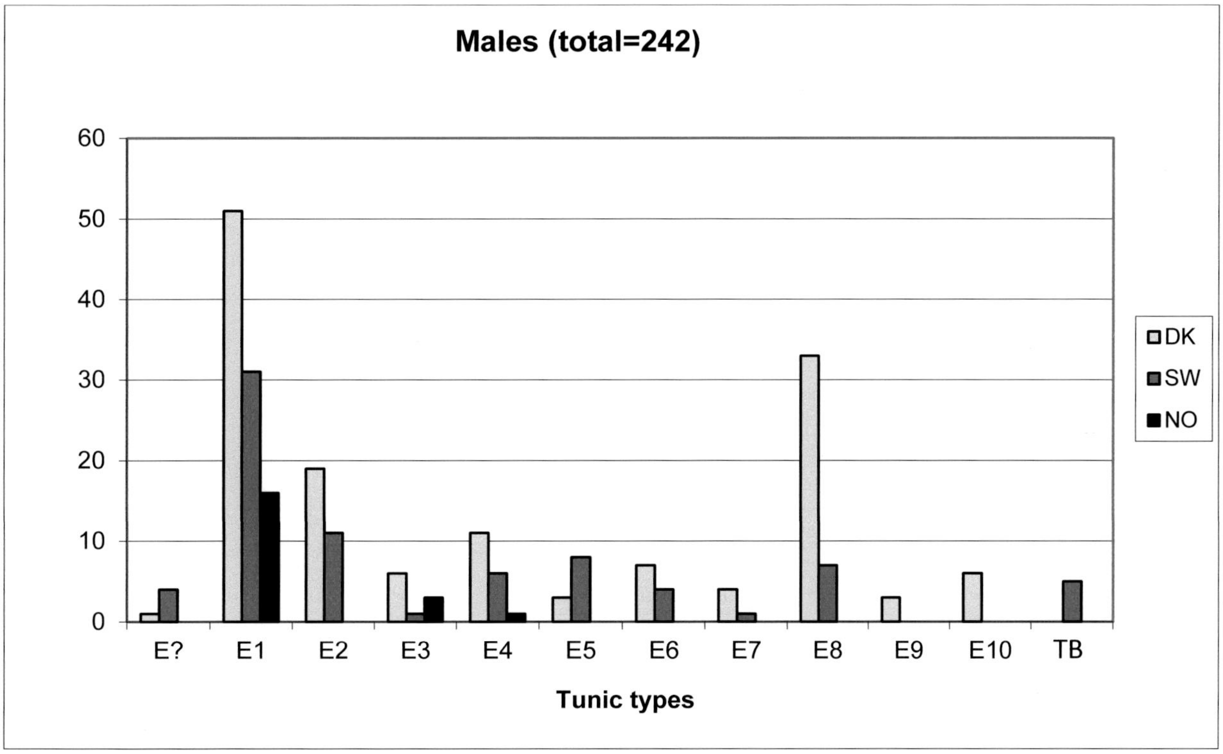

Fig. 2.36 Distribution of the male gold-foil figure tunic types in Denmark (DK), Sweden (SW) and Norway (NO).

Fig. 2.37 Distribution of the patterning of male gold-foil figure innerwear.

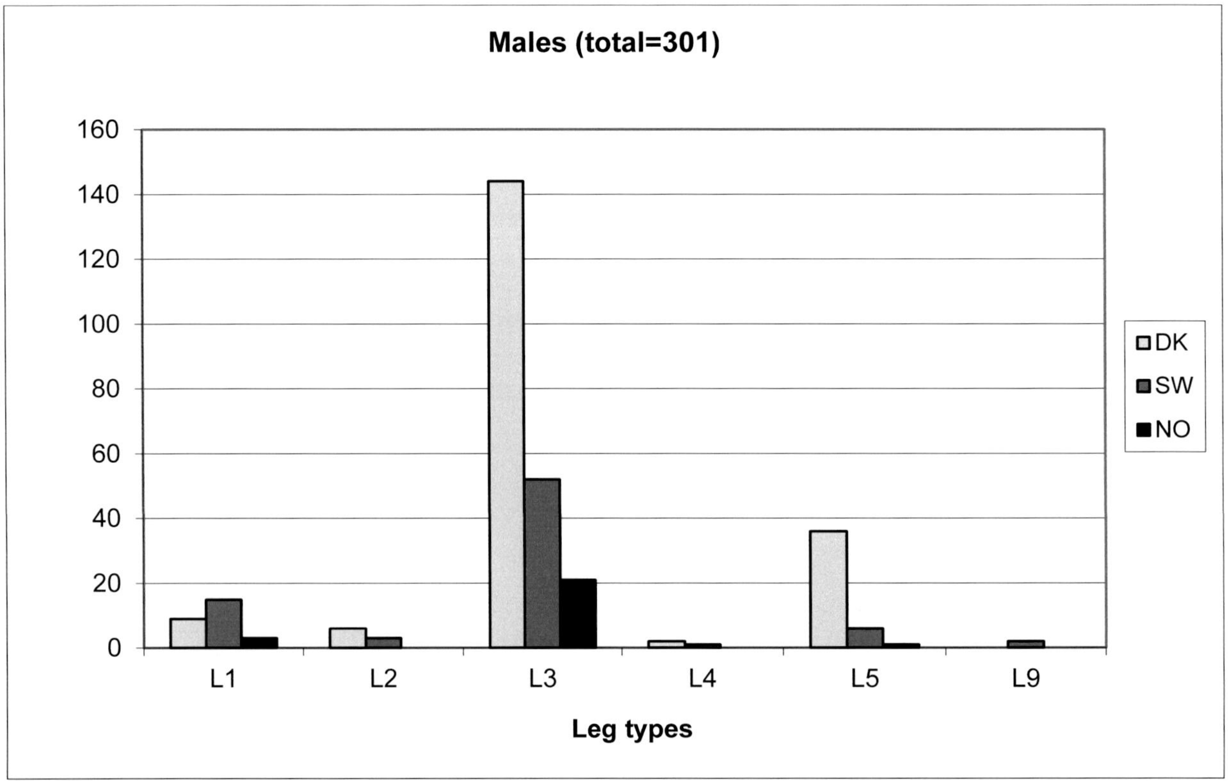

Fig. 2.38 Distribution of the male gold-foil figure leg types in Denmark (DK), Sweden (SW) and Norway (NO).

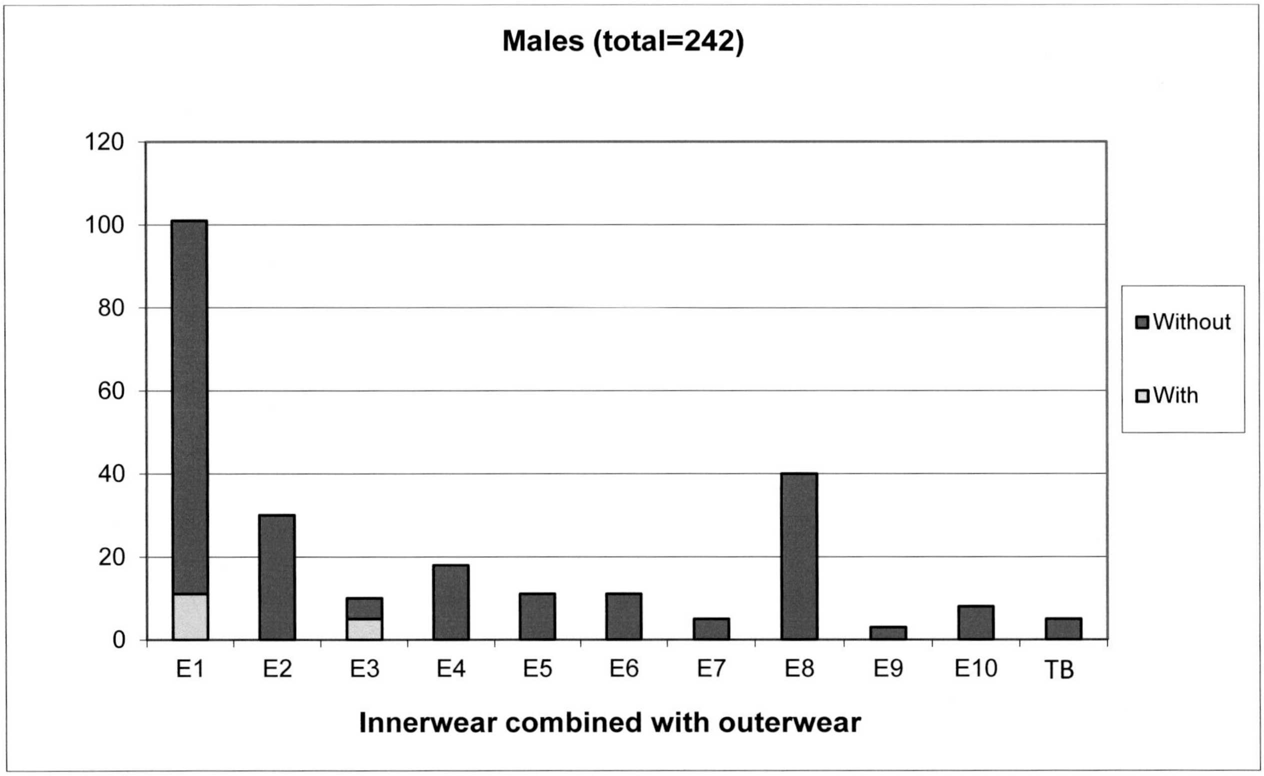

Fig. 2.39 Distribution of the male gold-foil figure innerwear combined with outerwear.

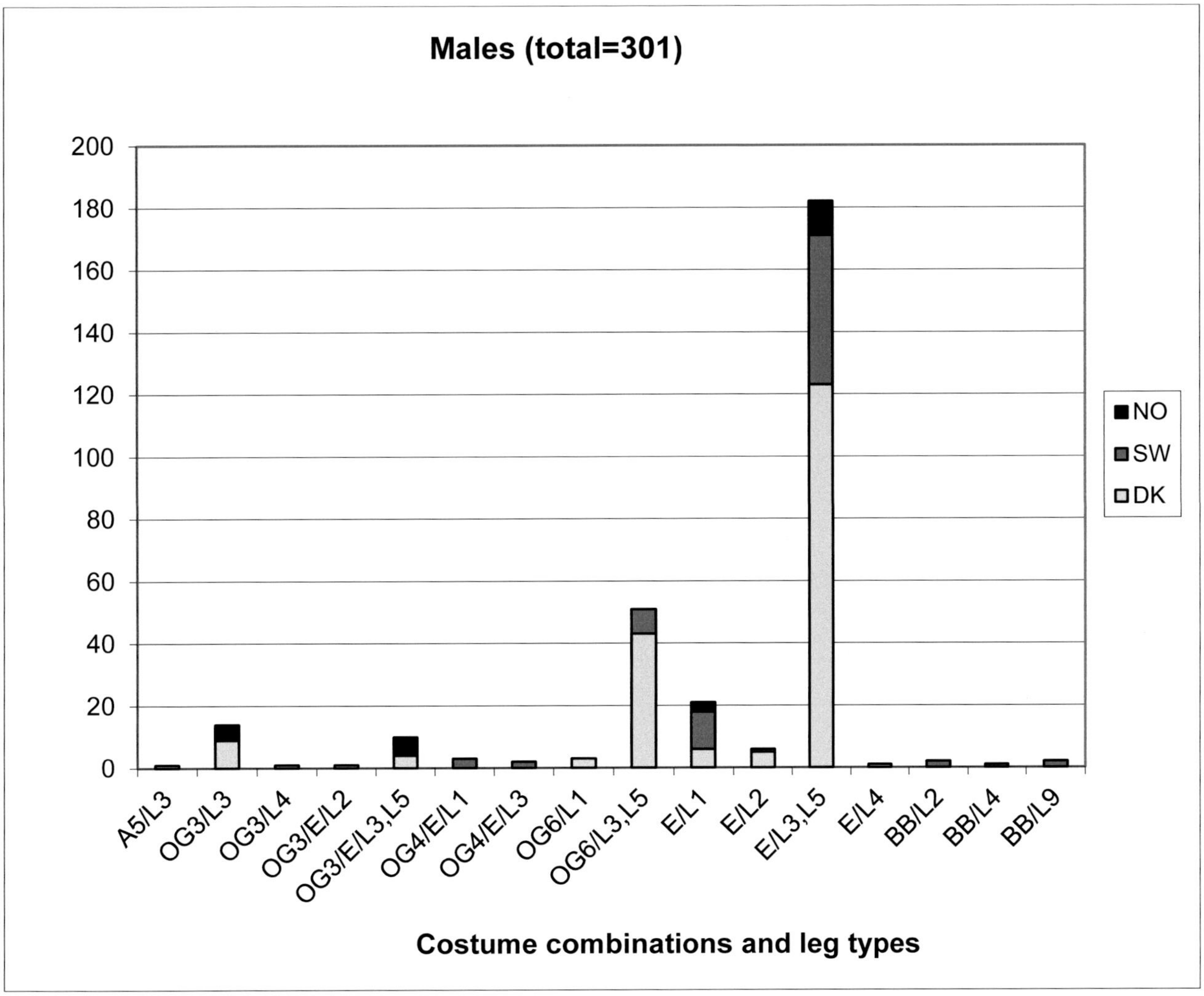

Fig. 2.40 Distribution of the male gold-foil figure costume combinations and leg types in Denmark (DK), Sweden (SW) and Norway (NO).

trouser types occur just as often in the blouse and trouser combinations or with jackets.

A more detailed picture of the male gold-foil figure costume items may be obtained by grouping the costume combinations into some more general categories. The most frequently occurring male costume is the tunic and trouser combination, and it occurs just as often on single as on double figures. This combination comprises 69% of the former and 75% of the latter groups (Table 2.15). The same picture is observed when looking at the Danish material alone where most of the singly represented figures originate from Sorte Muld, and the majority of the double figures from Lundeborg (Table 2.16). In contrast, the kaftan-clad males are found exclusively among the singly represented figures, of which most originate from Sorte Muld and a few from Uppåkra.

With the exception of the kaftan-clad males, outerwear is rare among the male costumes, and only constitutes 10% of the total number of costumes. Three costume combinations that have an uneven geographical distribution are involved here: cloak and tunic, cloak only, and jacket and tunic. The first two combinations are found in Norway and among the Danish examples that possibly originate from Norway. The other Danish examples of the cloaks only combination primarily originate from Lundeborg. Most of the cloak and tunic or cloak only combinations occur among double figures, except in Sorte Muld where they are also found on single figures, with one example of each.

In contrast, the jacket and tunic combination is only encountered among the double figures from Sweden, where they occur together with females in similar jackets. The blouse and trouser combination too, is exclusively found in Sweden on a few single figures, and slightly more on double figures.

The analysis of male gold-foil figure costumes demonstrates that they comprise two characteristic costume

Table 2.15 Male costume combinations on the gold-foil figures, die stamps and bronze-foil figures from Denmark (DK), Sweden (S) and Norway (N) in absolute and percentage terms.

Costume combination	*Males on single gold-foil figures*			*Males on double gold-foil figures*			*Males on single gold-foil figures (%)*			*Males on double gold-foil figures (%)*			*Total (%)*
Country	S	N	DK	S	N	DK	S	N	DK	S	N	DK	
Kaftan	7		47				18%		32%				17%
Cloak and tunic			1		6	4			1%		24%	7%	3%
Jacket and tunic				5						10%			2%
Tunic and trousers	30		96	40	14	46	79%		66%	82%	56%	76%	71%
Blouse and trousers	1			4			3%			8%			2%
Cloak			1		5	10			1%		20%	17%	5%
Total	38		145	49	25	60	100%		100%	100%	100%	100%	100%
Costumes in total		183			134								317

Table 2.16 Male costume combinations on the gold-foil figures, die stamps and bronze-foil figures from Denmark in absolute and percentage terms with a subdivision of the finds from Sorte Muld (SM), Lundeborg (LU) and the rest of Denmark (DK).

Costume combination	*Males on single gold-foil figures*			*Males on double gold-foil figures*			*Males on single gold-foil figures (%)*			*Males on double gold-foil figures (%)*			*Total (%)*
Country	DK	SM	LU	DK	SM	LU	DK	SM	LU	DK	SM	LU	
Kaftan	6	41					50%	31%					23%
Cloak and tunic		1		4				1%		27%			2%
Tunic and trousers	6	90		7	7	32	50%	67%		46%	100%	84%	69%
Cloak		1		4		6		1%		27%		16%	5%
Total	12	133		15	7	38	100%	100%		100%	100%	100%	100%
Costumes in total		145			60								205

types: either the kaftan as a specific item of outerwear or the combination of innerwear and outerwear. As far as the double figures are concerned, the closest similarities in costume combinations are found between Denmark and Norway, while the Swedish material has other combinations, such as jackets and the combination of blouse and trousers. Nevertheless, there are also great similarities between the costume selection seen on the single figures from Denmark and Sweden, and it is among this group of images that the kaftan predominates as outerwear.

Further scrutiny of the selection of various basic types and their patterning shows that the Norwegian material stands out from the other areas in terms of the selection of basic outerwear types and by primarily comprising plain innerwear. A similar picture emerges when the innerwear is examined, as the Norwegian finds are notably less varied, with a different distribution of innerwear types compared to the others. Most similarities are encountered between the Danish and Swedish finds, and, possibly due to the greater numbers, the Danish material is characterised by having the greatest variations of outer and innerwear and patterning, while the Swedish material distinguishes itself by having more special types of outerwear, such as jackets.

Combinations of male and female costumes

In the preceding sections, male and female gold-foil figures were analysed as separate entities without taking into consideration whether they were combined with another figure on the same object or not. A picture of characteristic costume types, costume combinations and costume patterning thus emerged indicating both similarities and differences in the material selected for this study. The next step in the analysis examined the manner in which all costume items and the accompanying jewellery are combined in the images. Due to the quantity of the material selected comprising 317 male costumes and 193 female costumes, a computer-based method of analysis, correspondence analysis[6] was employed to reveal similarities, differences and changes in the composition of data. This method has also been used on other large groups of archaeological material or finds, as for example analyses of grave materials or bracteates in order to create chronological sequences (Jensen and Høilund Nielsen 1997, 37; Axboe 1999; Barfod Carlsen 2002; Jørgensen and Jørgensen 1997).

The elements used in the correspondence analysis are the identified costume items with their subtypes (A, B, C, D, E, and L), and the jewellery. The length of leg for females and the leg/trouser type for males are also included in the analysis, and so are fibulae and brooches that are germane for dating purposes, *i.e.* penannular brooches, disc-on-bow fibulae, disc fibulae and plate fibulae. Two types of analysis were undertaken: analysis of the depictions as a whole and of the two genders separately. This method was chosen in order to relate the figures to their geographical or chronological context, and would not have been possible if the depictions only had been analysed from a gender perspective.

The correspondence analysis took place in several different steps, where the statements of the various elements and combination types were tested. The analysis presented here is thus the result of a lengthy testing process where an increasingly more accurate expression was sought for each test. Many different data combinations were employed in the course of the analytical process, such as the shape of the costume, patterning, geographical area and gender in order to ascertain which characteristics and groupings could optimally illustrate the aim of the analysis.

The result of the correspondence analysis is expressed graphically by either a seriation plot or a parabola (Jensen and Høilund Nielsen 1997, 33, 38–39). A normally distributed parabola thus reflects a uniform and sliding distribution of the different types of figures. However, as archaeological material rarely represents complete data, it is acceptable to work with less perfect parabolae (Jensen and Høilund Nielsen 1997, 49). Thus, in order to obtain a clearer picture, it may be necessary in the course of the analysis process to remove elements, which distort the parabola. In general, these units constitute elements that have few or no similarities with the other units. On a practical level, the process takes place on a step-by-step basis, and is completed when no further improvements can be discerned in the shape of the parabola.

The first correspondence analysis plots the distribution of double gold-foil figures from the entire Scandinavian area in relation to each other, and is in its final stage based on 88 figures, termed *units*,[7] and 24 costume items, termed *types*.[8] As may be seen from the seriation diagrams, the finds from the Central Swedish area, as well as Borg, Uppåkra, Sorte Muld and Lundeborg are grouped on the right side of the diagram (Fig. 2.41). This group is termed Costume Group 1 and is characterised by females with visible legs, trousers with narrow legs, female OG1, OG4, and dresses and tunics with checked patterning or vertical stripes, plain tunics with borders, or diamond or lozenge patterning. To the left of

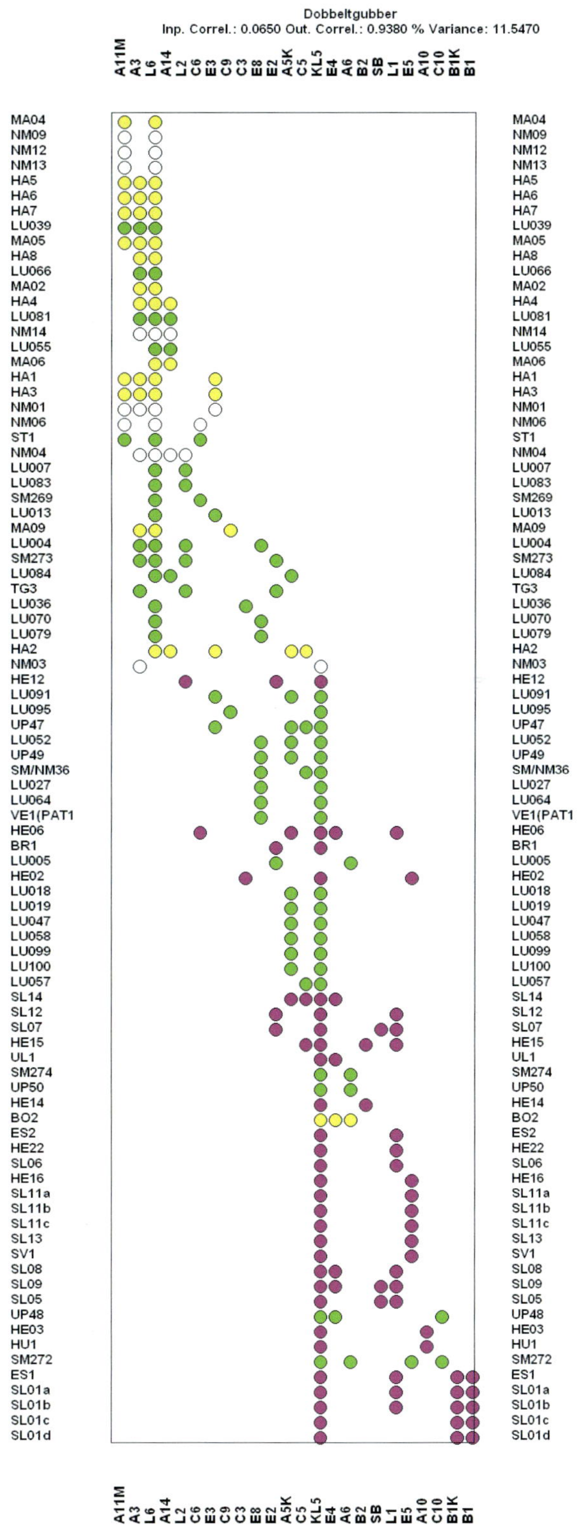

Fig. 2.41 The first correspondence analysis with the completed seriation diagram of double gold-foil figures from the entire Scandinavian area. Based on 88 units and 24 types. The colour key indicating the geographical distribution is as follows: Southern Scandinavia (green), Sweden (pink), Norway (yellow), Unknown provenance (white).

the diagram are the finds from Hauge, Mære, the National Museum of Denmark, Sorte Muld, Toftegård, Lundeborg and Stentinget. Termed Costume Group 2, this group is characterised by females without visible legs, wide-legged trousers, female OG2, male OG3, and horizontal-striped tunics and dresses. Thus, the southern Scandinavian finds are represented in both groups. It is further apparent that female fibula types primarily occur in Costume Group 1, while male fibula types occur in Costume Group 2.

In conclusion, Costume Group 1 is characterised by a great variety of shapes and patterning, many items of female outerwear, ankle-length dresses and female fibulae, while Costume Group 2 is characterised by a different and more limited range of male and female outerwear, floor-length dresses and male penannular brooches.

The second correspondence analysis plots single and double figures from the entire Scandinavian area in relation to each other and is in its final stage based on 314 units and 88 types excluding the jewellery types.[9] As is evident from the seriation diagram, the distribution moves from single female figures to double figures and then to single male figures (Fig. 2.42). Moreover, it appears that the southern Scandinavian finds (primarily Sorte Muld, Lundeborg and Uppåkra) are dispersed over the entire parabola, while the Swedish and Norwegian finds are clustered in the central part. When this cluster is examined further, it proves to contain a mixture of Costume Groups 1 and 2 from the first correspondence analysis. Furthermore, the kaftan types occur in a single cluster amongst the singly represented males followed by singly represented males in tunics, types E2, E4, E8 and E10, which characterise Costume Group 1. The singly represented females lie at the other end of the seriation diagram, and are characterised by females with visible legs, female cloaks A5, A15, A17, dress types C2, C8 and C10, which are types that are rarely represented in Costume Group 1.

In conclusion, the diagram highlights three features: single and double male figures are divided in two separate parts, Costume Groups 1 and 2 are mixed among the double figures, while the southern Scandinavian finds have an even geographical distribution over the entire diagram, compared to the Swedish and Norwegian gold-foil figures that cluster at the centre of the diagram.

The third correspondence analysis plots females on single and double figures in order to scrutinise the difference between these two groups of figures, and the diagram is based on 139 units and 25 types. As may be seen from the seriation diagram, one end consists primarily of females on double figures, while the other end of single female figures (Fig. 2.43). On closer examination, types A5, A15, A17, C2, C8 and C10 are primarily linked to singly represented females, which are termed Costume Group 3. Furthermore, it can be seen that, types A10, B1, B2, B3, C9 and SB are primarily connected to Costume Group 1, while A3, A4, and C6 are connected to Costume Group 2. Types A6, B4, C5 and C7 are common to singly represented females and Costume Group 1, while A1, A16 and KL5 are common to females in Costume Groups 1 and 2. Types C1 and C4 are general types, but are primarily found in Costume Groups 1 and 2, and only rarely among the single female figures in Costume Group 3.

The diagram thus emphasises a tendency towards a division into two parts as observed in the first correspondence analysis, but which in the second analysis was obscured by the presence of the male gold-foil figures. Concurrently, it is possible on the basis of this correspondence analysis to isolate more specific types belonging to single female figures in Costume Group 3. Furthermore, it may be concluded that, based on the first analysis using double figures and the third analysis with females only, it is possible to generate a seriation diagram that reflects a clear grouping of the identified costume items. In the second analysis, the groupings are further differentiated in terms of general characteristics, such as male leg types. The analysis thus provides an explicit formulation of the differences in the selection of costume categories (Tables 2.14 and 2.16).

In conclusion, Costume Group 1 is characterised by and comprises costume types A10, B1, B2, B3, C9, SB, E2, E4, E5, E8, E10, L1 and KL5. Costume Group 2 constitutes of costume types A3–A4, A11–A14, C6, E3, L2 and L6 and Costume Group 3 constitutes of A5, A15, A17, C2, C8, C10 and KL5, while costume types A1, A5, A6, A16, B4, C1, C3, C4, C5, C7, E1, ML3 and ML5 are general types that occur across the groupings. Furthermore, it is possible to constitute a fourth group, termed Costume Group 4 that incorporate singly represented kaftan-clad males, *i.e.* jacket types D1–D5.

Costume Group 1 is thus the group that displays the greatest variety of costume types, but this is primarily due to both male and female varieties being included here. Indeed, Costume Group 3 is the most varied, while Costume Group 2 appears more homogenous and has a limited number of costume types. If a geographical distribution of the units is incorporated, it appears that Costume Group 1 is a mixture of finds from Slöinge, Helgö, Eskiltuna, Ulltuna, Gullmarsberg, Svintuna, Brahekyrkan, Uppåkra, Järrested, Borg, Toftegård, Lundeborg and Sorte Muld. The singly represented males that are mainly found in Sorte Muld and Uppåkra also belong to this group. Costume Group 2 incorporates finds from Hauge, Mære, Toftegård, Lundeborg, Gudme, Stentinget, Sorte Muld and the National Museum of Denmark. Costume Group 3 comprises singly represented females primarily from Vester Egesborg, Rønne, Sylten, Neble, Tørring, Sorte Muld, Uppåkra and Eketorp, while Costume Group 4 consists of singly represented kaftan-clad male gold-foil figures from the National Museum of Denmark, Sorte Muld, Møllegård, Bornholm, Smørenge and Uppåkra.

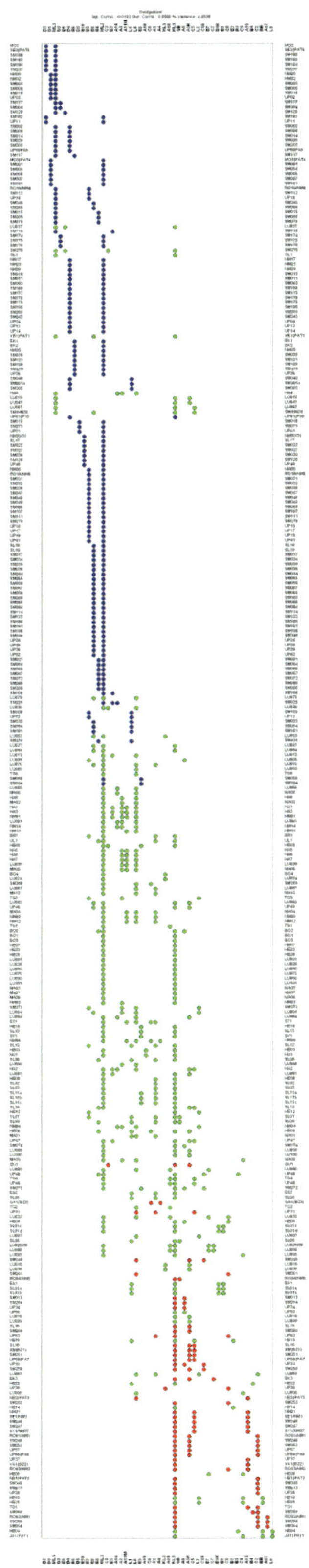

Fig. 2.42 The second correspondence analysis with the completed diagram of single and double gold-foil figures from the entire Scandinavian area. Based on 314 units and 88 types. The colour key is as follows: double gold-foil figures (green), males represented singly (blue) and females represented singly (red).

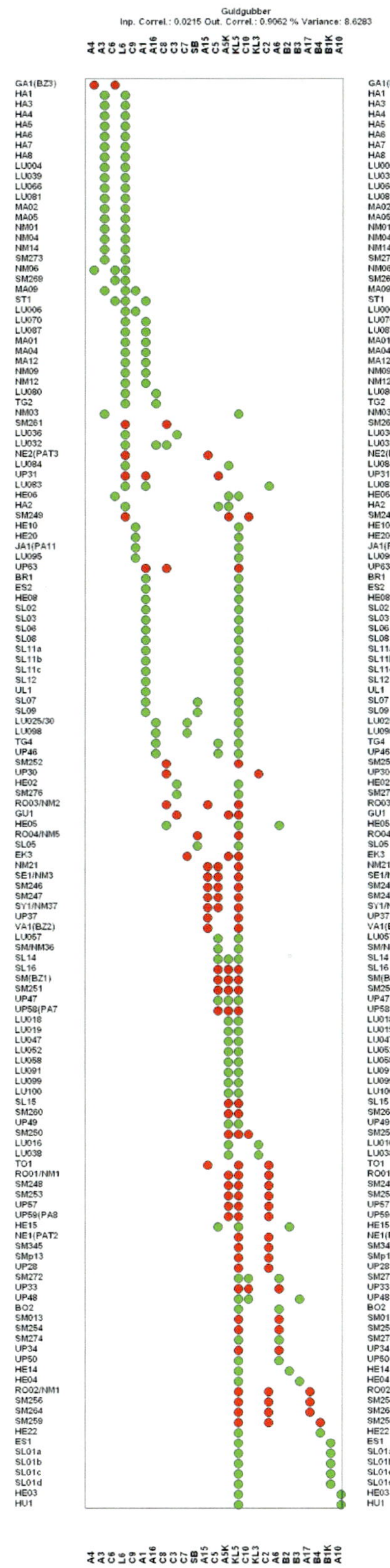

Fig. 2.43 The third correspondence analysis with the completed diagram of single and double female gold-foil figures from the entire Scandinavian area. Based on 139 units and 25 types. The colour key is as follows: single gold-foil figures (red) and double gold-foil figures (green).

Interpretation of the costume combinations

A significant aspect illustrated by the analysis is the manner in which the gold-foil figures that are geographically and possibly chronologically distant are closely related to one another iconographically. It is especially striking how the southern Scandinavian material, with an emphasis on Sorte Muld, Lundeborg and Uppåkra, is equally distributed in all three costume groups, while the Swedish material and Borg and the rest of the Norwegian material, *i.e.* Mære and Hauge, lie separate in the first two Costume Groups.

A method of anchoring the gold-foil figures to a known time period could be, for example to consider the various types of fibulae that accompany the figures. In the case of the female gold-foil figures, these are disc-on-bow fibulae, disc fibulae and plate fibulae that are well known in burials dated to the Late Germanic Iron Age (Jørgensen and Jørgensen 1997, Watt 2004, 215–217). The fibulae are encountered in both single and double female figures, primarily belonging to Costume Groups 1 and 3. However, a single female depiction from Group 2 (Mære 1) wears a plate fibula at the neck, a type which occurs most frequently in the 7th century AD (Jørgensen and Jørgensen 1997, 29, 43, and 59). In the case of male depictions, the penannular brooch is used in varying size and shape throughout the entire Late Iron Age. In the early part of the period, the penannular brooches are small, and worn by both men and women. However, some large and very special penannular brooches that resemble those depicted on the gold-foil figures have been recovered in several hoards dated to the Viking Age. Nevertheless, as the bulk of the fibulae depicted on the gold-foil figurines are oversized, this is not a valid argument for the hypothesis that this jewellery originates in the Viking Age. The penannular brooches are exclusively linked to the male figures on the double gold-foil figures from Costume Group 2, although in archaeological contexts, they are also found in female graves.

Thus, the female objects depicted here primarily link the gold-foil figures to the Late Germanic Iron Age, while male objects cannot be determined more closely than to the Late Iron Age, perhaps with an emphasis on the Viking Age. When this result is compared to the costume groups, it is likely that Costume Groups 1, 3 and 4 are linked to the Late Germanic Iron Age, while Costume Group 2 may perhaps be linked to the Viking Age.

Another method of anchoring the costume groups to a chronological sequence is to relate them to the overall chronological framework sketched in Table 2.1. As the survey indicates, the Scandinavian gold-foil figures are generally difficult to date, but if the overall dating framework is accepted, the bulk of the Scandinavian gold-foil figures must have been produced during the earliest part of the Late Iron Age. Watt dates the eastern south Scandinavian group, with Bornholm and Uppåkra as core examples, to the 6th century AD (Watt 2004, 216), whereas, *e.g.* Lundeborg is dated to the mid-7th century AD. Furthermore, Slöinge and Helgö are dated archeologically and iconographically to the 8th century AD, while Borg is dated to a period between the 8th and 10th centuries AD, and Mære and Hauge to the 9th-10th centuries AD (Lundquist 1996a, 15, 45–47: Lamm 2004, 132; Watt 2004, 216; Stamsø Munch *et al.*, 2003, 13, 252; Lidén 1969, 16).

The meticulous excavation of the more than seven building phases in the cult house at Uppåkra, however, emphasises the problems in dating the gold-foil figures on the basis of the other find material or connecting the gold-foil figures to specific constructions or housing phases. Excavations have demonstrated that the gold-foil figures were, more or less, deliberately deposited in the posthole fills or have ended in the landfill around the posts through the repeated transformations wrought in the various parts of the house (Larsson and Lenntorp 2004, 22–24, 42). With this in mind, it is necessary to review the proposed dating framework with a critical eye. Thus, it is possible that the dating of the gold-foil figures from Slöinge to AD 710–720 (Lundqvist 1996a, 15–17) may be considered a *terminus ante quem*, while in reality, no hint is given as to when they were produced. The same problem applies to the finds from Borg and most probably to those from Mære (Stamsø Munch *et al.*, 2003), where older material is present in both localities, although today, it is not considered significant for the dating of the gold-foil figures. Thus in Borg, directly beneath the hall that is dated to between AD 700–1000, the archaeologist excavated an almost corresponding house dated to the 5th–7th centuries AD (Stamsø Munch *et al.*, 2003, 13–15). Material indicating activity in the Germanic Iron Age was found in Mære as well (Lidén 1969, 13, 21). In an overall context, it is significant for the chronological anchoring of the costume groups that, Costume Groups 1 and 3 too, in this case, show a strong connection to the Late Germanic Iron Age and especially to the 7th and 8th centuries AD, while Costume Group 2 may belong to the Viking or the Late Germanic Iron Ages. Furthermore, it may be concluded that both the depicted objects and the chronological frameworks indicate that Costume Groups 1, 3 and 4 belong to the Germanic Iron Age, and specifically the 7th and 8th centuries AD. Although Borg can place Costume Group 1 further forward in time, the opposite is the most likely. Costume Group 2 which according to the chronological framework belongs to the Viking Age is less substantiated. It is contradicted by both the many gold-foil figures from Lundeborg and the rest of the southern Scandinavian sites, and the female depiction from Mære with the plate fibula from the 7th century (Mære 1, Fig. 2.5). Thus, there are several elements that place the gold-foil figures from Mære and Hauge further back in time together with the other southern Scandinavian sites than there are elements which place them in the latest period.

When the time frames for dating the gold-foil figures are subject to critical review, based on the costumes depicted, there are many indications for placing all costume groups in the Late Germanic Iron Age, presumably with an emphasis on the 7th century AD. Likewise, the correspondence analysis of the gold-foil figure costumes cannot reasonably be said to reflect any chronological sequence that stretches over a longer period. Rather, it is likely that the gold-foil figures were produced within a very short time period, and that the distribution pattern of the costume types that emerges, reflects, in reality, the differences in depiction strategies in the various regions. However, it should be noted that even though the costume analysis indicates that the production of the gold-foil figures may be squeezed into a period of 100–200 years compared to the previously assumed 500 years, it is still a long chronological period for an object without a clear line of development. Perhaps, by supplementing costume research with other elements, a finer calibration may emerge in the future. Even if it is maintained that the Norwegian gold-foil figures belong to a context of the later period of the Late Germanic Iron Age or Viking Age, it may be questioned whether the Norwegian gold-foil figures were actually produced in Norway. If this were not the case, it could explain the resemblance and die link that are observed between the Hauge finds and the double gold-foil figures without provenance in the National Museum of Denmark. The southern Scandinavian finds indicate that it was presumably the owner of the die stamp or patrix who travelled between the various cult and central places and produced the gold-foil figures (Watt 1999b, 188), but this does not exclude the possibility that after they were produced, the gold-foil figures were transported over a longer distance or used in other contexts than is evidenced by the archaeological data. When the distribution of the single and double gold-foil figure depictions in the diagrams is considered, a more obvious interpretation of the analysis results is that they are due to differences in the depiction strategies of the single and double figures. This conclusion is based on the fact that the costume groups are encountered across chronological and regional boundaries.

The different seriation diagrams demonstrate a striking difference in the image selection in the southern Scandinavian, Swedish and Norwegian material, which has an influence on the selection of the iconographical costume types in the various regions. In southern Scandinavia, both single and double gold-foil figures are encountered. Single figures, especially those representing males are, however, most conspicuous in the eastern area, *i.e.* Scania and Bornholm, while the double figures dominate in the western part, *i.e.* Zealand and Funen. The latter group also predominates in the other Swedish and Norwegian areas. When this result is compared to the costume groups, it is evident that the individual image types were subject to strict rules in terms of the careful selection of the content of the images.

The correspondence analysis clearly indicates that different costume types were prioritised and used in the different images, and that the Scandinavian gold-foil figures may be divided into four groups, with specific costume types and ways of decorating the costumes. For instance, the kaftan is only employed on an exclusive group of singly represented males (Costume Group 4). Vertically-striped cloaks, lozenge- or diamond-patterned dresses, checked dresses or dresses with borders characterise the singly represented females (Costume Group 3), while male and female jackets, lozenge or diamond shaped-patterned tunics, checked or vertically-striped tunics, and narrow trousers for men and plain or vertically-striped dresses and possibly plain cloaks for women characterise the double figures from Costume Group 1. Costume Group 2 is characterised by double figures with OG2 for women and OG3 for men, horizontally-striped dresses and tunics, and wide trousers for men.

The correspondence analysis further demonstrates that the groupings do not reflect a chronological sequence, most likely the Scandinavian gold-foil figures were produced in the course of a period of 100–200 years with an emphasis on the 7th century AD.

Conclusion

The examination of the iconographical costume types observed on the Scandinavian gold-foil figures demonstrates that, despite the abundant and geographically widely distributed material, it is possible to describe the figures with a relatively simple system of outerwear and innerwear and leg types. In total, the following costume items were recorded: 17 different cloak types divided into three outerwear groups, ten jacket types divided into three groups of which the third is termed a kaftan, ten dress types and a skirt and blouse combination and ten tunic types and a blouse and trouser combination, which altogether were combined with seven different leg and trouser types. This constitutes a total of 56 different costume types, of which 32 were specifically female and 27 specifically male items (Fig. 2.44).

Most costume types are gender specific, while some are recorded for both genders. Jacket type B1 is the only costume item that is common for both genders, while in a few other cases, a male is attired in a female cloak type (Lundeborg 87) or a female figure is attired in a male cloak type (Mære 6). As these are single occurrences, they were not registered as new and separate types. Furthermore, the analysis of the single elements for male and female figures

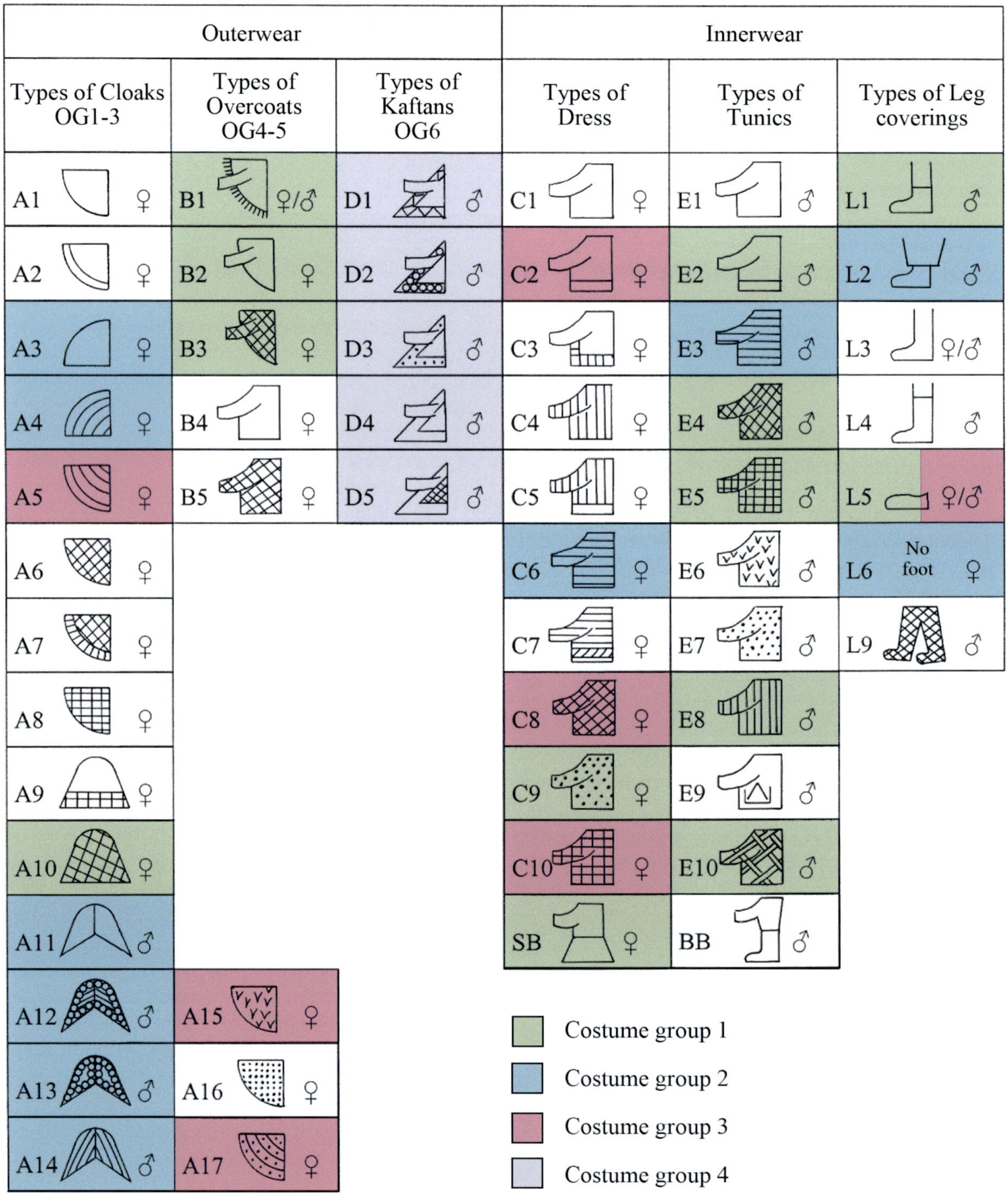

Fig. 2.44 Shape and patterning of the various gold-foil figure costume types with the specification of the Costume Group to which they belong.

respectively demonstrates that there was a specific way of decorating the costumes. The most common are the plain (non-patterned) costume types that sometimes have a border. Next are the patterned costume types that may be subdivided into vertically-striped, horizontally-striped, criss-crossed or checked types. Dotted types, types combining patterns or any of the other patterns are rare. Thus, there is no doubt that gold-foil figures were subjected to clear-cut conventions as regards the appearance of the costumes. This is attested both by the employment of a limited number of costume types closely linked to the two genders and a limited range of types of patterning employed for both genders.

Another important aspect demonstrated by the analysis is that different costume items and costume combinations are prioritised in different areas. In the case of female figures, the Norwegian depictions give priority to Outerwear Group 2 cloaks, and dress types C4, C1, C5 and C9 in that order, while the Danish material accords most importance to Outerwear Group 1a cloaks, and dresses of types C1, C4, C2 and C5 in that order. The Swedish find material is dominated by Outerwear Group 1a cloaks, while Outerwear Group 1b, 4 and 5 cloaks, too, are characteristic here. Concurrently, priority is given to dress types C1, C4, C5 and C9 in that order, followed by the skirt and blouse combination. In addition, the most conspicuous characteristic of the depictions from the Norwegian area is that there are no female figures with legs/feet.

In the case of the males, the Norwegian depictions give priority to Outerwear Group 3 cloaks and E1 and E3 type tunics in that order, while the Danish find material is dominated by Outerwear Groups 3 and 6 cloaks combined with all types of tunics (with the exception of the TBB type). In the Swedish find material, Outerwear Groups 4 and 6 cloaks are predominant together with tunics of type E1, E2, E4, E5, E8 and the trouser and blouse combination.

Generally, plain and vertically-striped patterned innerwear is common to both genders, as is plain innerwear with a lower border. In particular, the plain and vertically-patterned innerwear is so widespread that it falls outside the characterisation of the individual costume types, while the latter are more closely defined by the selection of patterned innerwear. Furthermore, the analysis of individual elements demonstrates that there are only a few similarities between male and female costumes, both in terms of outer and innerwear as well as in leg types. Thus, jacket type B1 of Outerwear Group 4 is the only costume worn by both genders. In contrast, Outerwear Group 6, *i.e.* the kaftan is exclusively worn by singly represented males. In other words, the kaftan only appears on images where there are no accompanying females.

In the case of males, generally the cloak is found primarily on the double gold-foil figures in the Norwegian and Danish material, while jackets are limited to the double figures from the Swedish area. The kaftan is mainly found on the single figures from Scania and Bornholm. The tunic and trouser category is common to both single and double figures in Denmark, Sweden and Norway, but is most homogeneous in the case of the single figures. The combination of a blouse and trousers is particular to the Swedish area.

In general, the open cloak is primarily found on female figures from the Norwegian and Danish area, while jackets are limited to the Swedish area. Other items of female outerwear occur in all areas, but especially on single figures. Furthermore, there is a preponderance of floor-length female dresses in the Norwegian and Danish find material. Thus, it is possible to emphasise both similarities and contrasts among the costumes depicted in the three countries.

Another striking difference between the costumes characterising the two genders, is the extent to which the figure is depicted with or without outerwear. Male costumes with outerwear constitute 27% of all male costume combinations and, of these, approximately two thirds are kaftans. Kaftans are exclusively found on males represented individually, while the other cloak combinations are to be found primarily on double male figures. Females with outerwear constitute on average 68% of the female costume combinations. The Danish double female figures, however, have a more even distribution of females with and without outerwear compared to the other areas, *i.e.* 46% without outerwear and 54% with outerwear.

In the correspondence analysis where male and female characteristics are kept together in the analysis, it was possible to divide the costume types on the single and double figures into four costume groups. This is interpreted as the core of a depiction strategy in relation to the contents of the images. This interpretation is substantiated by the fact that the costume groupings cut across both chronological and geographical divisions. Both single and double figures are found in the southern Scandinavian region. Male representations as single figures are, however, most conspicuous in the eastern region, *i.e.* Scania and Bornholm, while the double figures predominate in the western region, *i.e.* Zealand and Funen. The Swedish and Norwegian areas are dominated by the double figures. When the results of the correspondence analyses are compared to the geographical distribution, it is evident that the different costume types are prioritised in different images. The kaftan is exclusively reserved for males represented singly in Costume Group 4. The vertically-striped cloaks, lozenge or diamond patterned or checked dresses, or dresses with borders are characteristic for the females represented singly in Costume Group 3. Jackets for both genders, lozenge or diamond patterned, checked or vertically-striped tunics and narrow trousers for males and generally plain or vertically-striped dresses and possibly plain cloaks for females are all characteristic for the double figures belonging to Costume Group 1. Cloaks of Outerwear Group 2 for females and Outerwear Group 3 for males, as well as horizontally-striped dresses and tunics and wide trousers for males characterise the double figures from Costume Group 2.

On this basis, it can be concluded that the 317 male and 193 female costume combinations identified in the course of the analysis of the Scandinavian gold-foil figures were subject to strict rules in terms of image content, and especially in terms of the types of costume that were depicted on the various gold-foil figure types.

Notes

1 As of 2014, there were approx. 3100 gold-foil figures from 49 sites in contrast to the 2900 figures from 37 sites on which I based the research for my PhD thesis, pers. comm. Margrethe Watt 2015.

2 As of 2014, there were 24 known localities in Denmark, in contrast to the 15 on which I based the research for my PhD thesis, pers. comm. Margrethe Watt 2015.

3 At the time I wrote my thesis, this object was considered to have been found in Rønne, however, according to Margrethe Watt, these gold-foil figures should merely be ascribed to a location on Bornholm, Denmark and now be termed "von Melle", pers. comm., 2015.

4 As of 2014, gold-foil figures have been found in 18 different localities in Sweden, in contrast to the 13 on which I based the research for my PhD thesis, pers. comm. Margrethe Watt 2015.

5 As of 2014,18 die stamps have been found, in contrast to the 11 on which I based the research for my PhD, pers. comm. Margrethe Watt 2015.

6 The software used for the correspondence analysis, Winbasp from Bonn in Germany was downloaded in January 2006 from: http://www.uni-koeln.de/%7Ea1001/basp.html.

7 The omitted units are HE4, HE9, HE10, HE20, JA1, LU06, LU16, LU25/30, LU32, LU37, LU38, LU80, LU 98 and SM 276. They constitute cloak types A5 and A16 possibly combined with dress types C7–C9 and tunic types E7 and E9 that are somewhat rare.

8 The omitted types are fibulae, A1, C1, C4, E1, ML3, ML5 and MA01.

9 The omitted units are UP16, SM 162 and SMris and the types C1, C4 and E1. The male leg types ML3 and ML5 that were previously omitted in diagram 1 are not omitted here, as far too many singly represented male figures would thus be eliminated.

3

Bracteates

Bracteates are circular gold pendants impressed on one side with a matrix die (Figs 3.1–3.7). A few bracteates are made of silver or bronze. They are important from an iconographical perspective, as they are the earliest examples of highly detailed images from the Iron Age produced in Scandinavia. Bracteates are primarily found in Denmark, Norway and Sweden, but also occur in north and central Germany, the province of Friesland in the Netherlands, and England, as well as in a few other places in Europe. The vast majority of the bracteates are presumed to have been struck in Scandinavia, and are treated in this study as a uniform group despite their scattered provenance (Axboe 2007; Axboe and Kromann 1992, 278). A bracteate may measure up to 12 cm in diameter, its frame included. The centrally impressed motif normally measures 1.5–3.5 cm in diameter and comprises a combination of anthropomorphic figures, animals, symbols, imitation Roman inscriptions and/or runes.

Since the discovery of the first bracteates in the 17th century, research has centred on classification and chronology, and many different systems emphasising various characteristics relating to the composition of the images have been suggested over the years (Montelius 1869; Mackeprang 1952; Malmer 1963; Axboe 2007). The categorisation developed by Oscar Montelius and further elaborated by Mogens Mackeprang, in which bracteates were classified into categories A–F is still used today (Mackeprang 1952; Axboe and Kromann 1992, 278–279). Most bracteates discovered before 1988 are published by Karl Hauck and Morten Axboe (Hauck *et al.*, 1985a–c; 1986a–b; 1989; Hauck 2011). More recent finds have been published in several articles and, in general, this object group is easily accessible for research purposes (*e.g.* Lamm *et al.*, 2000; Axboe 2002; 2007; 2011). However, most of these studies only pay little attention to the costumes on the figures, primarily due to the costume being considered either problematic to interpret or peripheral to the interpretation of the image.

A key point of focus in the study of bracteates is the iconographical identification of the figures and the narrative portrayed in the image. Despite their obvious Roman prototypes, the interpretation of the figures has primarily been based on Old Norse mythology handed down in the written sources from the Scandinavian Middle Ages. Hauck especially successfully employed this complex coupling of text and image, and identified some of the persons depicted on the bracteates. Thus, Hauck's interpretations have been decisive for the understanding of the bracteates. According to him, the vast majority of the human figures that appear on the bracteates are male. Only a smaller number of figures are interpreted as female (Pesch 2002b and Simek 2002).

The majority of the approximately 900 bracteates were discovered in hoards or as stray finds and only a small number were found in other contexts, such as graves.[1] These often gender neutral contexts make it difficult to ascertain if bracteates are a male- or female-related item. In grave contexts, as for example in Norway and Kent in England, bracteates are often found in female graves (Axboe and Kromann 1992, 278). Thus, the gender of the owner does not seem to have any influence on the selection of images.

A- to D- and F-bracteates are dated to the Early Germanic Iron Age and more specifically to the period from AD 450 to *c.* 540 (Axboe 1999, 140–141). However, in order to take into account the long period in which the object was in use, it is crucial to differentiate between production and deposition (Barfod Carlsen 2002, 123). The stylistic-typological division of the A to D-bracteates were long considered to mirror a chronological sequence. However, such a sequence is impossible to document, and the different bracteate types appear to be, more or less, contemporaneous with the aforementioned period of production (Axboe 1999, 134; Barfod Carlsen 2002, 123; Lund Hansen 1976).

A-bracteates display a male head in profile. This is the most obviously Roman inspired type which presumably developed from ideals of military virtue, *i.e.* the emperor as warlord or victor. The heads displayed often bear a characteristic diadem and a circular fibula at the shoulder, presumably inspired by the equivalent imperial Roman symbol of dignity, as for instance portrayed in Roman coinage (Axboe and Kromann 1992, 288–289).

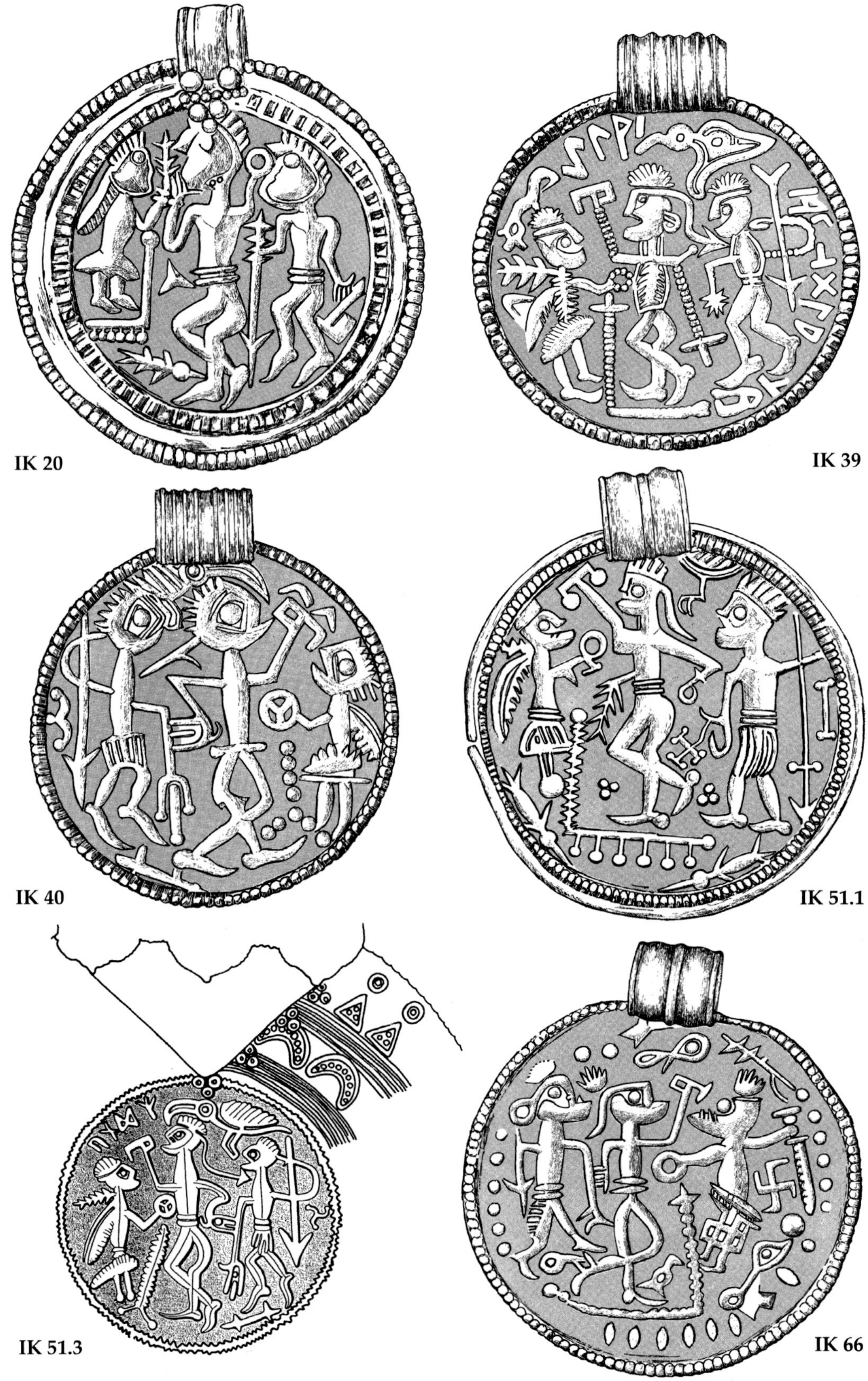

Fig. 3.1 Drawing of the "Drei Götter" Bracteates (IK20, IK39, IK40, IK51.1, IK51.3 and IK66) (after Hauck et al., 1985a–c; 1989b).

B-bracteates display one or more anthropomorphic full figures. The head is often seen in profile, while a smaller number are represented facing forward (see fig. 3.7). The body is predominantly displayed in profile with arms and legs. The figures may bear diadems, but are never seen with the so-called imperial fibula.

C-bracteates display a male head in profile above the body of a four-legged animal most often interpreted as a horse. Presumably, this type developed from ideals of civic virtue, *i.e.* of the emperor as a consul. The head often bears a diadem (Axboe and Kromann 1992, 279).

D-/F-bracteates display zoomorphic ornamentation in Animal Style 1 and their only anthropomorphic features are singly represented legs, hands or ears (Axboe and Kromann 1992, 278–279).

Of the A- to D- and F-bracteates discovered hitherto, *c.* 40% are C-bracteates, *c.* 35% are D-bracteates, while the rest are A- and B-bracteates respectively (Munksgaard 1978, 338). By discounting die-identical items, this represents *c.* 640 individual dies.

Medallion imitations are also included in the bracteate group. Similar to their Roman originals, they too are impressed on both sides, often with a head in profile on one side and a horseman on the other. The imitation medallions have been dated via grave finds to the 4th century AD and are thus older than the bracteates (Axboe 2007, 93–98; Axboe and Kromann 1992, 278). These are not included in the analysis, as there is some uncertainty about their costume sources. Imitation medallions only number 17 items from Scandinavia. Finally, a group of approximately 200 E-bracteates must be mentioned here. Their motifs comprise a zoomorphic triskelion under a semi-circular shape interpreted as a stylised human head. E-bracteates are primarily found on the island of Gotland in Sweden and are dated to the period around the end of the 7th century AD to around AD 900 (Gaimster 1998, 71, 93, 275). This type is not included in the costume analysis. In this study, the focus is on the A-, B- and C-bracteates with anthropomorphic figures whose prototype are medallions of Roman emperors and Roman coinage from the 3rd and 4th centuries AD (Axboe and Kromann 1992, 278 and 286). As a rule, A- and C-bracteates are inspired by the obverse side of the coin prototypes, while the B-bracteates are primarily connected to the reverse of the coin prototypes (Axboe and Kromann 1992, 281).

The costume analysis

In the costume analysis, both Roman and Scandinavian costume traditions are considered in the search for interpretation for the appearance of the figures, in order to ascertain the way in which the costumes on the bracteates have moved away from their original source of inspiration. That there is a temporal distance between Roman coinage and medallions and the product itself, however, plays a minor role, as what is crucial here is the composition of the costume, the costume traditions from which they originate, and how they can be understood.

On A-, B- and C-bracteates, the central motif is a human figure, sometimes together with an animal. In the image composition, the head is of primary focus, and where the status-denoting elements lie. In contrast, the costumes are simple and not prioritised in the images, and therefore, bracteates are difficult to study from a costume perspective. This means that bracteates have to be analysed using a slightly different method than with the gold-foil figures and any of the other object categories selected for this study. Bracteates with costume characteristics were selected for the analysis and divided into groups according to how much of the body is depicted. This also means that figures which would have been omitted in the other object categories, either because they are naked or because they have no clear costume characteristics, are included in this analysis. This was done in order to give a full description of the costumes displayed on the bracteates, even though this means that the analysis is not directly comparable with the results of the other object categories. The bracteates used in the costume analysis were selected from Hauck's publications (Hauck *et al.*, 1985a–c; 1986a–b; 1989). In cases where there are several figures on an image, they are termed "x", "y" and "z" figures in order from left to right. The costume terms employed here are the same as for the gold-foil figures, but in some instances, they are adapted to the special universe of the bracteates. As the strict composition of the motifs on the bracteates determines which parts of the human body are visible, the bracteates were first divided into two groups consisting of images with a human upper body and those with full figures. This division follows the classical iconographical division of the bracteates, whereby most of the A- and C- bracteates are separated from the B-bracteates. However, exceptions do occur, with for example full figures on A- and C- bracteates. In total, 133 bracteates with 151 figures were employed in the costume analysis, more or less with an equal number in each group (Table 3.1).

In the group constituting upper bodies, there are relatively fixed norms for the appearance of the images, and in this context, most attention is given to those bracteates where the figure comprises more than a head, for example where neck, upper body and/or arms are included. The homogenous style makes it possible to divide the upper body group into four subgroups on the basis of various ornamental features. The costume characteristics seen in the different subgroups are not necessarily different, but rather complementary, as they emphasise different details and parts of the body.

Table 3.1 Costume analysis of the bracteates selected for this study, male (M), female (F).

Find ID	*Type*	*Country*	*Location*	*Gender*	*Bracteate costume group*
IK5	C	Sweden	Åkarp	M	1.2
IK6	B	Denmark	Års	M	2.2
IK7	B	Denmark	Års	M	2.4
IK9	C	Denmark	Års	M	1.4
IK11	C	Sweden	Åsum	M	1.2
IK13	B	Denmark	Allesø	M	2.4
IK20	B	Russia	"Beresina" (Zagórzyn)	M/M/M	2.4,2.5
IK23	B	England	Bifrons	M	2.3
IK25	C	Norway	Bjørnsholm	M	1.4
IK33	C	England	Unknown (British Museum)	M	1.3
IK39	B	Denmark	Denmark X	M?/M/M	2.4
IK40	B	Denmark	Denmark	M/M/M?	2.4, 2.5
IK42	B	Denmark	Darum I	M	2.2
IK50	C	Denmark	Esrum Sø	M/M	1.3,2.1
IK 51.1	B	Denmark	Fakse	M/M/M	2.4, 2.5
IK 51.3	B	Denmark	Gudme II	M?/M/M	2.4,2.5
IK57.2	C	Sweden	Öster Ryftes, Gotland	M	1.2
IK58	C	Denmark	Fyn I	M	1.3
IK59	C	Denmark	Fyn	M	1.1
IK61	B	Denmark	Galsted	M	2.2
IK65	C	Norway	Gulbrandsdalen	M	2.3
IK66	B	Denmark	Gummerup	M/M/M?	2.4
IK 71	B	Germany	Hamburg	M	2.3
IK72.1	C	Norway	Hauge	M	1.4
IK73	C	Norway	Hauge, Klepp	M	1.4
IK74	B	Germany	Heide	M	2.3
IK75	C	Denmark	Hesselager	M	1.3
IK77	C	Denmark	Hjørlunde Mark	M	1.2
IK78	C	Denmark	Hjørlund Mark	M	1.4
IK79	C	Denmark	Hjørlunde Mark	M	2.4
IK 91	C	Denmark	Killerup	M	1.2
IK92	C	Denmark	Kitnæs I	M	2.2
IK93	C	Denmark	Kitnæs II	M	1.4
IK94.1	C	Denmark	Kitnæs III	M	1.3
IK95	C	Denmark	Kjøllergård	M	1.2
IK101	A	Denmark	Kongsvad Å	M/M	1.1, 2.2
IK104	B	Sweden	Lau Backa, Gotland	M	2.1
IK105	B	Denmark	Lellinge Kohave	M	2.2
IK112	C	Norway	Linnestad II	M	2.1
IK117.2	A	Norway	Hauge, Klepp	M	1.3
IK128	B	Germany	Nebenstedt I	M	2.4

(Continued)

Table 3.1 Costume analysis of the bracteates selected for this study, male (M), female (F). (Continued)

Find ID	*Type*	*Country*	*Location*	*Gender*	*Bracteate costume group*
IK129.1	B	Germany	Nebenstedt II	M	2.4
IK129.2	B	Denmark	Darum	M	2.4
IK131	B	Norway	Norway	M	2.1
IK132	B	Germany	Obermöllern	M	2.1
IK133	C	Sweden	Öjorna	M	1.4
IK135	C	Denmark	Ølst	M	1.4
IK141	B	Germany	Penzlin	M/M	2.4
IK143	B	Sweden	Ravlunda	M	2.1
IK145	A	Denmark	Revsgård	M	1.1
IK147	C	Denmark	Rynkebygård	M	1.3
IK148	B	Denmark	Sædding	M	2.4
IK149.1	B	Sweden	Skåne	M	2.2
IK150.2	C	Sweden	Tjurkö	M	1.4
IK154.2	C	Denmark	Overhornbæk	M	1.4
IK160	A	Sweden	Skättekärr	M	1.1
IK161	B	Denmark	Skodborghus	M	2.4
IK162.2	A	Denmark	Darum	M	1.1
IK163	C	Denmark	Skonager	M	1.4
IK165	B	Denmark	Skovsborg	M/M/M?	2.4, 2.5
IK166	B	Denmark	Skrydstrup	M	2.3
IK167	B	Norway	Sletner	M	2.3
IK168	B	Norway	Sletner	M	2.3
IK175	C	Denmark	Snesere	M	1.2
IK176	B	Sweden	Söderby	M	2.1
IK177	C	Norway	Søtvet	M	1.2
IK183	A	Sweden	Tjörkö	M	1.1
IK187	A	Sweden	Tossene	M	1.2
IK189	A	Sweden	Trollhättan	M	1.3
IK190	A/D	Sweden	Trollhättan	M, D-animal	-
IK194	C	Norway	Tveitane	M	1.2
IK195	B	Sweden	Ulvsunda	M	2.1
IK196	A	Denmark	Unknown	M	1.3
IK197	B	Denmark	Unknown VII	M/M	2.2
IK198	C	Denmark	Unknown VI	M	1.4
IK199	C	Denmark	Unknown III	M	1.2
IK206	B	Hungary	Várpalota	M	2.2
IK209	C	Norway	Vindingland	M	1.4
IK215	C	Denmark	Aversi	M	1.2
IK217	C	Russia	"Beresina" (Zagórzyn)	M	1.2
IK222	C	Sweden	Bostorp,Öland	M	1.4
IK225	A	Denmark	Broholm	M	1.1
IK230	C	Norway	Dalum	M	1.4

(Continued)

Table 3.1 Costume analysis of the bracteates selected for this study, male (M), female (F). (Continued)

Find ID	*Type*	*Country*	*Location*	*Gender*	*Bracteate costume group*
IK233	C	Sweden	Djupbruuns, Gotland	M	1.2
IK234	B	Sweden	Djurgårdsäng	M	2.1
IK236	C	Germany	Eckernförde	M	1.2
IK237	C	Germany	Eckenförde	M	1.2
IK238	C	Denmark	Ejby	M	1.2
IK240	A	Germany	Erin	M	1.1
IK244	C	Norway	Frederikstad	M	1.2
IK245.1	B	Germany	Freilaubersheim	M	2.1
IK245.2	B	Denmark	Vester Nebel	M	2.1
IK247	C	Sweden	Fridhem	M	1.4
IK249	C	Denmark	Fyn II	M	1.2
IK251	C	Denmark	Gammel Stenderup	M	1.2
IK254	A	Germany	Geltorf	M	1.1
IK257	C	Germany	Golm	M	1.2
IK258	C	Sweden	Gotland	M	1.2
IK259	B	Germany	Grossfahner	F	2.5
IK264	C	Sweden	Gurfiles, Gotland	M	1.2
IK267	C	Sweden	Hammenhög	M	1.2
IK282	A	Norway	Hov	M	1.1
IK291	C	Denmark	Lekkende Have	M	1.3
IK295	A	Denmark	Lundeborg	M	1.1
IK298	C	Denmark	Lynge Gyde	M	1.2
IK300	C	Denmark	Maglemose III/ Gummersmark	M	1.3
IK301	C	Denmark	Maglemose II/ Gummersmark	M	1.2
IK308	B	Germany	Nebenstedt	M	2.3
IK311	B	Germany	Oberwerschen	F	2.5
IK316	A	Norway	Rjoa	M	1.3
IK323	A	England	St. Giles	M	1.4
IK326	A	Sweden	Skåne VII	M	1.2
IK329	A	Denmark	Sjælland III	M	1.1
IK330	C	Denmark	Sjælland I	M	1.2
IK333	B	Germany	Sievern	M	2.3
IK335	A	Norway	Simonnes	M	1.4
IK337	B	Sweden	Sjöhagen	M	2.2
IK341	DB	Denmark	Sønder Rind	M	2.4
IK345	A	Norway	Store Salte	M	1.4
IK346	A	Denmark	Strangegården	M	1.1
IK347	B	Germany	Straubing-Bajuwarenstrasse	M	2.2
IK350	B	Germany	Südwestdeutschland	F	2.5
IK353	B	Denmark	Tønder	M	2.2

(Continued)

Table 3.1 Costume analysis of the bracteates selected for this study, male (M), female (F). (Continued)

Find ID	*Type*	*Country*	*Location*	*Gender*	*Bracteate costume group*
IK354	A	Sweden	Torpsgård, Senoren	M	1.1
IK357	C	Denmark	Tved	M	1.2
IK360	A	England	Unknown	M	1.1
IK361	B/D	Denmark	Unknown	M/M	-
IK362	B	Denmark	Unknown, Danmark VIII	M	-
IK364	C	Denmark	Unknown, Danmark II	M	1.2
IK365.2	C	Sweden	Almungs, Gotland	M	1.2
IK375	C	Hungary	Hungary	M	1.2
IK380	A	Norway	Vestly	M	1.1
IK385	C	Sweden	Visby Kungsladugård, Gotland	M	1.2
IK389	B	Germany	Welschingen	F	2.5
IK391	B	Denmark	Gudme II	F	2.5
IK595	B	Denmark	Sorte Muld	M/M/M	2.4, 2.5

All these figures, both in the Roman and the Scandinavian iconographical tradition, are interpreted as males. The various characteristics are grouped as follows:

Group 1.1 Upper body with a clear hint of outerwear, perhaps draped on the shoulder, and in a few cases with hints of innerwear. The figure often has a circular fibula on the left shoulder, a diadem on the head, and, on rare occasions, an arm is also visible. This category encompasses 15 A-bracteates and one C-bracteate (Fig. 3.2).

Group 1.2 Upper body depicted with a dotted outline and a diadem on the head. This category encompasses two A-bracteates, with arms, and 31 C-bracteates without arms (Fig. 3.3).

Group 1.3 Upper body depicted with horizontal stripes at the shoulder, chest or arm, and most often a diadem on the head. As a rule, an arm is also portrayed. This category encompasses four A-bracteates that are very different to one another and eight C-bracteates (Fig. 3.3 and 3.4).

Group 1.4 Outlined upper body without further details, and often a diadem on the head. An arm may be visible. This category comprises three A-bracteates and 16 C-bracteates (Fig. 3.4).

In the group of full figures, the head is often seen in profile, while the body as a rule is seen facing the front, or slightly turned. The legs are stretched or bent, and the arms are held at the side. However, exceptions occur where the figures are huddled up or separated into parts, are placed on horseback or seen in another context. This depends on the type of bracteate in question. Generally, B-bracteates portray full figures, although they also occur in A- and C-bracteates.

In general, it is difficult to find a consistent way of grouping the material, and the subgroupings depend on whether it is the entire/upper body or the leg covering that is in focus, and whether it is a male or female. In this group, the leg covering or the lower part of the costume is important, and is where the greatest variations are to be found. The various characteristics are grouped as follows:

Group 2.1 Long trousers demarcated at the foot, combined in several cases with a long-sleeved blouse or tunic and possibly a belt. This category encompasses nine B-bracteates and two C-bracteates (Fig. 3.4). Several of these have an upper body marked with horizontal stripes (IK 50, IK 104, IK 176, IK 195, and IK 234) (Figs 3.3 and 3.4). The upper bodies of IK 101x and IK 61 from Group 2.2 resemble them, although they are not demarcated at the foot (Fig. 3.2).

Group 2.2 Long trousers without demarcation at the foot, usually combined with a long-sleeved blouse and a belt. This category encompasses 11 B-bracteates and one C-bracteate (Figs 3.4 and 3.5). There may be a tunic on IK 92, but this cannot be definitely determined as the figure is on horseback. Again, the upper bodies of IK 101x and IK 61 are very similar to Group 2.1. On IK 61 the clothing on the upper body is defined as a tunic. IK 353 is special as the legs have an extra marking that possibly hint at trousers.

Group 2.3 Long trousers with a demarcation at the knee, usually combined with a long-sleeved blouse or bare upper body and a belt. This category comprises eight B-bracteates and one C-bracteate (Figs 3.5 and 3.6).

Group 2.4 Figures without costume details, but which are not naked in a physical sense. The body has been delineated very simply on these figures and at times

Fig. 3.2 Drawing of the "Drei Götter" Bracteates (IK165 and IK595) and bracteates depicting Costume Group 1.1 (IK59, IK160, IK183, IK101) (after Hauck et al., 1985a–c; 1989b; 2011).

IK 326

IK 187

IK 57.2

IK 117

IK 189

IK 50

Fig. 3.3 Drawing of bracteates depicting Costume Group 1.2 (IK326, IK187, IK57.2) and those depicting Costume Group 1.3 (IK117, IK189, IK50) (after Hauck et al., 1985a–c; 1986b; 1989b).

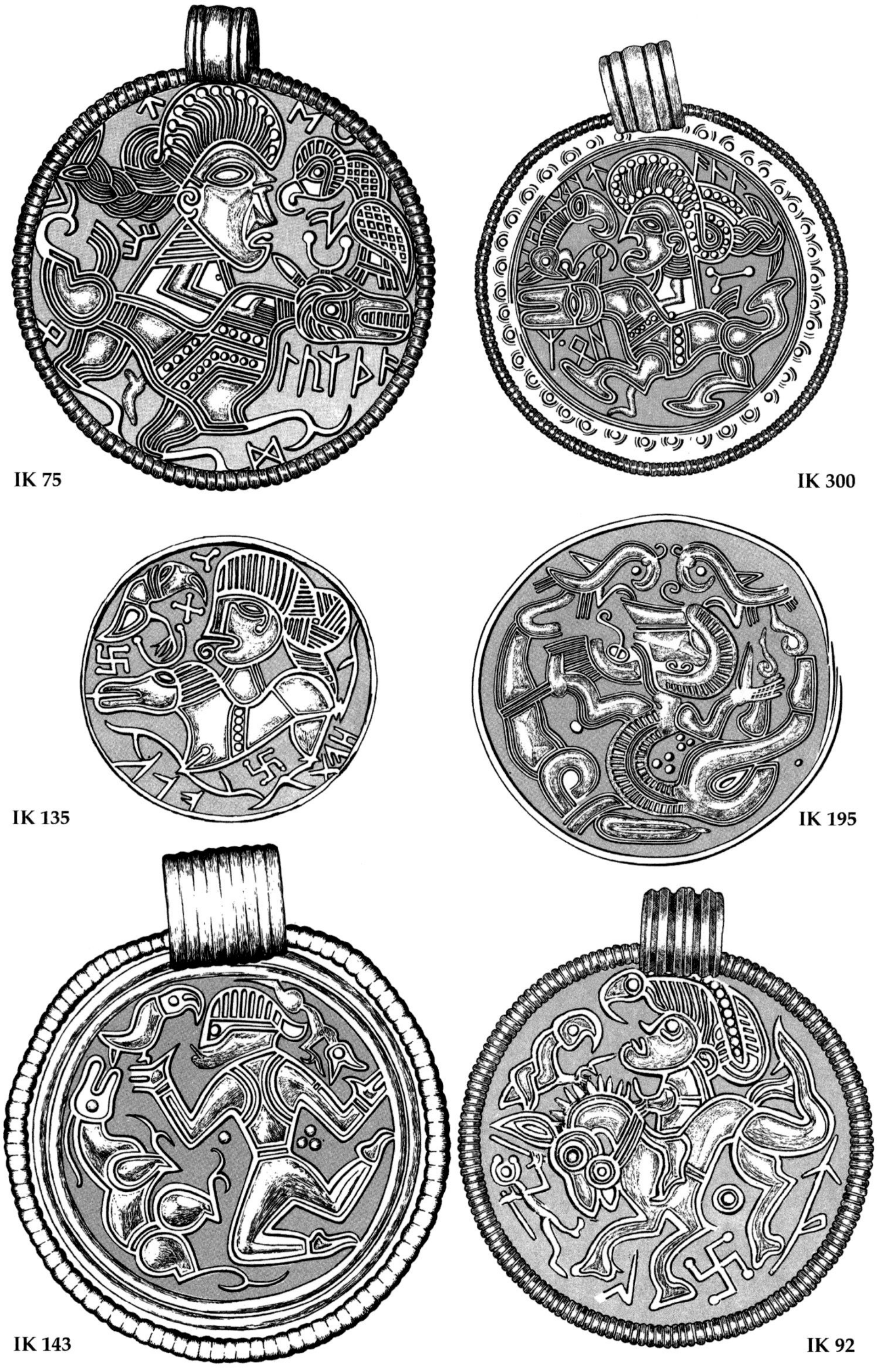

Fig. 3.4 Drawings of bracteates depicting Costume Group 1.3 (IK75 and IK300), Costume Group 1.4 (IK135), Costume Group 2.1 (IK195 and IK143) and Costume Group 2.2 (IK92) (after Hauck et al., 1985a–c; 1986b).

Fig. 3.5 Drawings of bracteates depicting Costume Group 2.2 (IK149, IK337, IK105 and IK6) and those depicting Costume Group 2.3 (IK65 and IK166) (after Hauck et al., 1985a–c; 1986b).

Fig. 3.6 Drawings of bracteates depicting Costume Group 2.3 (IK71, IK308 and IK167) and those depicting Costume Group 2.4 (IK148, IK341 and IK79) (after Hauck et al., 1985a–c; 1986b; 1989b).

Fig. 3.7 Drawings of bracteates depicting Costume Group 2.5 (IK350, IK389, IK391, IK311, IK259 and IK362) (after Hauck et al., 1986b; 1989b).

reduced almost to matchstick men. The figures can have arm rings and or foot rings and possibly a diadem on the head. This category comprises nine B-bracteates and one C-bracteate (Figs 3.6). A further 11 figures on the eight multiple-figure B-bracteates are included here. These are IK 20y, IK 20z, IK 39y, IK 39z, IK 40y, IK 51.1y, IK 51.3y, IK 66y, IK 165y, IK 595x and IK 595y (Figs 3.1 and 3.2).

Group 2.5 Long vertically-striped skirt without a clear upper section and possibly a belt. The figures are front-facing with arms at the side and short or invisible legs. The figure perhaps has short or upswept hair, with possibly a diadem. This category includes five B-bracteates that are all interpreted as females (Fig. 3.7). To this may be added a further eight figures on the eight multiple-figure B-bracteates that are recorded here as males. These are IK 20x, IK 40x, IK 51.1x, IK 51.1z, IK 51.3z, IK 66x, IK 165x, and IK 595z (Figs 3.1 and Fig. 3.2). A further five figures on five of the eight multiple-figure B-bracteates that were analysed could not be placed in any of the groups. These are IK 39x, IK 40z, IK 51.3x, IK 66z, and IK 165z (Figs 3.1 and 3.2).

Costumes in Group 1

Group 1.1. primarily contains the representation of a two-part costume. The outerwear constitutes a cloak, which in most cases is fastened at the shoulder, although it can, however, also be draped on the shoulder. More exact details cannot be deciphered. The innerwear constitutes a long-sleeved costume item, such as a blouse or a tunic, as portrayed on IK 160, IK 183 and possibly IK 145 (Fig. 3.2). These costume items fit both into the Roman and the Scandinavain clothing tradition (Croom 2000).

In the other three groups of upper bodies, Groups 1.2–1.4, only a few clear costume characteristics are discernible. In most cases, their form is ornamental rather than informative, and there is nothing to indicate that the representations portray more than a one-layered costume. Two A-bracteates (IK 187 and IK 326, Fig. 3.3) from Group 1.2 may, perhaps, be interpreted as a long-sleeved jacket with a dotted border along the front edge or a long-sleeved tunic/blouse with a dotted-patterned border on the chest. There are no further details for the C-bracteates, but the dotted patterning indicates that these belong to the same costume type. Thus, it is possible to consider the entire Group 1.2 as depicting a variation of the aforementioned jacket. Even fewer details are visible in Groups 1.3 and 1.4, and often in a highly stylised version. However, similar shapes are encountered among several of the figures belonging to Group 2.1, which may indicate that they represent more than a stylistic feature. Examples of this are IK 93 and IK 291 which possibly depict a costume detail, such as an added sleeve, also seen among several depictions of complete figures (IK 143 and IK 92, Fig. 3.4) and among figures in the jewellery category.

Finally, the existence of a male costume comprising an item of innerwear and, at times, also an item of outerwear may be extrapolated from the group of upper bodies. The outerwear is interpreted as a cloak that is typically fastened at the shoulder with a fibula. Most likely, the jacket is also represented (IK 187 and IK 326, Fig. 3.3), but it is uncertain if this is a kaftan type. Items of innerwear often consist of long-sleeved blouses or tunics.

Whether this outfit is rooted in a Roman or Scandinavian clothing tradition is difficult to determine at first glance, as it is primarily in the leg covering that the Roman male costume tradition differs from the Scandinavian. Kromann suggests that many figures on especially the A-bracteates have their origins in Roman coins, where the emperor is dressed as a general in a cloak (*paludamentum*), breastplate, circular fibula and diadem, which is consistent with the results in the present analysis. The C-bracteates, in contrast, primarily have their origins in coins of the emperor dressed as a consul in a toga (Axboe and Kromann 1992, 279). All in all, there are no details in the depictions that support the latter hypothesis.

Costumes in Group 2

This group encompasses both males and a few females. The costumes displayed are, generally, close fitting and consist primarily of a blouse and trouser combination or, in rare cases, a tunic and trouser ensemble. It is also possible that some of the figures in Group 2.4 wear trousers only with a bare upper body or that the figures are completely naked. No figures displaying genitals are seen, and especially the figures in Group 2.4 may be considered to be a more generic symbol of a human being or a man rather than a truly naked figure.

Generally, the types of trousers recorded correspond to those types recorded for the gold-foil figures, with roughly equal amounts of each type. Nevertheless, knee breeches are much more frequent on bracteates than in any of the other selected object categories. Leg coverings are often combined with a tight-fitting, long-sleeved blouse, and more rarely with a long-sleeved tunic. Markings at the waist indicate that the belt was an important component of the costume.

Group 2.5 is noteworthy, as some of the figures are interpreted as women. In terms of the costume, this identification is in accordance with the rest of the costume iconography employed for female figures (Fig. 3.7). All female figures are homogenously depicted in nearly floor-length, vertically-striped skirts, while the upper bodies are difficult to decipher, and could either depict a blouse or a bare upper body (IK2 59, Fig. 3.7). According to Hauck, this bracteate type represents an enthroned female figure, but a more specific suggestion is that they represent Odin's wife Frigg (Pesch 2002b, 51ff). These bracteates were found, and most likely also produced in continental Europe, with the exception of an item from Funen, Denmark, which could be an import (IK 301, Fig. 3.7). Most of the attributes the female figures hold in their hands are interpreted as symbols of power, while IK 311 is represented with a distaff, *i.e.* a textile tool. Especially striking is the small group of eight bracteates in Groups 2.4 and 2.5, the so-called "Drei-Götter" bracteates, that represent three persons in full figure (IK 20, IK 39, IK 40, IK 51.1, IK 51.3, IK 66, IK 165 and IK 595, Figs 3.1 and 3.2) (Hauck 1985a–c; 1986a–b; and 1989; Axboe 2002). The figures on this type of bracteate are interpreted as a Scandinavian variant of Roman coins depicting Victoria, the goddess of victory, who crowns the emperor while he receives the vanquished enemy (Axboe and Kromann 1992, 281, 284). In Old Norse mythology, this triad was translated into Loki who shoots Baldur with a mistletoe dart, while Odin watches. In the Roman version, the narrative consists of a female and two male figures, whereas in the Old Norse version, this becomes three male figures.

The figure at the centre of the images is interpreted in the Old Norse tradition as Baldur, whereas in the Roman model, it is the emperor. On the bracteates, this figure is seen frontally with the head in profile, arms at the side and legs turned towards one of the other figures. There are just as many that have their feet turned towards the right (IK 40y, IK 66y, IK 165y and IK 595y, Figs 3.1 and 3.2) as towards the left (IK 20y, IK 39y, IK 51.1y, IK 51.3y, Figs 3.1 and 3.2). This type of figure wears no easily recognisable costume, *i.e.* the body and leg are outlined in a continuous line, although separated at the waist with a hint of a belt. These matchstick men all belong to Group 2.4.

One of the other figures facing the central figure is interpreted as Loki and in the Roman prototype is synonymous with the goddess Victoria who crowns the emperor. These figures thus portray 14 different gendered figures in the Old Norse and Roman traditions respectively. The majority of these figures on the bracteates wear a costume comprising a bare upper body and a leaf-like lower part (IK 39x, IK 40z, IK 51.3x, IK 66z and IK 165z, Figs 3.1 and 3.2) and possibly a belt. Two figures wear a knee-length kilt and belt (IK 51.1x, and IK 595z, Figs 3.1 and 3.2), while one figure is attired in a knee-length costume with a tight-fitting upper part that widens at the bottom (IK 20x, Fig. 3.1).

The difference in the composition of the depicted figures makes it difficult to interpret the costume, but it is likely that it consists of mixed features due to the Roman origin in the form of a female goddess with a dress, wings, palm branch and a wreath in her hand with a presumed male figure with short hair and body that resembles the other simple male figures. As far as the leaf-like item of costume is concerned, it may not even be an item of costume, but possibly more of an attribute (a palm branch or a misteltoe) that has fused with the body contour. As far as the two figures in kilts are concerned, they probably wear identical types of costume, as described below, while IK 20x may be interpreted as a depiction of a dress (Fig. 3.1). As far as these figures are concerned, there is a clear mixture of male and female costume items, either because the model was misunderstood or it was more important to maintain the original features of the figures rather than stressing the gendered costume iconography.

The third figure that stands behind the central figure is interpreted in the Old Norse tradition as Odin who in the Roman tradition is synonymous with Mars, the god of war, with the inverted spear as his attribute (Axboe 2002, 300). On the Roman prototypes, this figure represents the enemy, but in these cases, the emperor turns towards the enemy and not away from him, as on the bracteates. In three cases, Odin's costume is remarkably similar to the central figure which is depicted without specific costume details and a belt at the waist (IK 20z, IK 39z and IK 595x, Figs 3.1 and 3.2). These three figures are without doubt males. In the other cases, the figures wear a costume consisting of a vertically-striped kilt, bare upper body and belt (IK 40, IK 51.1z, and IK 51.3z, Fig. 3.1), or a variant thereof (IK 66x and IK 165x, Figs 3.1 and 3.2).

The kilt that these figures wear is described as a "soldier's kilt" (Axboe 2002, 300), but with no further definition of what this entails. Roman soldiers' clothing comprised a flowing tunic that was approximately knee-length and held together at the waist by a belt. Over this, primarily officers wore a cuirass that was composed of metal or leather strips. Both items could be depicted with a vertically-striped effect, but was always associated with a clothed upper body (Sumner 2002, 2003; D'Amato 2005). Most likely, the outline of a Roman soldier's clothing was misunderstood or reinterpreted. Indeed, male kilts do exist among the Scandinavian archaeological textile and skin finds from the Early Bronze (Fig. 3.8) and Early Iron Ages (Fig. 3.9) (Broholm and Hald 1940; Hald 1980). It is evident that, in the case of the "Drei-Götter" bracteates, realism was overruled by the symbolic nature of the object, which clashed with the otherwise gender specific Scandinavian costume iconography.

Although the full figures on the bracteates are of varied expression and originate from very different prototypes, it is possible to extrapolate detailed information from their costumes. Even in cases where the costume is hardly noticeable, it is evident that the figures are not naked. Some can be described as partially naked, for instance with a naked upper body, but complete nudity with clearly delineated genitals is not seen on the bracteates. The male costume types comprise a blouse combined with various types of trousers or leg coverings, such as knee breeches, long trousers or a leg covering where no demarcation is drawn between leg and foot. The latter is often accompanied by a bare upper body. The blouse is tight-fitting, presumably with long sleeves and reaches the waist, which is marked with a belt. In a few cases, a knee-length tunic is also recorded. Females, in general, are represented in an almost floor-length, vertically-striped skirt, while the upper part of the figure is indecipherable. This could either be interpreted as a blouse or a bare upper body (IK 259, Fig. 3.7). Some of the "Drei-Götter" figures, that are interpreted as males, are depicted in a similar kind of outfit, here termed a kilt.

Fig. 3.8 A wool kilt and cloak found in the Early Bronze Age oak coffin (of a young man) from Borum Eshøj B in Denmark. The find is dated to 1345 BC (National Museum of Denmark).

Conclusion

Generally, the bracteates are iconographically close to their Roman prototypes, although the costumes are modified in several ways. As far as the upper body is concerned, there were evidently no problems of comprehension or translation from the Roman to the Scandinavian clothing tradition, even when dealing with costume shapes that presumably were unknown in the Scandinavian area, as for example the toga, armour and cuirass. The same applies to most of the full figures that, as far as may be ascertained, are depicted in typical Scandinavian costumes. In the few cases where there seem to be discrepancies in the costume identification between the Roman and the Scandinavian clothing tradition, these primarily occur in those instances where the genders on the Roman originals vary in relation to the suggested Scandinavian model, namely in the more complex "Drei-Götter" bracteates. Although it can be difficult to extrapolate precise costume characteristics from the group of upper bodies, often it is evident that the upper body was clothed in one or more layers of clothing. In most cases, this seems to be an item of outerwear.

While tunic or blouse and trouser combinations are only recorded in a few cases, the most characteristic male costume within the full figure group is the blouse and trouser combination. This predominance can hardly be a coincidence. Furthermore, no full figures attired in outerwear were encountered.

The representations of the female costume, though few, are entirely homogeneous and constitute skirts with little or no information on how the upper body was clothed.

Fig. 3.9 A skin kilt from Søgård in Denmark. The find is radiocarbon dated to AD 130–340 (scale = 10 cm) (National Museum of Denmark).

Moreover, bracteates possibly depict male figures clothed in an item of costume that, according to the definitions used here, has to be interpreted as a kind of kilt. On the basis of the costume iconography alone, the vast majority of the figures on the bracteates ought to be considered males, while only a few are females. The costume ensembles that may be extrapolated from the costume analysis are: cloaked males with a tunic or blouse; males in jackets; males in a blouse and long trousers; males in a blouse and knee breeches; males in a tunic and long trousers; males with a bare upper body and long trousers; males with a bare upper body and a kilt; and females with a long skirt and possibly a bare upper body.

When evaluated against the array of costumes of the gold-foil figures, it is evident that there are many similar costume ensembles. The clear costume iconographical links between the simplified gold-foil figures, from especially Sorte Muld and the slightly older bracteates, are even more accentuated when it is taken into account that most gold-foil figures with a correspondingly simple costume iconography were omitted from the analysis. These types of figures especially attest to an iconographical link between the two object categories. Other examples of iconographical similarities occur on the bracteates where the heads clearly resemble the more anonymous and stylised faces of the gold-foil figures (IK 361a, IK 361b and IK 362, Fig. 3.7). However, the analysis demonstrates that the human body plays another, and evidently less distinct role, in the bracteates than in the gold-foil figures, and that the composition of the images is different. Of the various body parts on the bracteates, it is the head that is in focus, and where the status-marking elements lie. The costume is indicated, but rarely depicted in great detail. Yet, the figures are not naked either, apart from the upper bodies that could be perceived as naked or bare. In all probability, these anonymous representations may be considered the epitome of human beings in general or men in particular. Furthermore, another crucial difference between the bracteates and the gold-foil figures is the clear correlation with various animals or zoomorphic figures, especially horses, a phenomenon unknown on the gold-foil figures.

Note

1 As of 2015 more than 1030 bracteates have been discovered, in contrast to the approx. 900 on which I based the research for my PhD thesis, pers. comm. Morten Axboe.

4

Helmet plates

Decorative plates on helmets were found in seven graves in Scandinavia, and in a stray find, several matrices were recovered. All originate from Sweden and are all made of bronze. In addition, a helmet from Sutton Hoo in England is included in the present analysis, as it is highly likely that the plates from this helmet are of Scandinavian origin (Evans 1986, 114–116, and fig. 27 and pl. iii) (see fig. 4.7). As shown in Table 4.1, the helmet plates are dated to the period AD 500–680. The material thus represents a well-defined time period and succeeds the bracteates, while they are more or less contemporaneous with the gold-foil figures and some of the jewellery.

Not all helmets or decorative panels are equally well preserved. The best preserved helmet with the most varied plates and the clearest image sequence is from the Valsgärde grave 7 in Sweden. No less than six different plates were found on this helmet. The remaining helmets have between one to three different plates each, but the number should be considered as a minimum (Table 4.2). In the present study, the different helmet plates are numbered with an abbreviation for the locality, a letter for the impress and a number for the figure.

It is quite common that the same plates occur several times on each helmet. Although it would have been both technically possible and simple to reuse the die stamps to create several similar helmets, each is unique. Similarly there are no preserved helmet plates produced from the known die stamps. Nevertheless, the motifs on the individual decorative panels have striking similarities, and it is obvious that they were subjected to a common style.

Table 4.1 The helmet plates selected for this study, male (M), female (F).

Find ID	*Object*	*Location*	*Country*	*Material*	*Gender*	*Dating*
VA7	Helmet plate	Valsgärde 7	Sweden	Bronze	M	AD 680
VA8	Helmet plate	Valsgärde 8	Sweden	Bronze	M	After AD 610/620
VE1	Helmet plate	Vendel 1	Sweden	Bronze	M	AD 600–630
VE11	Helmet plate	Vendel 11	Sweden	Bronze	M	AD 600
VE12	Helmet plate	Vendel 12	Sweden	Bronze	M	AD 610/620
VE14	Helmet plate	Vendel 14	Sweden	Bronze	M	AD 560/570
SH	Helmet plate	Sutton Hoo	England	Bronze	M	AD 625
UP	Decorative plates	Uppsala	Sweden	Bronze	M	AD 500
TO	Decorative plates	Torslunda	Sweden	Bronze	M	6th–7th century AD

Table 4.2 The various types of helmet plates and the total number recorded.

Type of helmet plate	*Va7A*	*Va7B*	*Va7C*	*Va7D*	*Va7E*	*Va7F*	*Va8A*	*Va8B*	*Ve1A*	*Ve1B*	*Ve1C*	*Ve11A*	*Ve11B*	*Ve12A*	*Ve14A*	*Ve14B*	*Ve14C*	*ShA*	*ShB*	*UpA*	*ToA*	*ToB*	*ToC*	*ToD*
Total	6	6	5	5	2	2	1	1	3	5	1	2	1	1	2	4	2	1	1	1	1	1	1	1

The material selected for this study is briefly described in the following.

Valsgärde 7: In 1933, the boat-grave field at Valsgärde in Sweden was excavated. Grave 7 turned out to be a richly equipped male grave, and among the objects were a helmet and the sheath belonging to sword I, both embellished with images. The grave is dated to *c*. AD 675 (Arwidsson 1977, 152; Nørgård Jørgensen 1999a, 186). The helmet cap is crafted of eight iron plates, which again are covered with bronze-sheet plates. Most of them contain human and zoomorphic images, while a few are equipped with interlaced patterns and animal ornamentation. The bronze plates are placed in four rows one above the other and the images in the three lowest rows are situated in such a manner that the figures appear to move towards the front. The bronze plates with human and zoomorphic depictions are stamped from six different square matrices of a similar size, approximately 4.7–5.2 × 4.7–5.2 cm, termed Types A–F (Arwidsson 1977, 116–118). In the bottom right corner of Figure 4.1 are five examples of Type A (I H2–6). In the second row, to the right is first an example of Type A (II H1), followed by four examples of Type C (II H2–5) and finally, an example of Type E (II H6). In the third row an example of Type C in section III H3 and an example of Type F in section III H6 are visible, while the rest are not preserved. In the bottom left corner, five examples of Type B (I V2–6) are visible. In the second row at left, four examples of Type D (II V2-5) are seen ending with an example of Type E (III V6). On the third row, an example each of Types B (III V3), D (III V4) and F (III V69) may be observed, while the remaining images are not preserved. Section I H1, I V1, III V5 and section IV H1–2, H4, H5 and V6-5 on row four are equipped with interlaced patterns and zoomorphic ornamentation (Arwidsson 1977; 21ff).

Type A (Fig. 4.2) depicts two identical right-facing helmeted men in profile, who stand behind each other, holding forth a shield in their left hand and a downturned spear held across the body in the other. Between their legs is a snake and in front of the foremost man is a stylised bird. Each man wears a long-sleeved, knee-length tunic (Va7A1 and Va7A2). The hem of the tunic is decorated with a border composed of three horizontal bands. Each sleeve is trimmed with a slightly different border. The tunic has a dotted pattern. Their legs appear to be bare, with no evidence of any footwear. The man in front has a triple marking around one of his ankles.

Type B (Fig. 4.2) depicts two identical left-facing, helmeted figures, standing one behind the other, holding a shield and sword in their right hand, and a downturned spear in the other (Va7B1 and Va7B2). From the preserved section, it may be deduced that, the men wear a long-sleeved, knee-length tunic, with a simple border. The tunic of the foremost figure (Va7B1) is ornamented with dots. The legs appear to be bare, with no evidence of any footwear.

Type C (Fig. 4.2) most likely represents a helmeted rider shown in profile on a right-facing horse (Va7C1). The rider holds a shield in his right hand. The left hand is raised as if to hold a spear, as in Type D. The horseman is not fully visible. He most likely wears a kaftan type jacket indicated by the V-border in the middle of the chest. On the upper part of one arm is a triple marking. The legs are either bare or he is wearing tight-fitting trousers ending at the ankle with a triple marking. There is no evidence of any footwear. Beneath the horse lies a person (Va7C2) shown in profile, whose sword is stuck into the ground, and who holds a shield in his left hand. This person wears a knee-length garment, which ends in a transverse-striped border. The garment itself is ornamented with dots. From the lower border to the arm appears a row of stripes, whose function is not clear. Given the ornamentation, it is most likely a tunic. The arm holding the shield has a marking at the wrist. The legs appear to be bare, with no evidence of any footwear. On the right hand side of this is a standing figure (Va7C3), holding a spear with his left hand, and the reins with his right. The head is not preserved, but given the position of the body, he is most likely seen from the front. Behind the horseman, and above the back of the horse, is a smaller figure (Va7C4), most likely in profile. This person wears a knee-length garment, which on the chest is marked with a dotted, V-shaped border. The garment has a horizontally-striped border at the armpit with a similar border below the waist. This barrel-shaped appearance is difficult to interpret. The arm which holds the horse has a marking at the wrist and the costume possibly has long sleeves. The legs are not fully preserved, and the legwear cannot be determined. With the left hand he appears to be holding on to the horseman's spear and with his right a slightly smaller spear or an arrow. The small figure has only slight hints of a costume, but it does not give the impression of being naked. Apart from a dotted, V-shaped border on the chest, the costume appears to be a type of close-fitting, one-piece garment with long sleeves.

Type D (Fig. 4.2) most probably depicts a helmeted rider (Va7D1) seen in profile on a left-facing horse. The right arm is bent upwards and appears to hold a forward-facing spear, while the left arm is turned downwards. On the chest, a sword handle can be observed. The costume of the horseman is not apparent. Beneath the horse lies a person (Va7D2), who has stuck a sword into the belly of the horse. The head and upper body of this figure is not preserved. This person wears a knee-length garment with a vertically-striped border at the lower hem. The rest of the costume is not preserved. The legs appear bare, with no evidence of any footwear. In front of the horse another faintly preserved kneeling figure (Va7D3) is glimpsed with a right-turned profile. The kneeling figure is very unclear, but is attired in a knee-length garment, with most likely a vertically-striped border along the lower hem. As the border continues upwards at an angle, it could be surmised

Valsgärde 7. Left side of helmet

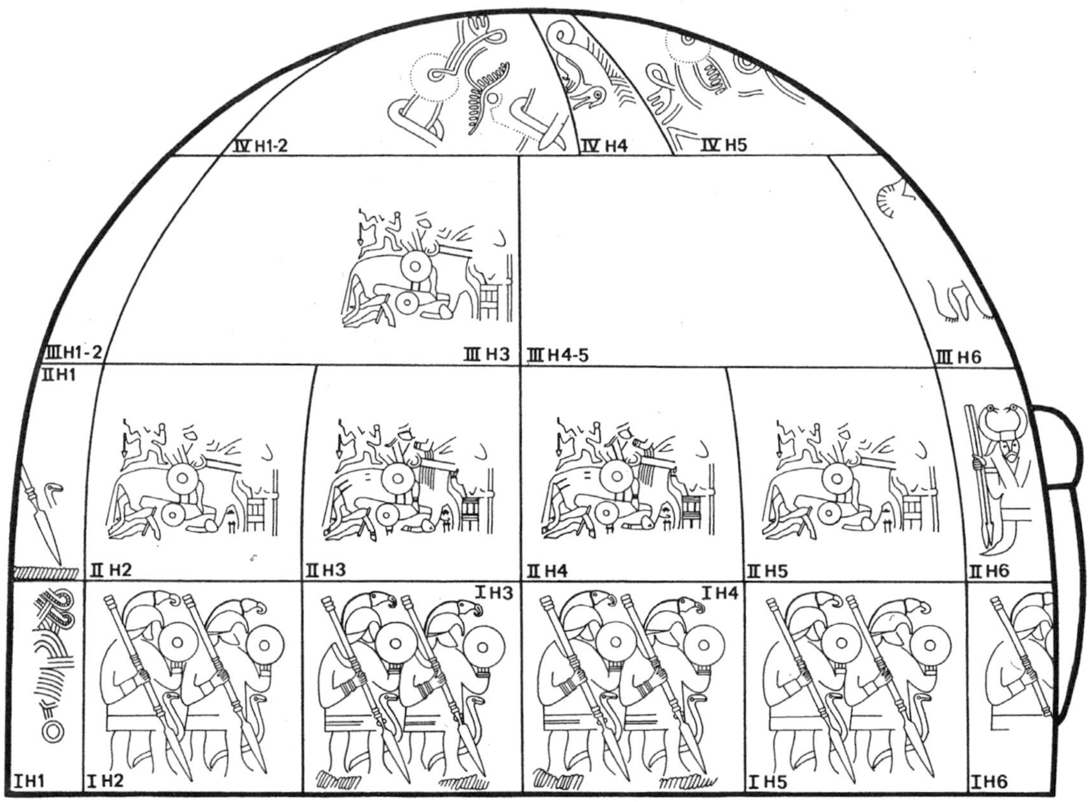

Valsgärde 7. Right side of helmet

Fig. 4.1 Drawings of both sides of the Valsgärde 7 helmet from Sweden (after Arwidsson, 1977, Abb. 26; drawing: Per-Olof Bohlin).

that this was a kaftan. The legs appear to be bare, with no evidence of any footwear. In the upper right-hand corner is depicted a smaller figure (Va7D4) with a right-faced profile who appears to be flying. The small flying figure displays no recognisable costume, apart from a horizontal border at the knee. The legs appear to be bare, with no evidence of any footwear. Behind the horseman, and above the back of the horse, is another small figure (Va7D5) with a horned helmet seen from the front. The small figure on the right is depicted in outline without any details. As the figure does not give the impression of being naked, it could be a kind of a one-piece outfit.

Type E (Fig. 4.3) depicts two identical men with horned helmets seen from the front (Va7E1 and Va7E2). They both

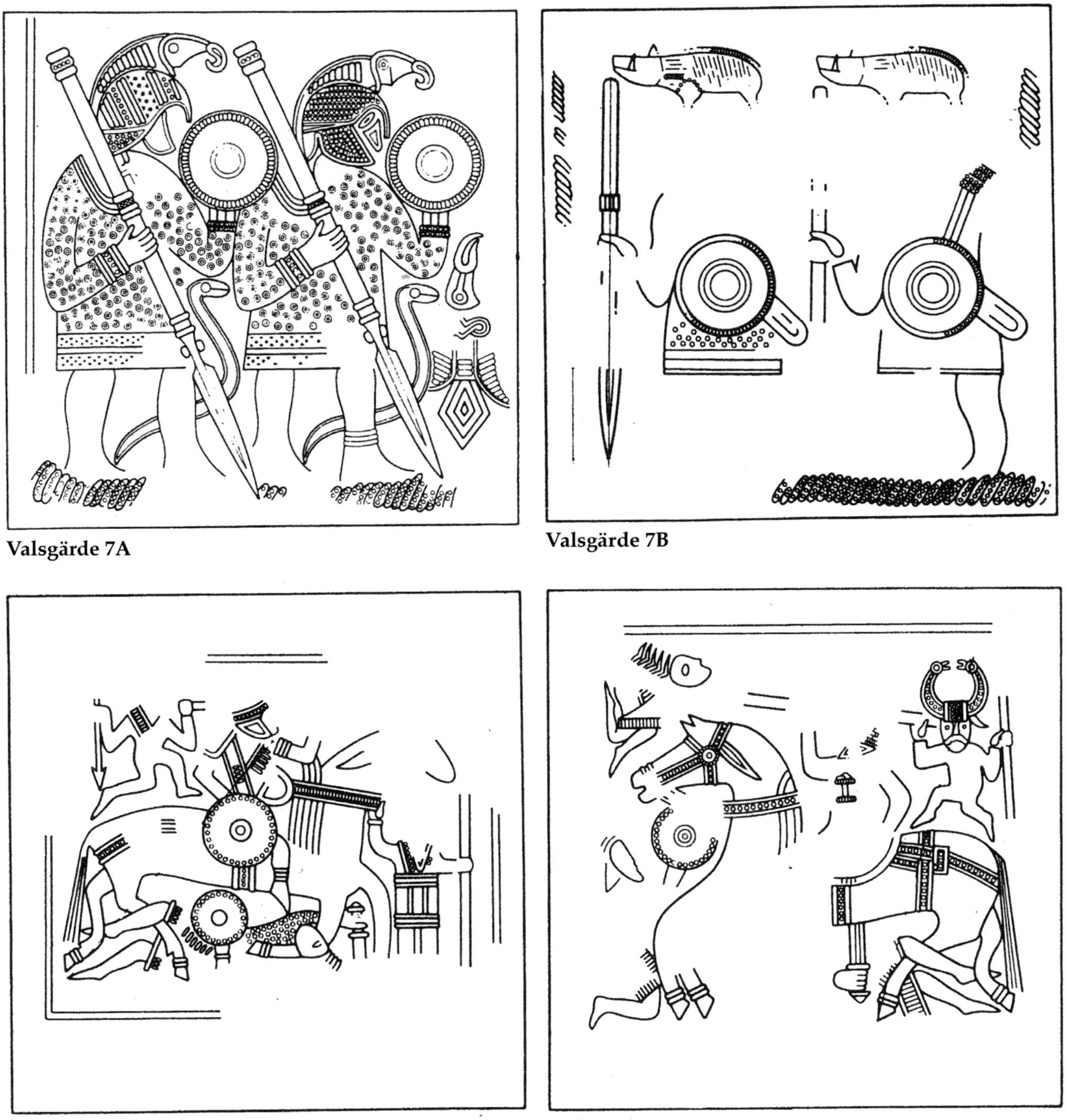

Valsgärde 7A **Valsgärde 7B**

Valsgärde 7C **Valsgärde 7D**

Fig. 4.2 Drawings of the Valsgärde 7A-D helmet motifs from Sweden (after Arwidsson 1977, Abb. 115, 120, 128 and 133; Drawing: Max Roosman and Per-Olof Bohlin).

hold two different spears in the outer hand and a short sword/knife in the inner. The most complete figure (Va7E1) to the left wears a long-sleeved, knee-length garment with a horizontal border at the wrist and lower hem. A similar border is seen in a V-shape on the chest, suggesting a kaftan type garment. The legs appear to be bare, with no evidence of any footwear.

Type F (Fig. 4.3) depicts a man seen from the front flanked by two animals, most likely bears. The bears are not completely preserved, but appear to be standing on their back legs. A ring placed around the neck of one of them could be a collar. Only the lower part of the man's attire is visible, suggesting a knee-length outfit, embellished with dots (Va7F1). The legs appear to be bare, with no evidence of any footwear.

Valsgärde 8: In grave 8 two different types of bronze plates, presumably belonging to a helmet were recovered. Both plates depict an equestrian scene similar to the bronze plates from Valsgärde 7, Vendel 1 and Sutton Hoo. Type A belongs to the left and Type B to the right side of the helmet. On the basis of the weapon equipment, the grave is dated to the period after AD 610/20 (Arrhenius 1983, 44; Nørgård Jørgensen 1999a, 186).

Type A (Fig. 4.3) depicts a helmeted rider seen in profile on a horse facing left (Va8A1). The rider holds a forward-facing spear in the right hand and a sword and shield in the left. This figure wears a garment, which is decorated with a ribbed pattern on the chest. It has long sleeves ending in a striped border. The arms are vertically striped. He wears tight-fitting trousers ending at the ankles, and embellished with narrow, vertical borders. Beneath the horse lies a figure facing the front, sticking a sword into the horse's belly with his right hand (Va8A2). This figure is attired in a knee-length garment which ends in a horizontal-striped border. It most likely has long, vertically-striped sleeves ending in a border. Across the stomach two more borders are visible, but whether they form part of a costume, denote a belt or a cross-belt is uncertain. On the legs, vertically-striped wrappings that end at the ankles can be seen. To the left, stands a figure seen from the front, holding the reins with the left hand and a spear in the other (Va8A3). This figure is attired in a knee-length garment, which at the chest has a hint of a V-shaped, vertically-striped border. The rest of the garment is embellished with narrow, vertically-striped bands. It has long sleeves and ends at the wrists in vertically-striped borders. It is uncertain if this is a tunic or a kaftan. The figure also wears tight-fitting trousers which end at the ankles and are embellished with narrow, vertically-striped borders. Behind the rider and above the horse's back, a small front-facing figure with a horned helmet can be seen (Va8A3). With his right hand, he holds on to the rider's spear and in his left hand, he holds a smaller spear. This small figure has only a faint hint of a costume, but does not give the impression of being naked. Presumably, it is a type of a tight-fitting, one-piece outfit with sleeves ending in a border. Around the waist he wears a broad belt. It is possible that from the right shoulder a narrow, cross-belt goes down to the waist belt. This depiction holds many similarities to Valsgärde 7, Type D, but the two are not identical.

Type B (Fig. 4.3) depicts a helmeted rider seen in profile on a right-facing horse (Va8B1). The rider holds a shield in the right hand and a forward-facing spear in the other. The rider most likely wears a kaftan type jacket hinted at by the vertically-striped V border on the chest. The costume has long sleeves ending in a vertically-striped border. The legs are either bare or he wears tight-fitting trousers ending at the ankles in a broad dotted border. There is no evidence of any footwear. Beneath the horse lies a figure facing the front whose right arm is stretched forward beneath the horse's foreleg (Va8B2). In his other hand, he holds a shield. This lying down figure wears a knee-length outfit which ends in a horizontally-striped border. On the chest a vertically-striped, V-shaped border is visible, suggesting a kaftan. The costume has long sleeves ending in a vertically-striped border. In front of the horse stands a figure holding a spear in one hand (Va8B3). The head is not preserved, but the body is seen in a right-facing profile. The standing figure wears a knee-length outfit with a dotted border composed of two similar patterned bands intersected horizontally by a third plain band at the lower hem. A similar border is visible at the waist, although this could be a belt. The centre of the costume is divided into three vertical sections. He most likely wears tight-fitting trousers ending at the ankles with a simple border. Behind the rider stands a small figure on the horse's back, facing the front (Va8B4). He holds on to the rider's spear with his left hand, and a smaller spear of his own in his right hand. The small figure behind the rider has only a faint hint of a costume, most likely a close-fitting one-piece outfit. Around the waist he wears a broad belt,

Table 4.3 The helmet plates divided into six image groups based on the content of the images.

Image group	*Description*	*Type of helmet plate*	*Total*
1	1–5 riders	Va7C, Va7D, Va8A, Va8B, Ve1A, Ve1B, ShB	7
2	Men in procession	Va7A, Va7B, Ve14A, Ve14B[1], ToC, ToD[2]	6
3	2 male combatants	Ve11B, Ve12A[3], Ve14C	3
4	2 front-facing men	Va7E, ShA, UpA?,	3
5	1 man with 1 animal	ToB, Ve11A, Ve1C	3
6	1 man with 2 animals	Va7F, ToA	2

[1]This panel deviates from the rest by depicting five figures following in a row; [2]This panel stands out from the rest by being a mixture of Categories 2 and 3; [3]This panel is only partially preserved and thus its placing in the group is uncertain.

from which a cross-belt goes up to his left shoulder. This depiction has many similarities to Valgärde 7, Type C, but they are not identical.

Vendel 1: The graveyard near Vendel in Sweden was discovered in 1881. The helmet which comprises an iron cap with a bronze crest at the top and a single row of images at the bottom, was found in grave 1 and was most likely intact when unearthed, but was damaged by the unskilled excavators. However, it was possible to reconstruct the helmet (Stolpe and Arne 1912, 13f). The burial is dated by its weapon equipment to AD 610/20–680 (Arrhenius 1983, 64–65; Nørgård Jørgensen 1999a, 186). On the helmet, at least three different images are depicted: one of a rider and a small man (Type A), another of a rider and a serpent (Type B), and a third of an animal and a standing male figure (Type C). At least three depictions of Type A, five depictions of Type B and at least one depiction of Type C have been found (Stolpe and Arne 1912, 14).

Type A (Fig. 4.4) depicts a helmeted rider seen in profile on a right-facing horse (Ve1A1). The rider holds a sword and a shield in the right hand and a forward-facing spear in the left. He wears a costume with long sleeves ending in a striped border. Across the chest, a vertically striped cross-belt is visible, but it is uncertain whether it belongs to the outfit itself or constitutes part of the weaponry. He also wears close-fitting trousers with a single marking at the ankle. In front of the horse stands a front-facing man, who holds the reins in his right hand and an upright spear in his left (Ve1A2). The standing figure wears a knee-length, long-sleeved kaftan, with a vertically-striped border along the lower hem and at the wrists, and a pair of close-fitting trousers with a single marking at the ankle. Around the rider's spear, fly three birds: two large and one small. This depiction has similarities to Valsgärde 7, Type C (Fig. 4.2), but contains only two figures.

Type B (Fig. 4.4) depicts a helmeted rider seen in profile on a left-facing horse (Ve1B1). The rider holds a shield in the left hand and a downturned spear in the other. In front of the horse is a serpent. On either side of the rider fly two large birds. The rider is attired in a long-sleeved costume, which on the front has a V-shaped border comprising two stripes. Whether this belongs to the garment or to the weaponry is uncertain. The rider's foot, sticks out from beneath the shield, and it is evident that he is wearing pointed-toed shoes.

Type C (Fig. 4.4) depicts a front-facing man, standing with an axe in his left hand holding an animal by a chain in the other. The animal, depicted on the left of the image is shown in profile sitting on its tail. The man is attired in a knee-length, long-sleeved kaftan with a herringbone-patterned border along the lower hem and a vertically-striped border at the wrists (Ve1C1). Two markings, whose function is unclear, are visible around the neck. He wears tight-fitting trousers with a single marking at the ankle. The animal has a dotted or lozenge-shaped embellishment on its belly and a plaited pattern on its legs. It is difficult to determine the animal from a zoological point of view. It could be a bear, but the ears appear to be a little too long, and thus is simply termed a beast. This depiction has many similarities to Vendel 11, Type A (see Fig. 4.5), but there, man and beast are placed facing opposite directions.

Vendel 11: Grave 11 was also excavated in 1881, but only a few fragments of an iron helmet, among them a fragment of a bronze crest and a few pieces of bronze plates with zoomorphic ornamentation, were described (Stolpe and Arne 1912, 42). Later, three more fragments of bronze plates with figural ornamentation were published (Arwidsson 1977, abb.148a–b) and the reconstruction of Types A and B were made on this basis (Hauck 1978). Type A exists in at least two examples, and Type B, in at least one example. The burial is dated to around AD 600 (Arrhenius 1983; Nørgård Jørgensen 1999a, 184).

Type A (Fig. 4.5) depicts a standing, and most likely helmeted, front-facing figure, holding an axe in the right hand and an animal by a chain in the other. The animal, which stands to the right of the image, is shown in profile, and obviously sits on its tail. Neither man nor beast is fully preserved. The man is clad in a knee-length kaftan with a double-striped border along the edges and most likely at the lower hem (Ve11A1). Around the neck are two markings whose function is uncertain. He wears tight-fitting trousers with a single marking at the ankle. Only the animal's head, a foreleg and a back leg are visible. This bronze plate highly resembles Vendel 1, Type C (Fig. 4.4), but man and beast are placed facing opposite directions.

Type B (Fig. 4.5) depicts two helmeted warriors fighting each other. They are shown in profile turning towards each other. The figure to the left is poorly preserved with no costume details discernible (Ve11B1). In his left-hand, he holds a shield with its inside turned towards the viewer, and in the right, most likely a sword directed at his opponent. The figure to the right holds a short sword in his right hand and a shield in his left through which a spear directed to the back, juts out (Ve11B2). This spear is most likely curved as in Vendel 14, Type C. Across the shoulder, a cross-belt which holds the sword/knife sheath is visible. He is clad in a long-sleeved, close-fitting, waist-length garment, with a V-shaped vertically-striped border on the chest and at the wrists. He wears a pair of dotted-patterned trousers, whose ornamentation continues over to the feet. A striped belt is visible around his waist. While there are many similarities with Vendel 14, Type C (Fig. 4.6), the costume is depicted differently.

Vendel 12: When this burial was uncovered in 1893, parts of a helmet resembling that from Vendel Grave 1 were recovered. It has at least two rows of bronze plates placed on top of each other. The majority of the bronze plates are decorated in Animal Style II, but above the left eyebrow, a depiction of two men is partially visible, suggesting that a similar one had been present above the right eyebrow

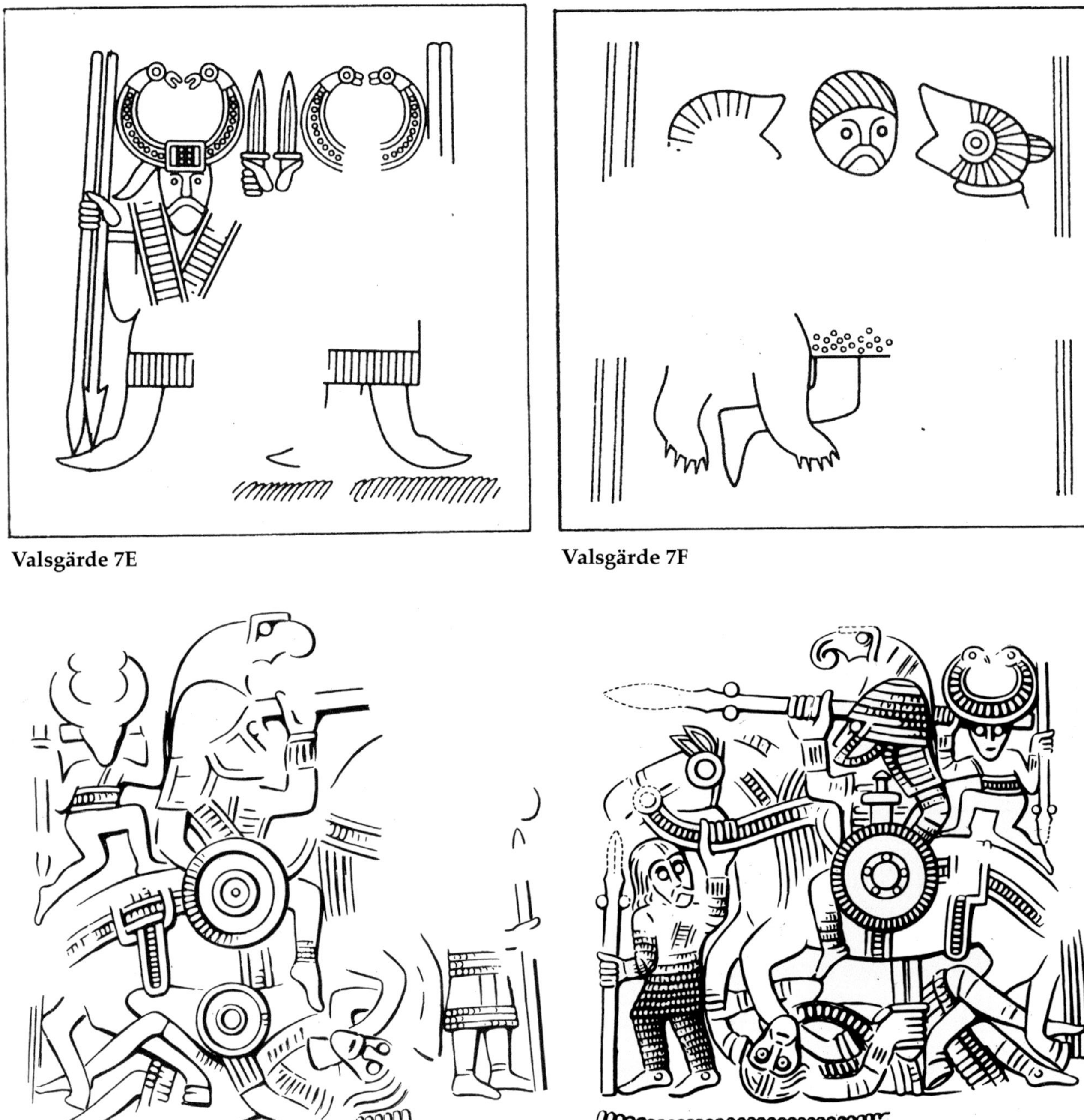

Valsgärde 7E

Valsgärde 7F

Valsgärde 8A

Valsgärde 8B

Fig. 4.3 Drawings of the Valsgärde 7E-F and 8A-B helmet motifs from Sweden (after Arwidsson, 1977, Abb. 115; drawing: Max Roosman and Per Olof Bohlin; and Hauck 1978, Abb. 20–1).

(Stolpe and Arne 1912, 46). On the basis of the weapon equipment, the burial is dated to AD 610/20 (Nørgård Jørgensen 1999a, 186).

Type A (Fig. 4.5) depicts two men in profile facing each other. In their outer hand, each holds a short sword, and towards the centre, each carries a shield with a spear stuck through it. The ends of the spears cross each other in front of the men's faces. The figure to the left has a simple marking on the upper right arm and presumably is bare-chested (Ve12A1). The rest of the figure is not visible. The figure to the right is bare-chested, with a dotted-patterned belt and dotted-patterned, tight-fitting trousers (Ve12A2). One of the legs crosses the other, as if he is about to take a step forward. The feet are not visible.

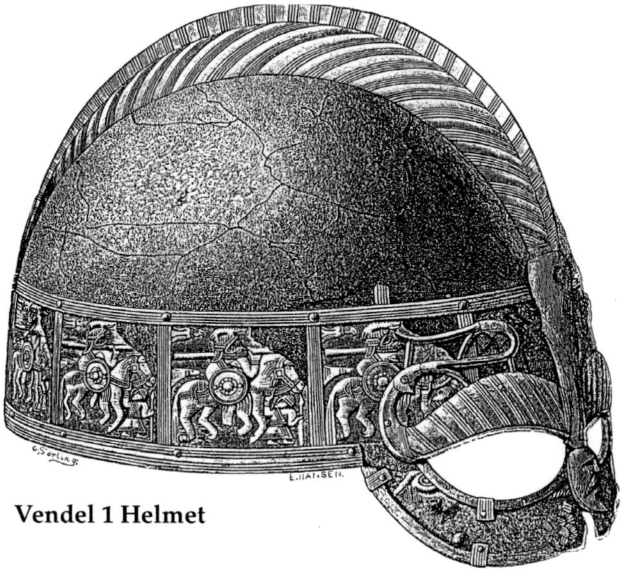

Vendel 1 Helmet

Vendel 1A

Vendel 1B

Vendel 1C

Fig. 4.4 Drawings of the Vendel 1 helmet and the Vendel 1A-C helmet motifs from Sweden (after Stolpe and Arne 1912. pl. v, fig.1; drawing: Olof Sörling and Evald Hansen; and Hauck 1978, Abb. 26, 29 and 12; drawing: W. Hilpert).

Vendel 14: When this grave was excavated in 1893, an iron helmet with bronze-plate figures was recovered. On the right side of the helmet are two consecutive plates, each with five warriors standing one behind the other (Type A). On the left side are four plates, each with two warriors standing one behind the other (Type B). Above the eyebrow ridge are two identical plates with two warriors facing each other (Type C). On the basis of the weapon equipment, the burial is dated to AD 560/70 (Nørgård Jørgensen 1999a, 185).

Type A (Fig. 4.6) depicts five identical helmeted warriors displayed in profile, all facing the right (Ve14A1-5). In their left hand, they hold a sheathed sword in front of themselves, and in their right, a downturned spear, held across their body. They wear short-sleeved, knee-length tunics ending in a simple border at the elbow and lower hem. The tunic is embellished with an irregular dotted pattern. Their legs are either bare or they are wearing tight-fitting trousers. This type has many similarities with Valsgärde 7, Type A (Fig. 4.2), but the latter only comprises two figures.

Type B (Fig. 4.6) depicts two identical helmeted warriors standing one behind the other (Ve14B1-2). They are seen in left-facing profile. They hold a downturned spear in their right hand and in their left, a shield. In a belt that goes from over their right shoulder to beneath their left arm, there hangs a sheathed sword. They wear short-sleeved, knee-length tunics ending in a simple border at the elbow and lower hem. The tunic is embellished with an irregular dotted pattern. Around the wrists and ankles, a simple marking is visible. The legs are either bare or they are clad in tight-fitting trousers. This type has many similarities to Valsgärde 7, Type B (Fig. 4.2).

Vendel 11A

Vendel 11B

Vendel 12 Helmet

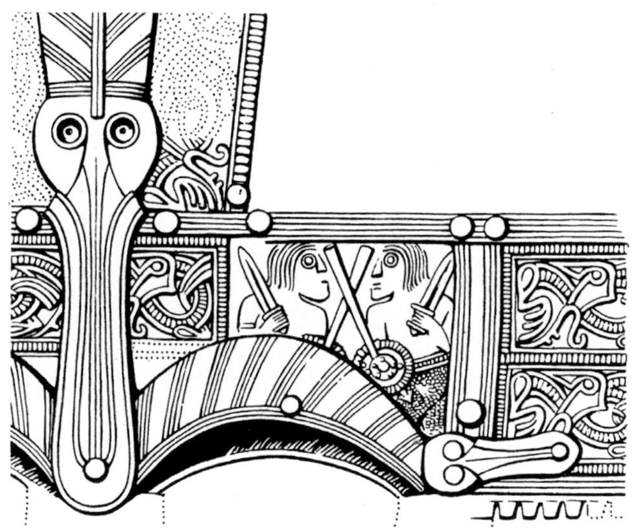

Vendel 12A

Fig. 4.5 Drawings of the Vendel 11A-B helmet motifs, the Vendel 12 helmet and the Vendel 12A helmet motif from Sweden (after Hauck 1978, Abb. 6a and Abb. 10a; drawings: I. Müller; after Stolpe and Arne 1912, xxxvi, fig. 4, Riksantikvarieämbetets arkiv ATA; drawing: Olof Sörling and Evand Hansen; and after Hauck 1978, 32, fig. 1; drawing: I. Müller).

Type C (Fig. 4.6) depicts two helmeted, battling warriors in profile, facing each other. The figure to the left holds a shield with its back facing the viewer in the left hand and a sword in the right aimed at his opponent (Ve14C1). The figure to the left is clad in a long-sleeved, knee-length kaftan with a triple border in a dotted pattern along the edges and a vertically-striped border at the wrists. The legs are either bare or he is clad in tight-fitting trousers, with a simple marking at the ankle. The figure to the right holds a short sword in the right hand and a shield in the left, and wears a short-sleeved, knee-length tunic which ends in a simple border at the elbow and lower hem (Ve14C2). The tunic is decorated with an irregular dotted pattern. Across the shoulder, a cross-belt, holding the sheath for a sword or knife, is visible. The legs are either bare or he is clad in tight-fitting trousers. A simple marking is seen at the ankle. Between the warriors' feet are seen two bent spears which pierce through the shield of the one and the costume of the other respectively. This holds many similarities to Vendel 11, Type B (Fig. 4.5), although the costumes are different.

Sutton Hoo: In 1938, privately sponsored archaeological investigations were begun in England at Sutton Hoo in East Anglia. In the first campaign, 15 burial mounds were recorded, of which three were excavated (burial mounds 2, 3 and 4). In the following year, it was decided to continue with burial mound 1, which turned out to be an undisturbed and very richly equipped boat grave – also later known as the Sutton Hoo chieftain's grave (Evans 1986, 16–18). The grave is dated by coin evidence to *c*. AD 625 and is possibly the burial of the East Anglian king Raedwald (Evans 1986, 16–22). In the grave was found a helmet, with tin overlaid bronze plates with figural motifs. The helmet was, more or less, deteriorated when it was recovered, but a careful reconstruction has shown that the iron helmet had a face mask, neck guard and cheek pieces. The design of the helmet, thus deviates from the Swedish helmets, while the iconography demonstrates close ties to Scandinavia (Evans 1986, 46–48). In addition to a number of bronze plates with zoomorphic and interlaced ribbon patterns, the helmet contains two different figural depictions. Of these, one is a scene of two standing men (Type A) and the other, a rider (Type B).

Type A (Fig. 4.7) depicts two helmeted, front facing men. They stand side by side, with their inner feet pointed towards each other. In their outer hand, they hold two downturned spears each, and in their inner hand, where the lower arms cross each other's, a short sword/knife. Beneath their crossed inner arms are two upturned spears that cross each other, with the lower part of each spear ending at the backward turning outer foot of each man. The clearest figure to the right wears a long-sleeved, knee-length kaftan (ShA2). Along the edge of the costume and at the wrists, a herringbone pattern similar to Vendel 1, Type C (Fig. 4.4) is visible. Around the waist is a vertical-striped belt. The legs appear bare and no marking is evident at the ankle. It is plausible that the figure to the left wears an identical costume (ShA1). This holds many similarities to Valsgärde 7, Type E (Fig. 4.3).

Type B (Fig. 4.7) depicts a rider seen in profile on a left-facing horse (ShB1). In his right hand, he holds a front-facing spear, while in his left, he holds a sword and shield. The rider is not completely preserved, but possibly wears a long-sleeved kaftan with a vertically-striped border. Around his waist he wears a vertically-striped belt. The legs are bare and no marking between leg and foot is visible. Underneath, in front of the horse, lies a figure that holds onto the reins with his left hand, while, with his right, sticks a sword into the horse's belly (ShB2). Beneath his right arm, the end of a sword sheath sticks out. This figure is wearing a long-sleeved, knee-length kaftan with a vertically-striped border. The kaftan is ornamented with a regular dotted pattern. Behind the rider, and above the horse's back, a small, left-profiled figure is visible (ShB3). He is depicted in outline, with few details. In his right hand, he holds on to the rider's spear and with the other, presumably holds a shield. On his chest, a hint of a V-shaped vertically-striped border and a similar one at the wrists can be seen, suggesting a kaftan. The legs appear bare and no marking is visible between leg and foot. This has many similarities to Valsgärde 7, Type C/D (Fig. 4.2).

Uppsala: In 1846, the eastern burial mound (østhøj) near Uppsala in Sweden was excavated. Among the finds were fragments of five bronze plates with figural depictions (Lindqvist 1936, 172–173). The bronze plates from the eastern burial mound most likely derive from a helmet and are dated to *c*. AD 500 (Lindqvist 1936, 171ff; Arwidsson 1977, 29–30).

Type A (Fig. 4.8) depicts the right side of a man (UpA1). In his right hand, he holds two downturned spears. It is not possible to ascertain if the man is facing the front or is depicted in profile. He wears a long-sleeved, knee-length garment with a vertically-striped border along the lower hem and at the wrists. Presumably, the marking at the waist denotes a belt. The arm and parts of the lap are embellished with a regular dotted/lozenge-shaped pattern. The legs appear bare and no marking between the leg and feet is visible. The rest of the image is not preserved. This has many similarities to Valsgärde 7, Type E and Sutton Hoo, Type A above (Figs 4.3 and 4.7) which could suggest that the man is depicted from the front.

Torslunda: The four bronze matrices from Torslunda on the island of Öland in Sweden were recovered in 1870 among a heap of stones. During the subsequent excavation, it became evident that these stones were the remains of a settlement, containing several houses and traces of a smithy (Hagberg 1976, 323–325, 331). A radiocarbon dating to the time around AD 400–500 supports a connection to the Swedish Vendel period (Hagberg 1976, 326, 349; Beck 1968, 237). The matrices are similar to the bronze-sheet plates, and most likely, they have been used to produce plates for helmets or other kinds of similar decorations.

Type A (Fig. 4.9) depicts a man seen from the front, flanked by two animals, presumably bears (Beck 1968, 238). With his right hand, he sticks a sword into the belly of the bear to the right, while holding up a knife in front of himself with his left hand (ToA1). The man wears a three-quarter sleeved, thigh-length tunic, decorated in a lozenge-striped or plaited pattern with a border at the lower hem. The legs appear bare, with no evidence of any footwear. The bears' fur is depicted with diagonal rows of vertical stripes. The bears hold the man's upper arms with their forepaws. The heads are ornamented with dots, and around the neck they presumably have a ring or collar. This resembles Valsgärde 7, Type F (Fig. 4.3) to a certain extent.

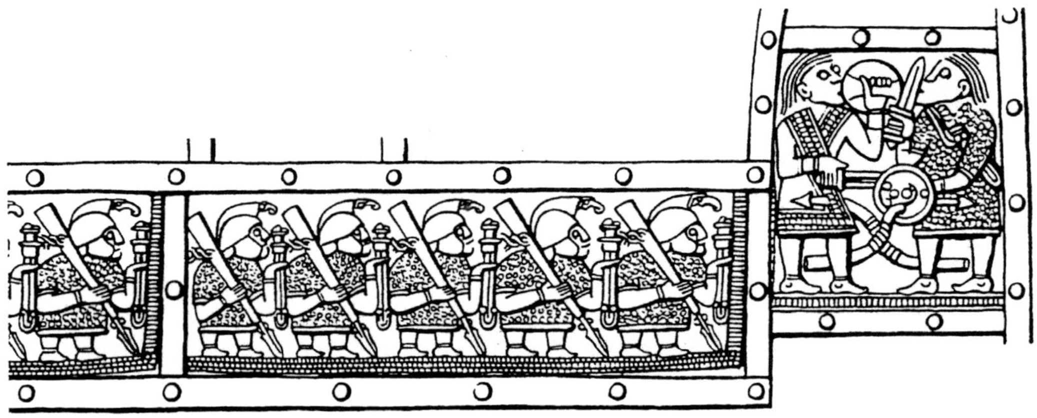

Vendel 14. Right side of helmet

Vendel 14. Left side of helmet

Vendel 14B

Vendel 14C

Fig. 4.6 Drawings of both sides of the Vendel 14 helmet, the helmet motifs from Vendel 14B and Vendel 14C from Sweden (after Hauck 1978, 34, fig. 2a and 35, fig. 2b; drawings: I. Müller; after Hauck 1978, 35, fig. 3; drawing: W. Hilpert; and after Hauck 1978, Tafel iv, Abb. 5a).

Sutton Hoo A

Sutton Hoo B

Fig. 4.7 Drawings of the Sutton Hoo A-B helmet motifs from England (after Evans 1986, 48; drawings: Carey Miller; © Trustees of the British Museum).

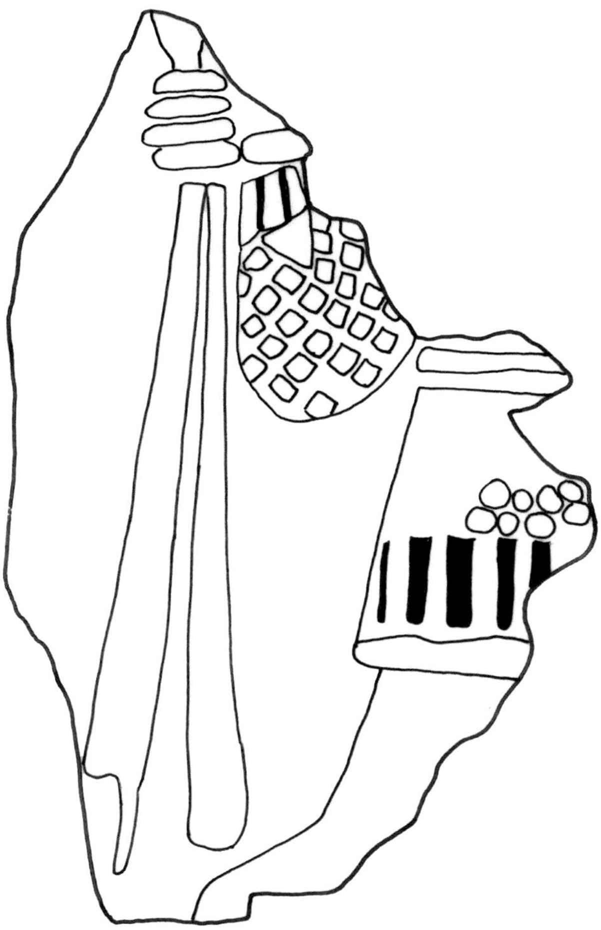

Fig. 4.8 Drawing of the bronze motif from Uppsala burial mound in Sweden (drawing: Pernille Foss; after Arwidsson 1977, Abb. 31).

Type B (Fig. 4.9) depicts a man standing facing the front with an axe in his right hand and holding an animal on a chain with the other (ToB1). The animal to the right is seen in profile and sits on its tail. The man is wearing long trousers with an irregularly dotted pattern, very similar to the trousers in Vendel 11, Type B2 and Vendel 12, Type A (Fig. 4.5). He is bare-chested and wears a vertically-patterned belt around his waist. The animal, which is difficult to determine zoologically, is often merely referred to as a beast, but it has been suggested that it is a feline type (Beck 1968, 239). This beast has a dotted pattern on its belly and its legs are decorated in a lozenge-striped or plaited pattern. This resembles Vendel 1, Type C and Vendel 11, Type A above (Figs 4.4 and 4.5) to a certain extent.

Type C (Fig. 4.9) depicts two identical warriors standing one behind the other (ToC1 and ToC2). Their faces are shown in left-facing profile. They both wear a helmet with a neck guard, and are displayed holding, in front of themselves, a sheathed sword in their left hand and a downturned spear in their right hand. The foremost warrior's sword-handle has a ring knob (ToC1). The warriors wear a plain, short-sleeved, knee-length tunic ending in a simple border at the lower hem and at the elbows. The border on the foremost warrior is composed of vertical and horizontal-striped rectangles, and on the other an S-twisted pattern. Their legs are either bare or they wear close-fitting trousers without any markings at the ankle. They have many similarities to Valsgärde 7, Types A and B (Fig. 4.2).

Type D (Fig. 4.9) depicts a helmeted warrior and a "wolf" warrior. The warrior standing to the left is seen from the front, with one leg stretched in front of him and the other bent

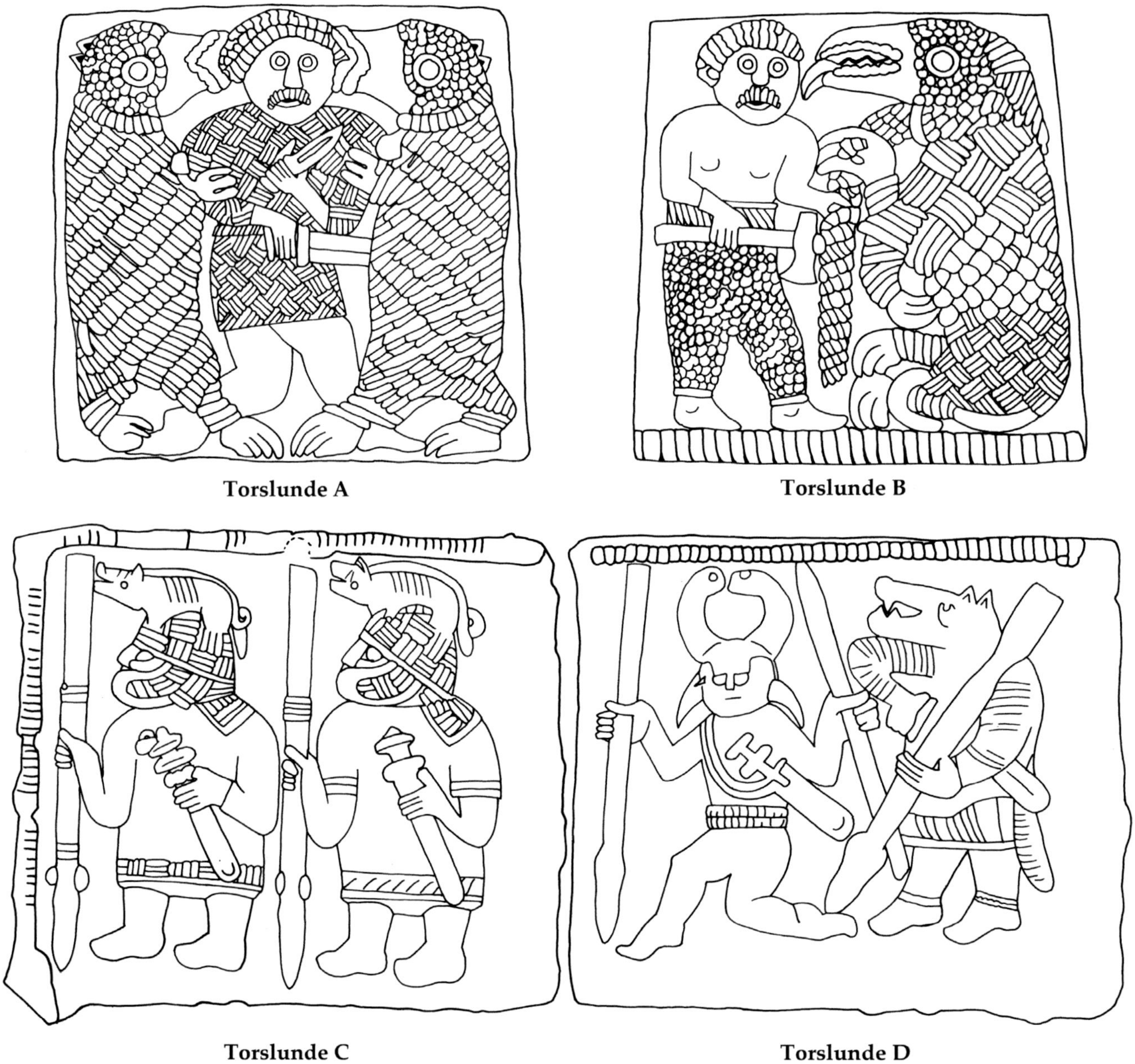

Fig. 4.9 Drawings of the Torslunda A-D bronze matrices from Sweden (drawing: Pernille Foss; after Hauck 1978).

behind, as if he was in the process of moving (ToD1). It is, by now, a well-accepted notion that this warrior is one-eyed (Beck 1986, 239). In his right hand, he holds a downturned spear in front of himself, and in his left, an upturned spear. In a cross-belt around his chest hangs a sheathed sword. He most likely wears a close-fitting blouse and trousers with feet, and, around his waist, a belt composed of two dotted rows is visible. The figure with the wolf's head holds a spear pointing downwards across the body with the left hand, and with the right hand bent near its chest, it unsheathes a sword (ToD2). The figure wears a long-sleeved, knee-length tunic. It is embellished with sections of horizontal stripes, and the sleeves appear to be flowing. Behind this figure, beneath the sword sheath, a similarly striped "tail" sticks down midway to the legs. It may possibly be a tail fixed onto a costume or perhaps the figure is wearing a complete wolfskin, with the head and tail intact.

In conclusion, the images on the helmet plates depict men in combat or other military situations, or men fighting with wild animals. No depictions of females are encountered in this category. Moreover, the similarities among the motifs of the plates enable the 24 different recorded images to be condensed into six groups. The grouping is based on the overall appearance and the elements in the image (Table 4.3).

The most common types of helmet plates belong to Groups 1 and 2 that incorporate equestrian images and men in procession respectively. The equestrian images comprise a representation of a centrally positioned rider in full figure. As a rule, this figure is heavily armed with a helmet, shield, lance or spear and sword. In addition, a group of up to four other men, each with their different posture and function are depicted. Underneath the horse, the full figure of a fallen man is about to stab a sword into the horse's belly. This figure also bears a shield in several cases and is represented in five panels, although not in Vendel 1A or Vendel 1B. In front of the horse, there may be yet another full figure standing, holding the horse. He only carries a spear and is represented on five panels, although not on Sutton Hoo B or Vendel 1B. Sutton Hoo B (Fig. 4.7) depicts a fallen figure that both holds the horse and sticks his sword into it. In addition, there are one to two smaller figures that are most often placed behind the rider above the back of the horse, or in front of its head. One of them seems to be flying, while the others are almost kneeling. These figures are helmeted in most cases, and carry spears. The tiny figures are represented on five panels, although not on Vendel 1A or Vendel 1B. Instead, on Vendel 1A and Vendel 1B (Fig. 4.4), flying birds are portrayed in the same position. The images of men in procession most often depict two men in full figure standing one behind the other. One image, however, has five figures in a row (Vendel 14B) (Fig.4.6). Both their appearance and armaments are entirely homogenous with helmet, sword, spear and possibly a shield. Torslunda D deviates from the other images in this group by depicting two figures of different appearance that seem to be in motion, rather than standing still (Fig. 4.9). The other groups are less common, and have only half as many examples of each as in Groups 1 and 2. Group 3 comprises two male warriors in full figure facing each other possibly in combat. Their weaponry constitutes of a shield, spear and sword, and the weapons are aimed at each other. Group 4 also comprises two men in full figure, but they are seen from the front and seem to be more static. Their armaments constitute of a helmet, 2–3 spears and a sword. Overall, there is a clear similarity among the images in Groups 2, 3 and 4 depicting two warriors engaged in various activities.

Groups 5 and 6 are depictions of men and animals. Group 5 depicts a man in full figure who holds a chained animal or beast. In his other hand, he holds an axe. Group 6 shows a man in full figure flanked by two bears. On Torslunda A (Fig. 4.9), the man sticks a sword into the belly of the bear and holds a knife in the other hand. These details are not preserved on the other example belonging to this group. Animals in combination with men are thus found in three groups (1, 5 and 6) with 12 different types of helmet plates in all. In addition, a further panel from Group 2 (Valsgärde 7A) (Fig. 4.2) also depicts two serpents and a bird. The animals depicted are horses, serpents, birds, bears and an unidentified beast with feline or canine traits. Apart from the serpents and birds, the other animals play a clear function in the image either as animals to ride on or as adversaries. It is possible to isolate and identify 54 males from the 24 helmet plates. Of these, 18 turn to the right, another 16 turn towards the left, 18 are seen from the front, while the direction of two cannot be determined. Thus, there does not seem to be any preferred angle of depicting the figures on the decorative panels, and most likely, it is the composition of the image and its position on the helmet that play a decisive role in the direction the figures face (Table 4.4).

Most of the men stand upright, while only a few figures either ride, kneel, or lie fallen. A single figure may even be considered to be flying (Valsgärde 7D5) (Fig. 4.2). The upright figures are primarily depicted on the images from Groups 2–6, while the riding, fallen and kneeling figures belong to the more complex images in Group 1. On the images in the former, the figures take up the entire area and are of equal size, while the figures appearing on the latter images are of varying sizes. The kneeling figures that are often seen both behind and in front of the mounted figures are smaller in size than the rest.

As the helmet plates depict various combat situations, the figures often carry different requisites or weapons. The weaponry on the helmet plates is comparatively homogenous with various combinations of helmets, shields, lances, *spathae* (double-edged swords), knives and axes. Helmets are represented in all groups, except Group 6 and constitute four different types: simple helmets, helmets with crests shaped like a bird or a boar and horned helmets ending in animal heads. All the shields are round with a circular shield boss, sometimes with the rivets depicted. Shields occur only in Groups 1–3. With the exception of the tiny figures, most of the men carry swords, and as far as may be ascertained, in most cases, it is a *spatha*. The length of the swords varies greatly, and, in some cases, it is perhaps only a long knife. On Torslunda A (Fig. 4.9), the man bears a sword in the one hand and a single-edged knife in the other. Often the sword is sheathed, and only in a few cases is the sword or knife drawn for combat (*e.g.* Torslunda A, Sutton Hoo A, Valsgärde 7E, Vendel 14C, Vendel 11B, Vendel 12A and Valsgärde 7C, Valsgärde7D, Valsgärde 8A, Sutton Hoo B) (Figs 4.2–4.3, 4.5–4.7 and 4.9). On several panels, a baldric or cross belt that the scabbard is attached to over the shoulder may be discerned. On two decorative panels, the swords are furnished with the highly discussed and prominent status marker: the hilt ring (Vendel 14A1-5 and Torslunda A1) (Figs 4.6 and 4.9) (Steur 1987; Nørgård Jørgensen 1999a). It has also been suggested that Vendel l4B1 has a hilt ring (Nørgård Jørgensen 1999a, 198), but in my opinion, it is more likely that the ring in question belongs to the embellishment of the costume (Fig. 4.6). Furthermore, it has been suggested that the scabbard fittings can be divided into orders of rank, with three rings denoting the highest rank, and the lowest rank having no rings (Nørgård Jørgensen 1999a, 198). The costumes that these figures wear are, however, homogeneous and therefore cannot support this hypothesis.

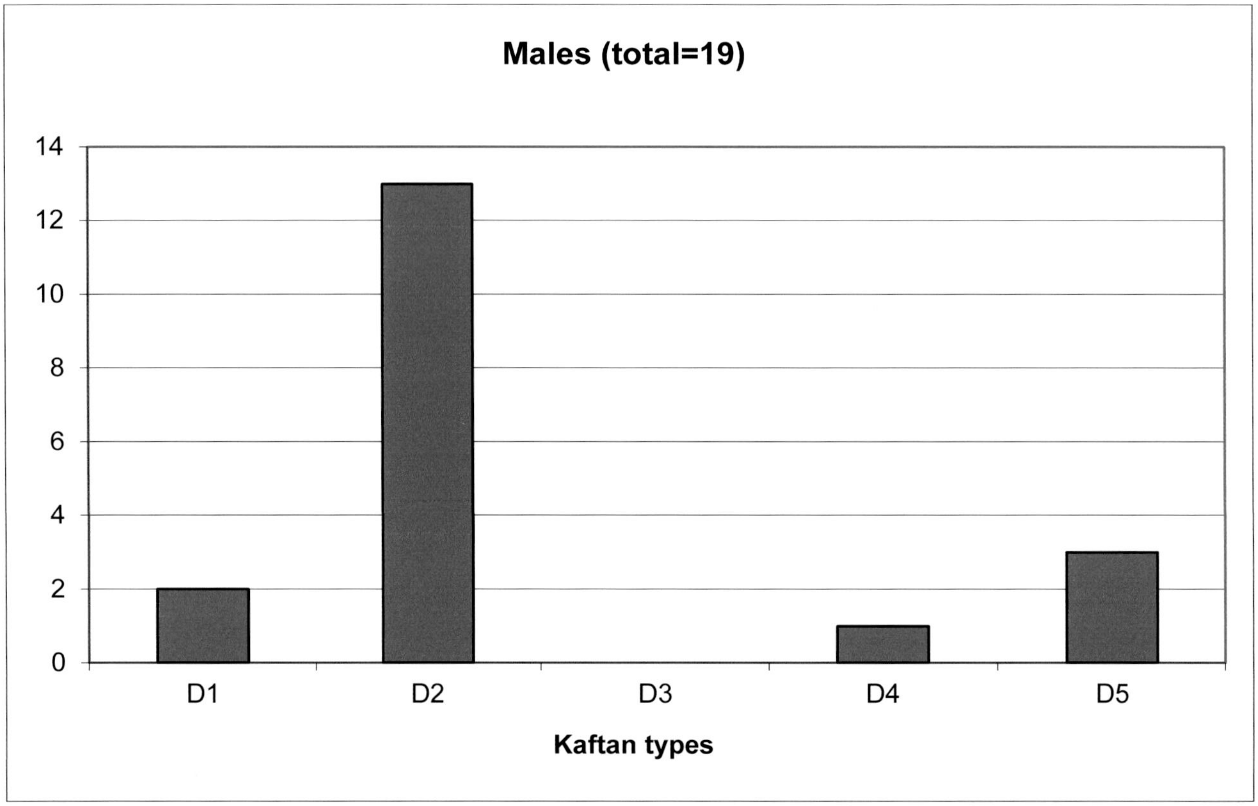

Fig. 4.10 Distribution of kaftan types on helmet plates.

Most men in Groups 1–4 carry lances, and several of them carry more than one, while no lances are seen in Groups 5 and 6. Axes occur in three decorative panels, all from Groups 5, and this type of weapon seems to be linked to exactly this kind of representation.

Male outerwear and innerwear

Approximately half of the depicted male figures wear Outerwear Group 6, also termed a kaftan, which is the only type of outerwear portrayed in this object category (Table 4.4). The kaftan possesses the same characteristics as that recorded for the gold-foil figures: long sleeves and wide borders along the slanted, closed front edge and lower edge, and occasionally, also borders at the wrist. The kaftan on the gold-foil figures are, however, longer than these specimens that generally reach to the knee. Most of the kaftans are devoid of embellishment. However, all-over patterning occurs on a few occasions (*e.g.* Sutton Hoo B2 (Fig. 4.7), and possibly Uppsala A1 (Fig. 4.8) and Valsgärde 8A2 (Fig. 4.3)). The bands are primarily of type D2. It is thus the same types, but with a markedly different distribution than is the case with the gold-foil figures (Fig 4.10) (see Fig. 2.34). The kaftan is primarily represented among the images in Group 1. It occurs less frequently in the other groups, and there are no kaftan-clad figures among the men in procession in Group 2. On the mounted figures, the kaftan is indicated by a V-shaped border or edging on the chest, but this costume is difficult to identify, as in many instances, the body is concealed behind the shield. In at least one case (Vendel 1A1) (Fig. 4.4), it could also be interpreted as a baldric. The kaftans also predominate among the fallen figures. On Valsgärde 7C and 7D (Fig. 4.2), the figures are so damaged that identification is uncertain. However, these figures do have an identical all-over patterning as the kaftan-clad figures in Valsgärde 7C2 and Sutton Hoo B2 (Figs 4.2 and 4.7), thereby strengthening the identification of the costume.

Almost half of the depicted male figures wear a tunic without accompanying outerwear (Fig. 4.11). The tunic is primarily of the dotted-patterned type (E7) that can have a lower border as a supplement (E7x) (Table 4.4). It is possible to interpret the closely dotted regular pattern as depicting chain mail. On the gold-foil figures, this patterning is relatively rare and only seen on figures from Helgö in Sweden and Lundeborg in Denmark, but in these cases, the dots are more spread out and give no obvious associations to chain mail. In addition, a braided-patterned tunic (E10) and a vertically-striped tunic with a border (E11) have been recorded. The latter is unique for this object category. A further two plain tunics (E2) have borders. Thus, they are not entirely plain tunic types, as is common among the gold-foil figures. All tunics recorded are knee-length, with either long or short sleeves, and resemble the corresponding gold-foil figure costumes. The figures in tunics occur mostly in

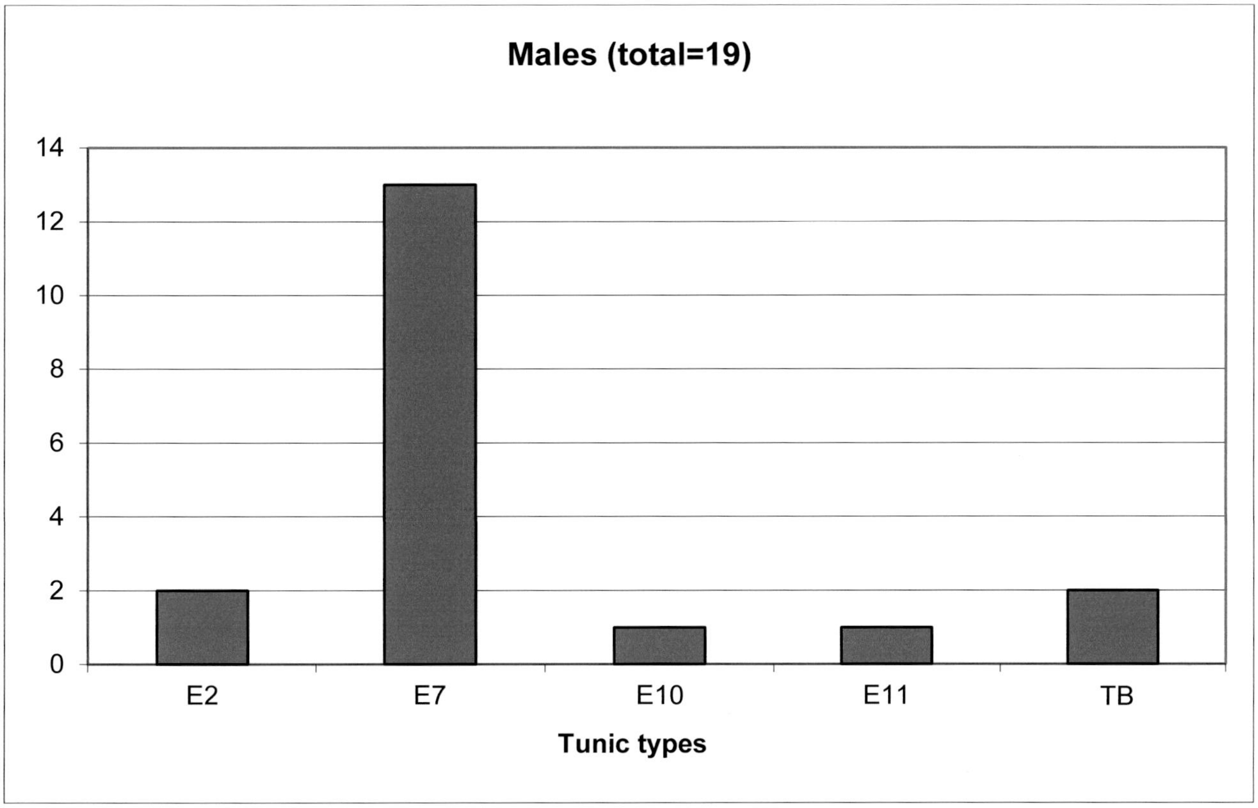

Fig. 4.11 Distribution of tunic types on helmet plates.

Groups 2 and 6, and are only represented by few specimens in the other groups. The identification of the costumes of those figures standing in front of the horse in the Group 1 images is less certain (Valsgärde 7C3, Valsgärde 7D3, Valsgärde 8A3, Valsgärde 8B3 and Vendel 1A2) (Fig. 4.2-4.4). One is recorded as a tunic with matching trousers and a baldric (Valsgärde 8A3) (Fig. 4.3). Another possibility is that this is a kaftan, as matching borders are displayed at the wrist. The other figures with a similar function seem to wear kaftans, but this too is not completely certain. Thus, the costume seen on Valsgärde 7C3 and Valsgärde 8B3 (Figs 4.2 and 4.3) is ambiguous, with mixed attributes of both tunic and kaftan. The most difficult of all to identify is the barrel-shaped figure with the V-shaped border on the chest. Whether it is a kind of belt or an unrecognizable costume is uncertain. Moreover, the costume items seen on Valsgärde 8A1 and Vendel 11B2 respectively (Figs 4.3 and 4.5), too, deviate on several points from the remaining representations, and have to be defined rather as blouse and trouser ensembles. The riding figure, Valsgärde 8A1 (Fig. 4.3), is exceptional in that the chest is ornamented with a wide horizontal area, and it is the only occurrence of this type of decoration among the entire selected material. Due to the presence of the shield, it is impossible to discern either the length of this upper part, or whether it is open or closed in front. A less likely possibility is that it depicts a kind of breastplate. Furthermore, the patterned trousers are indicative of this hardly being a kaftan. The upper part of the costume on Vendel 11B2 (Fig. 4.5) is reminiscent of a kaftan (V-necked border on the chest and cuffs at the wrist), but combined with trousers. This is a mixture of characteristics not usually seen together. The tiny kneeling or flying figures, positioned either in front of or behind the riders, are rarely depicted in a detailed costume, and often only portrayed in outline (*e.g.* Valsgärde 7D4) (Fig. 4.2). Some of these figures (Valsgärde 7C4, Valsgärde 7D5, Valsgärde 8A4 and Valsgärde 8B4) have a band placed diagonally across the chest (Figs 4.2 and 4.3), and as a rule a belt, which can be interpreted as a kind of baldric. The figures never carry a sword, only a spear. Furthermore, it is difficult to discern if the figures wear a tight-fitting blouse and trousers or no specific costume at all. It has obviously not been important to specify the costume for identification purposes or function.

Male costume combinations

Leg coverings comprise the following types: plain lower legs (L3), plain lower legs with a demarcation at the foot (L1), plain trousers (L8) and patterned trousers (L9) (Table 4.4). On one figure, the patterning may possibly be interpreted as leg wrappings (Valsgärde 8A2) (Fig. 4.3).

The group of tunic and kaftan-clad men are nearly equally distributed between leg coverings consisting of a plain lower leg and trousers with narrow legs (Fig. 4.12). In the latter case, a border may be perceived at the bottom

Table 4.4 Costume analysis of the helmet plates selected for this study.

Find ID	*Image group*	*Costume not identified*	*Outerwear*	*Tunic*	*Long sleeves*	*Short sleeves*	*Leg type*	*Belt*	*Head seen from R=right L=left F=front*	*S=standing R=riding L=lying, K=kneeling F=flying*	*X=helmet 2=bird 3=boar 4=horn*	*Shield*	*Spear*	*Sword + hilt ring (ring sword)*	*Baldric*	*Knife*	*Axe*	*H=horse S=serpent F=fowl/bird B=bear U=unidentified animal/ beast*
Va7D1	1	X							L	R	X	?	?	X				H
Va8A1	1			TB?	X		L9		L	R	2	X	X	X				H
Va7C1	1		D2				L1		R	R	X	X	?					H
Va8B1	1		D2		X		L1		R	R	2	X	X					H
Ve1A1	1		D2?		X		L1		R	R	3	X	X	X	?			H3F
Ve1B1	1		D4?				L1		L	R	2	X	X		?			HS2F
ShB1	1	X	D2?		X		L3?		L	R		X	X	X				H
ShB2	1		D5		X		L3	X	R	L				X				
Va7C2	1		D5?				L3		L	L		X		X				
Va7D2	1	X	D2?				L3		R	L				X				
Va8A2	1		D2/5		X		L9?	X?	F	L				X	?			
Va8B2	1		D2		X		L3		F	L		X		?				
Va7C3	1		D2?		X		L?	X?	F?	S			X		?			
Va7D3	1	X	D2?				L3		R	K								
Va8A3	1			E7	X		L9	X?	F	S			X		?			
Va8B3	1		D3?	(E2?)			L1	X?	?	S			X					
Ve1A2	1		D2		X		L1	X	F	S			X					
Va7C4	1			Full body	X		L8		R	K	?		X		?			
Va7D4	1			Full body	X		L8		F	K	4		X					
Va8A4	1			Full body	X		L8	X	F	K	4		X		?			
Va8B4	1			Full body			L8	X	F	K	4		X		?			
Va7D5	1	X		Full body			L3?		R	F					?			
ShB3	1	X		Full body	X		L3		L	K		?						
Va7A1	2			E7x	X		L3		R	S	2	X	X					S
Va7A2	2			E7x	X		L3		R	S	2	X	X					SF
Va7B1	2			E7x			L?		L	S	3	X	X	X				
Va7B2	2	X	D?	E?	X		L3		L	S	3	X	X	X	?			
Ve14A1	2			E7x		X	L1		R	S	2		X	X+				
Ve14A2	2			E7x		X	L1		R	S	2		X	X+				
Ve14A3	2			E7x		X	L1		R	S	2		X	X+				
Ve14A4	2			E7x		X	L1		R	S	2		X	X+				
Ve14A5	2			E7x		X	L1		R	S	2		X	X+				
Ve14B1	2			E7x		X	L1		L	S	2	X	X	X	X			
Ve14B2	2			E7x		X	L1		L	S	2	X	X	X	X			
ToC1	2			E2		X	L3		L	S	3		X	X+				
ToC2	2			E2		X	L3		L	S	3		X	X				

(Continued)

Table 4.4 Costume analysis of the helmet plates selected for this study. (Continued)

Find ID	Image group	Costume not identified	Outerwear	Tunic	Long sleeves	Short sleeves	Leg type	Belt	Head seen from R=right L=left F=front	S=standing R=riding L=lying, K=kneeling F=flying	X=helmet 2=bird 3=boar 4=horn	Shield	Spear	Sword + hilt ring (ring sword)	Baldric	Knife	Axe	H=horse S=serpent F=fowl/bird B=bear U=unidentified animal/ beast
ToD1	2/3			Bare upper body?			L8	X	F	S	4		2	X	X			
ToD2	2/3			E11	X		L3		L	S			X	X				
Ve11B1	3	X							R	S	X	X	X	X				
Ve11B2	3			BB?	X		L9	X	L	S	X	X	X	X	X			
Ve12A1	3	X		Bare upper body?					R	S		X	X	X				
Ve12A2	3			Bare upper body?			L9	X	L	S		X	X	X				
Ve14C1	3		D2/3		X		L1		R	S		X	X	X				
Ve14C2	3			E7		X	L1		L	S		X	X	X	X			
Va7E1	4		D2		X		L3		F	S	4		2	X				
Va7E2	4	X	D2?				L3		F	S	4		2	X				
ShA1	4	X							F	S	4		3	X				
ShA2	4		D1?		X		L3	X	F	S	4		3	X				
UpA1	4		D5?	E7x?	X		L3	X	L	S			2					
Ve1C1	5		D1		X		L1		F	S							X	U
Ve11A1	5		D5?				L1		F	S	X						X	U
ToB1	5			Bare upper body			L9	X	F	S							X	U
Va7F1	6	X		E7?			L3		F	S								2B
ToA1	6			E10		X	L3		F	S				X		X		2B

of the trousers, and especially the very wide edging on Valsgärde 7C1 and Valsgärde 8B1 support this interpretation (Figs 4.2 and 4.3). Seen in relation to the dating of the helmet plates to the Late Germanic Iron Age, this could possibly be an archaic element, as trousers with tablet-woven cuffs and hook clasps are an element in the archaeological record belonging to the Early Germanic Iron Age (Ræder Knudsen and Mannering 2007). Yet another possible interpretation is that it is a demarcation of the foot, with a bare lower leg.

On Valsgärde 8A3 (Fig. 4.3), both the standing and tunic-clad figures wear patterned trousers. The remaining representations of patterned trousers may be found on two depictions of men in combat (Group 3) and a single representation of a man and an animal (Group 5). Generally, the patterned trousers portrayed on the images have a high resemblance to one another with their closely-dotted patterning. This pattern is very similar to the dotted-patterned tunics or kaftans, and it possibly depicts a kind of leg protection.

A single figure wears plain trousers with no patterning (Torslunda D1, Fig. 4.9), but otherwise this is primarily a characteristic of the tiny figures in Group 1 (Valsgärde 7C3, Valsgärde 7D3, Valsgärde 8A3 and Valsgärde 8B3) (Figs 4.2 and 4.3). Many of the aforementioned trousers are of a type with a clearly delineated crotch, and often portrayed in conjunction with a bare upper body (Vendel 11B2, Vendel 12A2, Torslunda B1 and Torslunda D1) (Figs 4.5 and 4.9). A further two blouse and trouser ensembles are included here. Belts are found only among a few kaftan-clad men, while this is a regular feature among the blouse and trouser ensembles. In one instance, a belt is shown together with a tunic, but otherwise, there is nothing to indicate that the belt belongs together with the tunic.

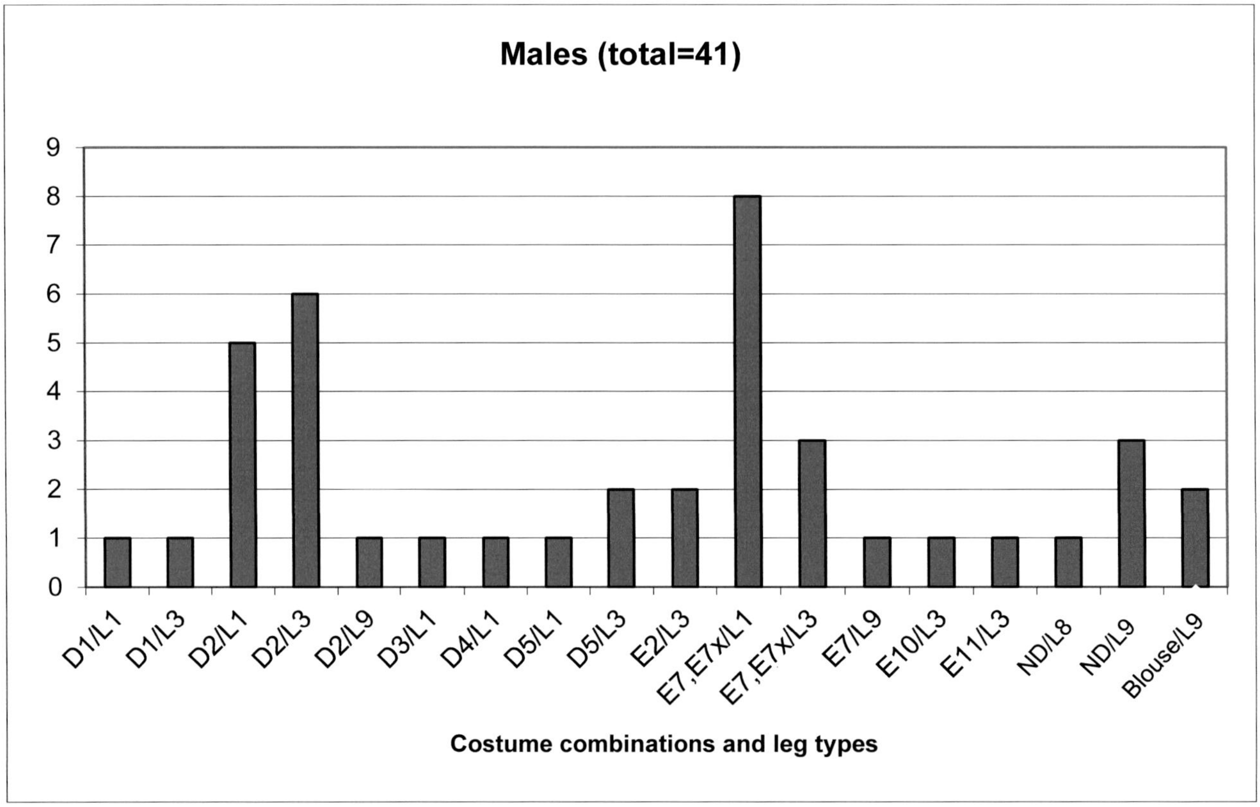

Fig. 4.12 Distribution of costume combinations and leg types on helmet plates.

Conclusion

The costumes depicted on the helmet plates are similar in many respects. Men in tunics and kaftans are, more or less, equally distributed, while no females feature in this object category (Table 4.4). The costumes depicted may be summarised in the following groups: kaftans combined with close-fitting trousers with cuffs, bare legs or leg wrappings; tunics combined with close-fitting trousers with cuffs, or bare-legs; trousers combined with a bare upper body or a blouse (Table 4.5). Kaftans predominate among the mounted warriors in Group 1, as well as among those posing in Groups 4 and 5. Compared to the gold-foil figures, the kaftan-clad figures are clearly predominant on the decorative helmet plates. The tunic is primarily linked to the men in procession (Group 2) and the representations of a man with two animals (Group 6). Blouses or bare upper bodies combined with trousers mostly occur in Group 3 (warriors/fighting men), but are also found in an almost identical version in Group 2 (Torslunda D1 is an anomaly here) and Group 5 (Torslunda B1). Bodies in silhouette or one-piece costume are reserved for the tiny figures in Group 1. Only a few tunic patterns are used, and primarily a regular closely-dotted patterning that gives associations of chain mail. Both costumes and the many weapons clearly link the men in this object category to the military. In contrast to what is seen on the gold-foil figures, the kaftan-clad men in this object category may be combined with men in other costumes. Thus, it may be concluded that the kaftan is not a costume that is reserved for single representations, but rather has to be considered a costume type that appears in an exclusively male milieu. Seen from this perspective, it is plausible that the kaftan-clad figures on the gold-foil figures that are never accompanied by females are also affiliated with military functions. With the dating of the helmet plate finds to the Late Germanic Iron Age, there is further evidence for the existence of the kaftans in this period, which accords well with the results from the gold-foil figures. Depictions of kaftans on the gold-foil figures are found in contexts primarily belonging to the Early and Late Germanic Iron Age (Watt 2004, 215).

Table 4.5 Male costume combinations on the helmet plates.

Costume combination	*Total*
Kaftan and trousers or kaftan and bare legs	19
Tunic and trousers or tunic and bare legs	17
Blouse and trousers or bare upper body	6
Body in silhouette (blouse and trousers), tiny figures	6
Costumes in total	48*

**Six costumes could not be identified.*

5

Jewellery

Jewellery with figural representations constitutes a mixed object category, which in most cases were used as costume fasteners, fibulae or pendants. Several scholars have presented various parts of the material under diverse themes or perspectives (*e.g.* Mackeprang 1935; Holmqvist 1960; Pesch 2002b; Simek 2002; Capelle and Fischer 2005). However, the costumes on the figures are often described at random, and with little focus on the costume itself and its significance. An exception is Flemming Bau's article from 1981, where several of the objects analysed here were included in a proposal for a reconstruction of Viking Age female clothing (Bau 1981). Twenty-nine items of jewellery with clothed figural representations were collected for the present analysis (Table 5.1). Fibulae or garment fasteners constitute approximately a third, while the majority of these objects comprise pendants.

The material selected for this study is briefly described in the following. *Tåstrup (1)*: In 1951, a fragment of a relief fibula was recovered in a burial near Tåstrup in Denmark. The fibula is crafted in the Nydam style and dated to the early Germanic Iron Age (Haseloff 1986, 68–69; Voss 1955, 177). On the fibula can be seen parts of a seated, right-profiled figure with two legs and an arm stretched out in front (Fig. 5.1). Its hand and feet are not preserved. Only the back of the head is visible, but this shows no special details. The body is shown in simple outline. The upper body is embellished with zigzag lines and punched with dots, while the legs are ornamented with stripes and dots. Most likely, the figure depicts a male wearing a blouse and trouser ensemble.

Hol (2): The remains of a gilded silver relief brooch were recovered in 1910 in a burial in Hol in Norway. The fibula is 15.4 cm long, shaped in the Nydam style and dated to the early Germanic Iron Age (Hougen 1936, 17–18, 36, pl. 10; Haseloff 1981, 11; 1968, 68). On the right side of the foot-plate of the fibula is depicted a full figure (Fig. 5.1). It is 1.8 cm long, facing the right with a leg, an arm and a hand, on which the thumb is turned towards the face. The body is simply delineated, while the head is more detailed, although without any indication of hair. The upper body is embellished with different diagonal lines, dots and circles, while the legs are ornamented with circles. Around the waist are seen two double horizontal stripes, suggesting a belt. Around the wrist is a marking. Most likely, the figure depicts a male clad in a blouse, trousers and belt ensemble.

Gummersmark (3): A large silver gilded relief fibula was unearthed amongst other finds during peat cutting in 1852 near Gummersmark in Denmark (Mackeprang 1952, 144–145). The fibula is 15.9 cm long and ornamented in Animal style IA (Haseloff 1981, 174–176, Taf. 23). The find is dated to the Early Germanic Iron Age (Fig. 5.1). On the foot-plate, the two small upright full figures are depicted in profile. The right figure is 1.7 cm high and the left figure, 1.8 cm high. Both figures are 0.7 cm wide. The left figure (3a) is facing right and the head appears very similar to the manner in which the gold-sheet figures are depicted. The hair is represented as a short, dotted border, but may also be interpreted as a kind of headgear or diadem. The figure has two arms, folded across the belly. Around the wrists is a simple marking. The garment has long sleeves and is knee-length, with a zigzag border at the wrists and lower hem. Possibly the figure also wears trousers. The right figure (3b) faces left, and is, more or less, similar to the above-mentioned, but has only one hand, enabling a bigger part of the vertically-striped costume to be visible. This costume is thigh-length, with a zigzag border at the hem. It could either be a tunic or a cloak and tunic combination. Most likely the figures are wearing trousers, as suggested by the marking visible at the ankle. On the basis of the hair-styles and the lengths of the outfits, the figures are most likely male. A third figure (3c), approximately 1. 8 cm long and 0.6 cm wide is placed on the foot-plate. The figure is similar to the two above-mentioned, but simpler. The figure is seen in right profile, but gives the impression of crawling. Only one arm with a marking at the wrist and one leg are depicted. This figure is not included in the costume analysis.

Ekeby (4): An equal-armed, silver relief fibula with traces of gilt was recovered in 1888 in Ekeby in Sweden (Fig. 5.1).

Table 5.1 The jewellery material selected for this study. Early Germanic Iron Age (EGIA), Late Germanic Iron Age (LGIA) and Viking Age (VIK), male (M), female (F).

Find ID	*Object*	*Location*	*Country*	*Material*	*Gender*	*Dating*	*Museum no.*
1	Relief brooch	Tåstrup	Denmark	Gilded silver	M	EGIA	NM615/51
2	Relief brooch	Hol	Norway	Gilded silver	M	EGIA	T9822
3	Relief brooch	Gummersmark	Denmark	Gilded silver	M	EGIA	NM12524
4	Relief brooch	Ekeby	Sweden	Gilded silver	M	EGIA	SHM8284
5	Costume fastener	Vester Egesborg	Denmark	Gilded bronze	M	7th century AD	
6	Oval brooch	Lousgård	Denmark	Bronze	M	LGIA 2A	NMC5587a-b
7	Valkyrie fibula	Stentinget	Denmark	Gilded bronze	F/F	Early VIK	NMI6917/89
8	Valkyrie fibula	Ribe	Denmark	Bronze	F/F	Early VIK	ASR9x476
9	Valkyrie fibulae	Tissø	Denmark	Gilded silver	F/F	VIK	KN-0855
10	Valkyrie fibulae	Hedeby	N. Germany	Bronze	M?/F	VIK	LMS/HS11918
11	Pendant	Norsborg	Sweden	Bronze	F/F	8th–9th century AD	
12	Pendant	Roskilde	Denmark	Gold	M/F	6th–8th century AD	
13	Pendant	Hagebyhöga	Sweden	Gilded silver	F	Pre-AD 850	SHM16429/16560
14	Figure pendant	Birka 825	Sweden	Silver	F	9th century AD	
15	Figure pendant	Birka 968	Sweden	Silver	F	10th century AD	
16	Figure pendant	Grödinge	Sweden	Silver	F	–	SHM20672
17	Figure pendant	Klinta	Sweden	Silver	F	–	SHM129
18	Figure pendant	Tuna	Sweden	Silver	F	–	GR. III, 23
19	Figure pendant	Gamla Uppsala	Sweden	Bronze	F	9th century AD	GR. 36/188
20	Figure pendant?	Hjorthammar	Sweden	Silver	F	–	
21	Figure pendant	Nygård	Denmark	Gilded bronze	F	VIK	
22	Figure pendant	Öland	Sweden	Bronze	F	VIK	
23	Figure pendant	Tissø	Denmark	Gilded silver	F	VIK	FB-0209/FG-2234
24	Figure pendant	Tissø	Denmark	Gilded silver	F	VIK	FG-3589
25	Figure pendant	Tissø	Denmark	Gilded silver	F	VIK	KN-1591
26	Figure pendant	Birka 571	Sweden	Silver	M	10th century AD	
27	Figure pendant	Ekhammar	Sweden	Bronze	M	Early VIK	
28	Figure pendant	Ekhammar	Sweden	Bronze	M	Early VIK	
29	Figure pendant	Uppåkra	Sweden	Silver	M	VIK	

The fibula is *c*. 14.3 cm long, crafted in Animal Style IB, and is dated to the Early Germanic Iron Age (Haseloff 1981, 186; Arrhenius 1986, 144–145). The decoration on the fibula is highly intricate and comprises both zoomorphic and human body parts, such as arms/hands, legs/feet and faces. Due to the compressed style, it is difficult to determine individual figures, but they seem to be wearing a costume consisting of a long-sleeved blouse, long narrow trousers and shoes. On the basis of this costume, all the figures are considered to be male.

Vester Egesborg (5): A unique, openwork, gilded, bronze brooch was recovered in the settlement of Vester Egesborg in Denmark (Fig. 5.1). Its motif is symmetrical and comprises a central figure flanked by two large and two smaller prehensile beasts. The brooch measures 5.8 × 3.8 cm and is dated to *c*. AD 600, *i.e.* the Late Germanic Iron Age (Gärtner and Ulriksen 1997, 17–19; 2000, 5–7). The figure is depicted in right profile, with the arms held out on either side of the body. The head resembles that of the gold-foil figures, and the hair is depicted in a short style. The figure is clad in a thigh-length costume, with a triple marking on the chest and a double marking immediately above the lower hem, and wears trousers or has bare legs. No marking is visible at the ankle. On the basis of the hair-style and costume, it is most likely a male. The item is interpreted as a representation of

Fig. 5.1 Depictions of males from relief fibulae from Tåstrup, Hol, Gummersmark in Denmark and Ekeby in Sweden, brooch from Vester Egesborg, and oval brooch from Lousgård in Denmark (drawings: Pernille Foss; after Haseloff 1986; Gärtner and Ulriksen 2000; Capelle 1968).

the biblical story of Daniel in the lions' den (Gärtner and Ulriksen 2000, 7).

Lousgård (6): Two identical oval brooches were recovered in 1886 in Lousgård grave 3 on the island of Bornholm in Denmark, each depicting a full figure (Vedel 1886, 413; Lyngstrøm 1989, 150) (Fig. 5.1). The burial is dated to the Late Germanic Iron Age period 2A (Jørgensen 1990, 141f). The figure is seen from the front holding both arms upwards, and with two legs. The head is seen in profile turning to the left. The figure is clad in a costume comprising a long-sleeved blouse with markings at the wrist and elbow, knee breeches and shoes. The lower legs are plain, suggesting that the legs are bare (Klindt-Jensen and Wilson 1965, 9–10). On the basis of the costume, the figure is most likely interpreted as a male.

Stentinget (7), Ribe (8), Tissø 1 (9), Haithabu (10): Five very similar Valkyrie brooches have been found in four locations in Denmark and Germany. One of these, a gilded bronze fibula was unearthed in 1991 in the topsoil near Stentinget in Denmark (Fig. 5.2) (Petersen 1992, 217; Nilsson 1994, 68–69). It measures *c.* 3.2 × 2.2 cm. As this was found using a metal detector, it is roughly dated to the Viking Age. A similar bronze fibula was recovered in 1991 during excavations in Ribe in Denmark (Fig. 5.2). It was found in a cultural layer belonging to the early Viking Age. Furthermore, two identical bronze fibulae dated to the Viking Age were unearthed during excavations in Tissø in Denmark in 1997 (Fig. 5.2) (Jørgensen *et al.*, 2004) and a fifth bronze fibula was recovered in Haithabu in north Germany (Fig. 5.2) (Capelle 1968, 107, Taf. 30, 12). This fibula is dated to the Viking Age and measures *c.* 4.4 × 3.2 cm.

The following description applies to all five fibulae: To the left of the image a rider in profile is seen on a horse facing right (7a-10a). Beneath the figure's thigh, a forward-facing spear sticks out. The rider holds a sword in its right hand. The figure is clad in a cloak which ends in a point at the back. The rider's hair is knotted at the nape of the neck. Also at the neck, a wide marking is visible. The broad thigh and the narrow lower leg are perhaps indicative of a type of wide knee breeches. Another possibility is that, it illustrates how a dress would look like, when its wearer is seated on a horse. In front of the horse stands a left-facing figure (7b-10b) holding a circular shield. Above the shield is seen an arm holding out a drinking horn. This figure wears a floor-length costume divided into backward tapering vertical stripes (Petersen 1993b, 43).

The fibula from Haithabu (10) deviates slightly in terms of detail from the others, for example the rider (10) has no hair knot nor is it possible to see a spear or sword, and the figure wears a costume without a cloak, but with a belt around the waist. A shield is held by the figure standing in front of the horse. This figure wears a costume with many narrow backward tapering stripes and probably also wears a helmet.

The hair knot and the backward flowing cloak belonging to the figure on the fibulae from Stentinget (7) and Ribe (8) suggest that, this could be a female, although its gender is open to question. On the basis of the costume, the rider on the Haithabu fibula (10) is most likely a male.

In all five cases, the standing figure is interpreted as a female, wearing a long dress. This feature is consistent with the dating of the fibula to the Viking Age. The heads of the figures have clear gender markings, and they most likely wear a helmet (Petersen 1993b, 43–44), although this is not evident in the Ribe fibula. Conflicting gender indicators are apparent on the fibulae too, as weapons and helmets are generally connected to males. The checked square seen beneath the horse is interpreted as the fabric of destiny, the cloth continuously woven by the Valkyries from human entrails throughout an individual's lifetime.

Norsborg (11): A square bronze pendant depicting two figures facing each other was recovered in the 1930s in a burial in Norsborg in Sweden (Fig. 5.2). The burial is dated to the 8th or early 9th century AD (Holmqvist 1960, 110–111). The pendent measures 1.9 × 2.0 cm, and including its loop, it is *c.* 2.5 cm high. The figures are depicted in full profile, and the depiction is edged with a simple frame. The figure to the left (11b) has finely combed back, and possibly, braided hair gathered in a knot at the nape of the neck, ending in a shoulder-length ponytail. No details of the face are given. The costume consists of an item of floor-length, triangular outerwear, which most likely has long sleeves and is fastened at the front. This jacket ends in a train, depicted with three transverse lines, which are marked above with two vertical stripes. Beneath this, a plain ankle-length dress with a simple border at the lower hem is visible. The arm of the figure is outstretched and holds onto the other figure's upper arm. Beneath the dress, the feet are visible. The figure to the right (11a) wears an almost identical costume, but on the jacket, there are no markings above the train. The hair is neatly arranged, similar to the person to the left, but the face has more details, such as eyes and a nose. The arms are similarly stretched out, and rest beneath the other figure's upper arm. On the basis of the hair-style and costume, the two figures are most likely females.

The composition of this depiction is very similar to that of the gold-foil figures, but is unusual in that the two figures are of the same gender. In contrast to other depictions, the two females appear unusually buxom.

Roskilde (12): An elongated gold pendant was unearthed in 1942 near Roskilde in Denmark (Fig. 5.2). The pendant measures 1.2 × 1.7 cm, and including the loop, it is 2.5 cm high (Mackeprang 1943, 69). The find is stylistically dated to the 6th–8th centuries AD. The two figures standing in profile facing each other are similar, yet small differences exist. The figure to the right (12b) has waist-length hair and wears a plain, long-sleeved, ankle-length tunic, with the feet visible beneath it. The outstretched arm has a marking at the wrist and rests on the hip of the figure opposite. The figure to the left (12a) has shoulder-length hair and is attired in a similar garment, although it is not quite as long. The feet and parts of the lower legs are visible beneath a costume. The faces of the figures are depicted simply with very large eyes and pointed chins. On the basis of the hair and length of the costumes, it is plausible that the figure to the left is a male and that to the right a female.

Hagebyhöga (13): A round silver pendant with traces of gilt was unearthed in 1929 in the disturbed barrow of female grave 1 near Hagebyhöga in Sweden (Arne 1932; Arrhenius 1962). The pendent depicts a figure seen from the front, inside a ring, which is 3 cm in diameter, with a

Fig. 5.2 Depictions of females on pendants from Norsborg and Roskilde in Denmark, Valkyrie fibulae from Stentinget, Ribe and Tissø 1 in Denmark and Haithabu in North Germany, and pendant from Hagebyhöga in Sweden (drawings: Pernille Foss; after Holmqvist 1960; Axboe 1986; Petersen 1993b; Jørgensen et al., 2004).

total height of 4 cm (Fig. 5.2). A ring with which to hang the pendant had once sat on the top of the head. Altogether, the objects found in the grave range over a period of more than 150 years, with the pendant being among the oldest. Most likely, the grave was constructed around AD 975 (Arne 1932). The figure has coiffured hair and possibly a diadem on the forehead. Around the neck is an over-sized disc-on-bow fibula. The costume appears more decorative than instructive, but most likely comprises a flowing cloak, open in the front, and a dress. The arms, decorated with dots, are folded beneath the breast in a bow shape. The four rows of dots around the neck possibly indicate a necklace. The upper part of the dress is plain, while the lower part is divided into several sections, with dotted and half-bow shapes respectively. At the hem is a circle-shaped border. The feet are not visible. On the basis of the costume, this is most likely a female.

Birka 825 (14): A silver pendant in the shape of a human figure measuring 2.7 cm in height and some beads were recovered in the 1870s in Birka grave 825 on the island of Björkö in Sweden (Fig. 5.3). The burial is dated to the 9th century AD (Arbman 1943, 298–300; Geijer 1938, 167). The face of the figure is turned towards the left, and the hair is coiffured into a knot, which ends in a ponytail. The left arm sticks out at the side of the costume and is bent upwards, so that the hand rests on the chest. The costume constitutes of an ankle-length outfit, divided into three sections. The foremost part has three vertical grooves whose top end is concealed by the arm. The rearmost part is similarly separated into three vertical sections. The arm sticks out from between the first and the second groove and only the two rearmost grooves continue all the way up to the shoulder. Both the foremost and rearmost parts of the costume have irregular marked horizontal stripes, suggesting that these parts belong together. The middlemost part comprises a large blank area with a triple horizontal-striped border at the lower hem. The costume flares out a little at the back, and the feet are visible beneath the costume. Based on the hair-style and costume, it is most likely a female.

Birka 968 (15): A silver pendant in the shape of a human figure, 2.6 cm high, suspended from a necklace was recovered in the 1870s in Birka grave 968 in Sweden (Fig. 5.3). The burial is dated to the 9th century AD (Arbman 1943, 394–396; Geijer 1938, 172). The figure's face is turned to the right and the hair is coiffured into a knot which ends in a ponytail. The right arm sticks out from the side of the costume and is bent upwards so that the hand rests on the chest. The costume constitutes of an ankle-length outfit divided into three sections, very similar to the above-mentioned one from grave 825. The foremost part is marked off into vertical grooves which at the very top are covered by the arm. The rearmost one is similarly divided into three vertical grooves. The arm sticks out between the first and second groove, and it is only the two groves at the very back that continue all the way up to the shoulder. Both the foremost and rearmost parts of the costume have irregular, horizontal stripes, suggesting that these parts belong together. The middlemost part comprises a large blank area with a double checked border at the lower hem. The costume flares out a little at the back, and thereby the rearmost part acquires several folds. The feet are visible beneath the costume. On the basis of the hair-style and costume, it is most likely a female.

Grödinge (16): A silver pendant, 3.4 cm high, was recovered in a cremation grave in Sibble in Sweden (Tillväxten 1934, xxxiii). Based on stylistic similarities, it is dated to the Viking Age (Fig. 5.3). The figure is seen in profile facing left. The hair is coiffured into a knot which ends in a ponytail. The costume comprises a rectangular cloak, which tapers down into a point at the side (*i.e.* the centre of the actual figure). It is embellished with a dotted pattern. Underneath and to the back of the cloak, a backward tapering costume, divided into a few diagonal stripes can be seen, and in front, the horizontally-striped part of the same dress is visible. The front and lower borders are embellished with fuzzy dots. On the chest, the border edge broadens itself into almond-shaped striping. To what extent this illustrates the dress beneath or the figure's hands, is uncertain. From the chin to the thighs another marking is visible, which could be the other side of the cloak. The feet are not visible. On the basis of the hair-style and costume, it is most likely a female.

Klinta (17): A silver pendant, 2.8 cm high, was found among a silver treasure from Klinta on the island of Öland in Sweden (Fig. 5.3). Based on stylistic similarities, it is dated to the Viking Age. The figure is seen in profile, facing left. The hair is coiffured into a knot at the neck. The arm is outstretched and holds a drinking cup. Around the wrist are two markings. The costume is either two-piece, comprising a blouse and a dress, or it is a dress with a belt at the waist. The lowermost part of the costume is striped, and tapers backwards into a train, while the front is straighter. The majority of the stripes that the dress is divided into are diagonal, which, especially in the lower part of the dress, give a herringbone-like pattern. The upper part of the dress has a plain centre. The outline border continues all the way around the figure, but as the inner back line is interrupted at the belt, it could also indicate the presence of a cloak. The feet are visible beneath the dress. On the basis of the hair-style and costume, it is most likely a female.

Tuna (18): A gilded bronze pendant measuring 3.4 cm was unearthed in 1893 in the damaged boat grave III in Tuna/Alsike in Sweden (Fig. 5.3). It has a loop on the back. Based on the stylistic features, it is dated to the Viking Age (Arne 1934, 26–28, Taf. vii). The figure is seen in profile facing left. The hair is coiffured into a knot at the neck, and ends in a long ponytail. The costume comprises a plain, triangular cloak that flares out at the back. Moreover, a vertically-striped train is visible at the back. At the front is seen a plain dress, embellished with two bands with crisscross patterning. Around the neck lies a circle interpreted as a disc fibula. In front of the chest a row of four circles placed above one another consecutively can be interpreted as a necklace. The interpretation of the two parallel stripes hanging down from the necklace is uncertain. The feet are visible beneath the dress. On the basis of the hair-style and costume, it is most likely a female.

Gamla Uppsala (19): A bronze pendant *c.* 3 cm high was unearthed in 1973 in a boat grave near Gamle Uppsala in

Fig. 5.3 Depictions of females on pendants from Birka 825, Birka 968, Grödinge, Klinta, Tuna, Gamla Uppsala, Hjorthammar and Öland in Sweden; Nygård, Tissø 2, Tissø 3 and Tissø 4 in Denmark (drawings: Pernille Foss; after Arbman 1940; Bau 1981; Nordahl 2001; Holmqvist 1960; Watt 1994; Hildebrand 1880; Jørgensen 1997; Jørgensen et al., 2004).

Sweden (Fig. 5.3). The burial is dated to the 9th century AD (Nordahl 2001, 46–52, 58). The figure is depicted in profile, facing left. The hair is unclear, but ends in a long ponytail. The left arm is outstretched, marked with a stripe and holds a large drinking horn in the hand. Around the wrist are two markings. The figure is somewhat blurred, but the costume comprises at least two layers. The outermost layer is presumably a cloak which tapers down into a point at the side (*i.e.* the centre of the actual figure). The cloak is plain, but perhaps the triangular striped area on the shoulder belongs to this costume item too. A striped dress is visible on either side of the cloak giving it a train-like end at the back and a patterned front. No feet are visible. On the basis of the hair-style and costume, it is most likely a female.

Hjorthammar (20): The lower part of a 1.6 cm high silver pendant, was recovered in a burial near Hjorthammar in Sweden (Fig. 5.3). Based on stylistic similarities, it is dated to the Viking Age (Holmqvist 1960). The figure is seen in profile facing right. The costume constitutes of a triangular cloak decorated with a dotted pattern. The right hand holding a drinking cup sticks out from the front edge of the cloak. A dress ending in a train and marked with stripes along the back is visible beneath the cloak. The feet are visible beneath the dress. On the basis of the costume, it is most likely a female.

Nygård (21): A gilded bronze pendant *c.* 3 cm high, was unearthed with a metal detector in 1992 near Nygård on the island of Bornholm in Denmark (Fig. 5.3). Based on stylistic similarities, it is dated to the Viking Age (Petersen 1992, 230; Watt 1994, 81). The figure is seen in profile, facing left. The hair is coiffured in a knot high up on the back of the head and ends in a short ponytail. The figure holds out a drinking cup in the left hand. The costume constitutes of a long, triangular cloak with a pattern of bows or half-circles. In the curved front edge of the cloak, a diagonally-striped border is visible, suggesting a fringe. Underneath the cloak, a plain dress can be seen. The costume flares a little out to the back, but does not end in a train. The feet are visible beneath the dress. On the basis of the hair-style and costume, it is most likely a female.

Öland (22): A 3 cm high bronze pendant, similar to that from Nygård, was recovered as a stray find in 1880 on Öland in Sweden (Fig. 5.3). It is dated to the Viking Age (Hildebrand 1880; Petersen 1992, 230; Watt 1994, 81). The figure is seen in profile facing left. The hair is coiffured into a knot high up on the back of the head and ends in a short ponytail. The figure holds out a drinking cup in the left hand. The costume constitutes of a long garment that flares a little at the back. It is decorated with a dotted border that goes all the way around the garment. At the centre are four rhombi of different sizes, placed one on top of the other, with the narrowest at the top and the widest at the bottom, separated by horizontal bands. It is difficult to determine if this is an item of innerwear or outerwear. No feet are visible. On the basis of the hair-style and costume, it is most likely a female.

Tissø 2 (23): A gilded silver figure with inlaid enamel work. The upper part was unearthed in 1995 during an excavation at Tissø in Denmark (Fig. 5.3). Continued excavation revealed the missing lower half in 1997. The figure is dated to the Viking Age (Jørgensen 1997). The figure is seen in profile facing left. The hair is coiffured in a knot at the neck, and ends in a short ponytail. The outer edge of the figure divides the body into two parts: a bell-shaped upper part and a crescent-shaped lower part. In the upper half, the left arm is depicted interlaced with the borders on the chest. The right arm emerges from below the shoulder and is bent upwards and holds on to the front opening of the costume. The left hand emerges at the point of the chest. Beneath the left arm, a trefoil knot comprising two parallel bands is visible. This decoration gives the body and the costume an ornamental, rather than realistic, character. The lower part of the costume is ornamented with slightly curved horizontal stripes suggesting that the costume ends in a train. This costume can most likely be interpreted as a cloak and dress, although the position of the arms and the embellishment of the cloak make the identification uncertain. The feet are visible beneath the dress. On the basis of the hair-style and costume, it is most likely a female.

Tissø 3 (24): A gilded silver pendant was unearthed at Tissø in Denmark in 1999. It is dated to the Viking Age (Jørgensen *et al.*, 2004, 58). The figure is seen from the front (Fig. 5.3). The hair is coiffured into two knots ending in ponytails. These are grasped by the two upstretched arms. The costume is difficult to interpret, but comprises an item of innerwear and outerwear respectively, most likely, a dress and a cloak. On the chest is possibly depicted a large jewel. No feet are visible. On the basis of the hair-style and costume, it is most likely a female.

Tissø 4 (25): A gilded silver pendant dated to the Viking Age was unearthed at Tissø in Denmark (Jørgensen *et al.*, 2004, 58) in 1999. The figure is seen in profile, facing left (Fig. 5.3). The face is stylised and the figure most likely bears a helmet. The hair is not visible. The upper body is covered by a shield. The left hand is placed below the cheek. The costume is vertically-striped in two different directions. It is difficult to determine, but constitutes of an item of inner and outerwear respectively, most likely a dress and cloak. No feet are visible. On the basis of the costume, it is most likely a female. The figure has many similarities to the upright figure on the Valkyrie fibulae (7b-10b).

Birka 571 (26): A 2.9 cm high silver pendant, shaped as a human figure, was recovered in the 1870s in Birka grave 571 in Sweden (Fig. 5.4). The grave is dated to the beginning of the 10th century AD (Arbman 1943, 185–186; Ringquist 1969, 288, 295). The figure is seen from the front, wearing a horned helmet. The arms are held out at the sides, and in its right hand, the figure holds a staff, and in the left, a sword.

The feet are turned towards the left (Arwidsson 1989b, 7). The figure wears an ankle-length costume that narrows down at the waist, ornamented with eight circles whose placement emphasises the shape of the costume. On the basis of the costume, it is most likely a male wearing a long tunic.

Ekhammar (27 and 28): Two small bronze figures were recovered in 1968 in cremation grave 6 near Ekhammar in Sweden (Ringquist 1969, 286). The grave is dated to the early Viking Age (Ringquist 1969, 295–296). The first figure (27) is *c.* 3 cm high and depicts a horned helmeted figure facing the front (Fig. 5.4). The arms are held out at the sides, with a staff in the right hand and two crossed staffs or similar in the left. The feet are turned towards the left. The figure wears a thigh-length costume, with a three-grooved belt at the waist. On the basis of the equipment and the costume, it is most likely a male wearing a tunic, and possibly trousers. The other figure (28) is similarly *c.* 3 cm high and depicted in profile, facing right. The face depicts a wolf's head with fangs. The figure seems to be biting a long wormlike object, which it holds out in front of itself. The rest of the body resembles a human being wearing a thigh-length costume encircled by a diagonal-striped border. The legs are either bare or the figure is wearing tight-fitting trousers. Due to the preservation conditions, it is not possible to see if there are any markings at the ankle. On the basis of the costume, it most likely depicts a wolf warrior in a short tunic and perhaps trousers.

Uppåkra (29): A 3.4 cm high silver pendant was unearthed with a metal detector in 1996 at Uppåkra in Sweden (Fig. 5.4). The item is dated to the Viking Age (Larsson and Hårdh 1998a, 140–144). The figure is seen in profile facing right and has an arm and two legs. The face has no details, and the topmost area flattens out into a snout-like shape, which is difficult to decipher. The right hand is outstretched and holds a crescent-shaped object which could be a drinking horn. The costume most likely constitutes a long-sleeved blouse. Around the upper body is a horizontal V-shape, divided into two diagonal-grooves, which could either be a type of cross-belt or some other kind of decoration. The figure is wearing baggy knee breeches with many folds. The lower legs are marked at the ankle. On the basis of the costume, it is most likely a male.

In conclusion, the selected jewellery is found mainly in Sweden and Denmark, while only a single find comes from Norway and another from Haithabu, in north Germany (Table 5.1). Around half are found in graves, while the rest are stray finds, metal detector finds, settlement finds and a single treasure hoard. Approximately a third was found in the 19th century, and the majority before 1980. The selected material covers the period from the Early Germanic Iron Age to the Viking Age. The relief brooches are among the early finds from the Early Germanic Iron Age, while the Late Germanic Iron Age is lightly represented by an oval brooch and a costume fastener respectively. The remaining finds, Valkyrie brooches and pendants are primarily attributed to the Viking Age. Only in a few instances have the find contexts given a precise dating. However, it is generally accepted that the representations reflect the times in which they were deposited. The grave from Hagebyhöga in Sweden constructed around AD 975 is, however, an exception (Arne 1932, 112). When the gender distribution is compared to the dating, male representations are perceived to cover the entire period under discussion, while female representations primarily originate from the Viking Age. This implies that male representations overlap with all other object categories, while female representations are primarily contemporaneous with the tapestry category. Generally, the quality of the selected material is of high craftsmanship. The silver and bronze specimens are, more or less, equally distributed,

26. Birka 571 **27. Ekhammar** **28. Ekhammar** **29. Uppåkra**

Fig. 5.4 Depictions of males on pendants from Birka 571, Ekhammar and Uppåkra in Sweden (drawings: Pernille Foss; after Arbman 1940; Ringquist 1969; Larsson and Hårdh 1998a).

although more of the silver jewellery is gilded in comparison to the bronze jewellery. One pendant is made exclusively of gold (12, Fig. 5.2).

Even though great variations in the size of the jewellery occur, the bulk of the figures are about 3 cm high, *i.e.* about double the size of the gold-foil figures. The grade of detail is, however, not correspondingly high, and often the jewels contain fewer details than on the gold-foil figures. With the exception of the pendants, which are primarily made as clothed figures, other figural representations form part of a decorative whole, circumscribed partly by style and partly by technique. This phenomenon is especially evident on the figural representations with Animal Style ornamentation, as is, for example the case with the relief brooches (1–4, Fig. 5.1).

The bulk of the jewellery figures comprise single depictions of either men or women, while only six items contain depictions of two figures. These comprise two pendants with double representations whose composition closely resembles that of the gold-foil figures (11 and 12, Table 5.2) and Fig. 5.2). One of them, moreover, is special in that it depicts two figures of the same gender. The five nearly identical fibulae displaying more action-oriented images are the so-called Valkyrie brooches. Four of them depict two women (7–9, Table 5.2 and Fig. 5.2), while the fifth (10) most likely portrays a male and a female.

All in all, 23 females and 14 males are recorded in the 29 selected jewels. The gender is identified partly through the appearance of the costume, and partly through other gender characteristics, such as hair-style and jewellery. Among the female representations, a slight majority turn their profiles towards the left, while only two are seen from the front (13 and 24, Table 5.2, Figs 5.2 and 5.3). Most of the females are depicted with only one arm. Around half have feet or shoes, and most have their hair in a knot or a long ponytail, while none are depicted with a cap or hair-net. Incidentally, hair is an important indicator of gender, and thus one of the characteristics on which scholars have focused a great deal. The vast majority of the female figures stand upright, while the five Valkyrie fibulae display the only mounted female figures in the entire selected material (7a-10, Table 5.2 and Fig. 5.2). In the tapestry category females are depicted, as passengers in horse-drawn wagons or sledges, but never riding. Males, in contrast, are often portrayed riding and/or in other connections with horses on the bracteates, helmet plates and tapestries. On this basis, female riders may be ascribed special status and position. Moreover, on the five Valkyrie brooches and a singly represented female (25, Fig. 5.3), the figures bear various combinations of helmets, shields, swords, spears and drinking horns respectively. A further six singly represented females hold a drinking horn or cup stretched out in front of them (17, 19–22, Fig. 5.3). Valkyries, according to Old Norse mythology, are the god Odin's handmaidens who help select the warriors who are to be slain in battle. Their attributes are weapons and drinking horns or cups with which they welcome the fallen warriors to Valhalla (Petersen 1993b, 41). It is thus likely that the figures on these brooches do indeed depict Valkyries. The male rider (10a, Fig. 5.2), however, if the gender is correctly identified, may be interpreted as a warrior being welcomed to Valhalla. It is uncertain if all depictions of females with drinking horns should be interpreted as Valkyries, as has been suggested earlier (Simek 2002, 110). The drinking horn may also be linked to the role and privilege of the lady of the house (Enright 1988), and it has further been suggested that women carrying drinking horns could represent venerable female ancestors, the *disir* who were venerated as semi-goddesses (Simek 2002, 114–116).

Jewellery is portrayed only on three different occasions: a disc-on-bow fibula in combination with a necklace (13, Fig. 5.2); a disc fibula combined with a necklace (18, Fig. 5.3); and a fibula of indeterminate type (24, Fig. 5.3). Thus, jewellery is not a prominent element on these depictions. In percentage terms, this corresponds to 13% of the females, which is somewhat less than for the gold-foil figures. The earlier mentioned representations are all attributed to the Viking Age through their find context. The disc-on-bow fibula seen on the Hagebyhöga pendant, however, may suggest a link to the Late Germanic Iron Age, where this type of fibula predominates (Jørgensen and Jørgensen 1997, 29, 41–42). The disc fibula has a somewhat broader time frame and is consistent with the Viking Age, as the specimen from Tuna (18) also confirms. Necklaces were an important part of women's jewellery collections throughout the entire Iron Age and cannot be linked to any specific period. The female riders on the Valkyrie fibulae also have a demarcation at the neck that can either be interpreted as a neck ring or the neckline of the costume. As both possibilities are equally viable, this element was not included in the analysis.

In contrast to the females, the vast majority of the male representations have a right-facing profile, while two representations are seen from the front (26 and 27, Fig. 5.4). Most of the men stand upright and have two feet or legs visible, while an equal number of men are portrayed with one or two arms. Only in a few cases can the gender be determined exclusively on the basis of the hair-style, but, in general, male hair is short and well groomed. Two male figures (26 and 27) also bear helmets, but in contrast to the female helmets, these have horns. Their weaponry is a combination of sword and staff. The male figure from Uppåkra (29, Fig. 5.4) holds a curved object in his hand that greatly resembles a drinking horn. However, this identification is uncertain as the drinking horn is a typical female object. Finally, many males have markings at the wrist and at the transition to the foot, which, as in the female case, can be interpreted as a part or a demarcation of the costume.

Table 5.2 Costume analysis of the jewellery material selected for this study.

Find ID	*Costume not identified*	*Gender*	*Outerwear*	*Dress*	*Tunic*	*L=long sleeves, S=short sleeves*	*Leg type*	*Train*	*Belt*	*Head seen from R=right L=left F=front*	*R=rider S=standing L=lying/sitting*	*H=hair knot S=short hair L=long hair*	*Number of arms*	*Number of legs*	*Helmet=X; 4=horn*	*S=shield P=spear W=sword T=staff*	*B=disc-on-bow fibula D=disc fibula; =necklace*	*Drinking horn*
1		M			TB	L?	L8			R	L	?	1	2				
2		M			TB	L	L8		X?	R	L	%	1	1				
3a		M			E2	L	L2			R	S	?	2	2				
3b		M			E11	?	L1			L	S	?	1	2				
3c	X	M			TB		L8?			R	L	%	1	1				
4		M			TB	L	L8			RL	?	S	2	2				
5		M			TB	L?	L8			R	S	S	2	2				
6		M			TB	L?	L4		?	L	S	S	2	2				
7a		F?	A1	C1?			L5?			R	R	H	1	1		PW		
7b	X	F		C4			L6			L	S		1	0	X?	S		X
8a		F?	A1	C1?			L5?			R	R	H	1	1		PW		
8b		F?		C?			L?	X		L	S		1	0	X?	S		X
9a		F	A1	C1?			L5?			R	R	H	1	1		P?W?		
9b		F		C4			L6	X		L	S		1	0	X	S		X?
10b		F		C4			L6	X		L	S		1?	0	X?	S		X
10a	X	M?			E1?		L3?		X	R	R	?	2?	1?				
11a		F	B2	C2			L5	XJ		R	S	H	1	2				
11b		F	B2	C2			L5	XJ		L	S	H	1	2				
12a		M			E1	L?	L3			R	S	S	1	2				
12b		F		C1		L	L5			L	S	L	1	2				
13		F	A5	C2/ SB		L?	L6	?		F	S	?	2	0			NB	
14		F	A18	C2		L	L5	X		L	S	H	1	2				
15		F	A18	C2		L	L5	X		R	S	H	1	2				
16		F	A21	C2			L6	X		L	S	H	1?	0				
17		F	A5	C4			L5	X	?	L	S	H	1	2				X
18		F	A1	C2			L5	X		L	S	H	0	2			ND	
19	X	K	A20	C6			L6	X		L	S	L	1	0				X
20		K	A16	C4			L5	X		R?	S	?	1	2				X
21		K	A15	C1?			L5			L	S	H	1	2				X
22		K		C9?			L6			L	S	H	1	0				X
23		K	A1?	C4			L5	X		L	S	H	2	2				
24		K	A5?	C5F?			L6	X?		F	S	H	2	0			?	
25		K?	A5	C4			L6	X		L	S		1	0	X?	SW?		X?
26		M			E7	L	L3			F	S		2	2	X4	WT		
27		M			E1	L?	L3		X	F	S		2	2	X4	W 2T		
28		M			E1	L?	L3			R	S	%	1	2				
29		M			TB	L?	L7		X	R	S	?	1	2				?

Female outerwear and innerwear

Of the 23 female figures, over half are dressed in outerwear. Of these, the cloak appears most frequently, while two representations on the same object (11a and 11b) wear jackets (Fig. 5.2 and Table 5.2).

Cloaks are found in both short and long versions, but most often in the latter. The vast majority belong to Outerwear Group 1a (OG1a) that is fastened at the centre of the chest, in two cases with a clearly depicted fibula (13 and 18) (Fig. 5.5). Two female figures have been recorded with an A18 type cloak (14 and 15) belonging to OG3, which is only worn by males in the gold-foil figure category (Fig. 5.6). This garment covers the back and chest and is open at the side allowing a glimpse of the innerwear. One possibility is that the garment was fashioned as a cloak, and that the sleeves belong to the innerwear. Another possibility is that the garment only rests on the shoulders as in OG2, and that the foremost part belongs to the dress. In this case, the depiction may be interpreted as an example of a dress and apron ensemble, as seen on several gold-foil figures. On the basis of the patterning, the first possibility is most likely. Cloaks A20 and A21 are also special types, unknown among the gold-foil figures. However, this type occurs frequently among female figures on the tapestries, and is recorded as OG1c, as it belongs to the group of closed cloaks. In contrast to the gold-foil figures, the jacket seen on the two female figures, portrayed on the same object (11), is floor-length and ends in a clearly delineated train (Figs 5.2 and 5.6). Although the jacket on the gold-foil figures is worn by both genders, there can be no doubt in this case that, both figures are female. Their characteristic hair-styles with a knot and ponytail, the rounded chests and the long costumes clearly support this interpretation. Generally, outerwear is nearly equally distributed between plain and patterned types, of which the vertically-striped cloak is the most common.

The innerwear is dominated by dresses (Table 5.2). The costume from the Hagebyhöga pendant (13, Fig. 5.2) may also possibly be a skirt and blouse combination, but as this interpretation is ambiguous, therefore, it is recorded as a dress. Long sleeves are clearly visible in a few cases (12-15, Figs 5.2 and 5.3), corresponding to the dresses seen on the gold-foil figures, but such details are rare. A little more than a third of the dresses have various borders on the lower edge (C2 and C5) that often comprise a simple border, although sometimes more intricate ones also occur (Fig. 5.7). Plain and patterned dresses are equally distributed, the vertically-striped being the most common among the latter. A few unusual types, such as the horizontally-striped (C6) and the dotted (C9) types are also recorded. In many ways,

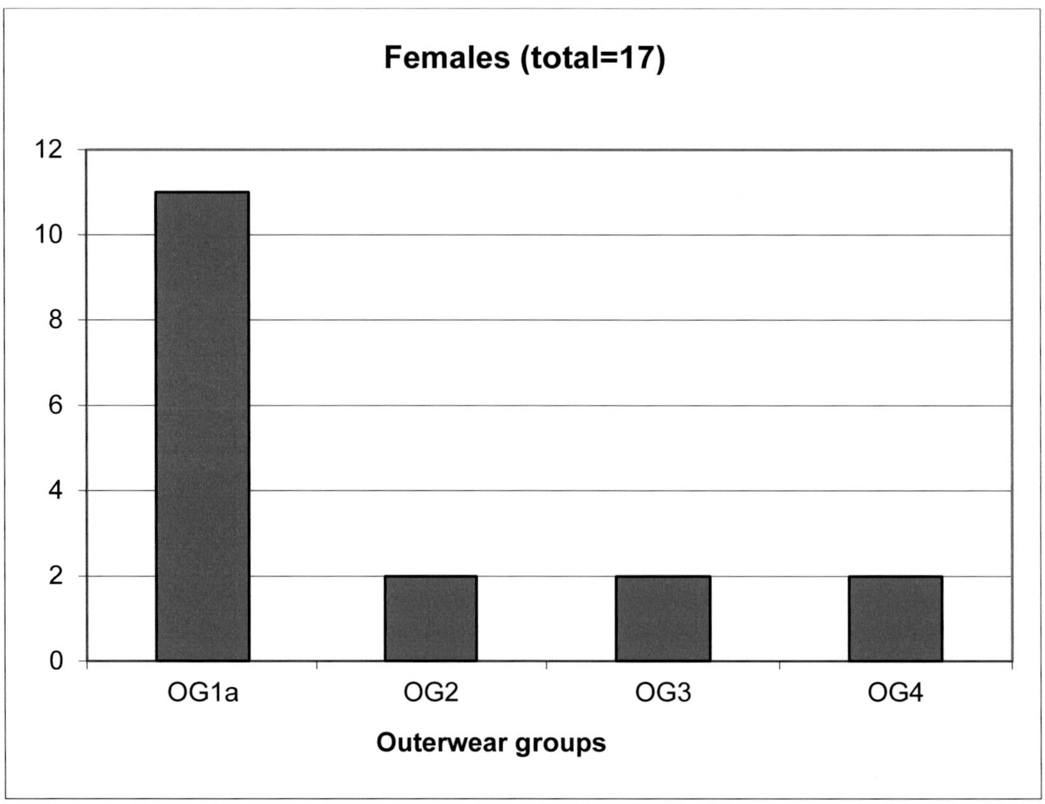

Fig. 5.5 Distribution of female outerwear groups in the jewellery.

this distribution corresponds to that recorded among the gold-foil figures.

Over half of the dresses are furnished with a train, which is most often depicted as an integral part of the dress. This characteristic is depicted with a marking at the back of the costume that ends in a downward pointing triangle. The type of train seen on the gold-foil figures is highly similar to the figure from Tuna (18, Fig. 5.3). Moreover, the figures from Hagebyhöga, Norsborg (11, Fig. 5.2) and Klinta (17, Fig. 5.3), have outerwear that clearly include a train, which is unusual.

The innerwear worn by the three female riders (7a-9a, Fig. 5.2) deviates from the other representations and is difficult to interpret. At first glance, it appears as if the women are depicted in a kind of baggy innerwear reminiscent of the wide breeches seen on several male representations. It is rare to see gender specific costumes used by the opposite gender. Written sources do not reveal whether the Valkyries wore typical male garments or not. Thus, it is possible that the lower parts of the costumes depict how female riders wearing long dresses would look like. This is an example of how female and male appearances could have been blended, thereby creating a new meaning. The other figures that are also interpreted as Valkyries wear manifestly female costume, with matching male attributes, such as shields and swords.

Most female representations wear a combination of inner and outerwear (Fig. 5.8). No feet are visible in the case of a little less than half of them, but there are just as many without feet wearing a combination of inner and outerwear, as those merely wearing a dress. However, no feet tend to be connected with patterned dresses and cloak types like OG1c, but it is difficult to identify general combinations, on the basis of the limited material. Gold-foil figures with no feet displayed are a characteristic that is primarily seen among the female figures from Norway and attributed to Costume Group 2. In comparison, floor-length costumes are thus more common in the jewellery category, as well as the tapestry category.

Male innerwear

No male figures in outerwear are found in this object category. The male costume is evenly distributed among tunics and blouse and trouser ensembles (Table 5.3) (Fig. 5.9). The tunics are always long-sleeved and can either be thigh- or knee-length. These two types are more or less equally distributed and both types are worn together with long trousers. The majority of the tunics are plain or plain with a border (E1 and E2), while a few are patterned. The horizontally-striped tunic type (E11) is only found in this object category. The tunic is most often combined with a plain lower leg (L3) although trousers also occur.

In the case of blouse and trouser ensembles, the trousers are generally long and tight-fitting (L8) (Table 5.3 and

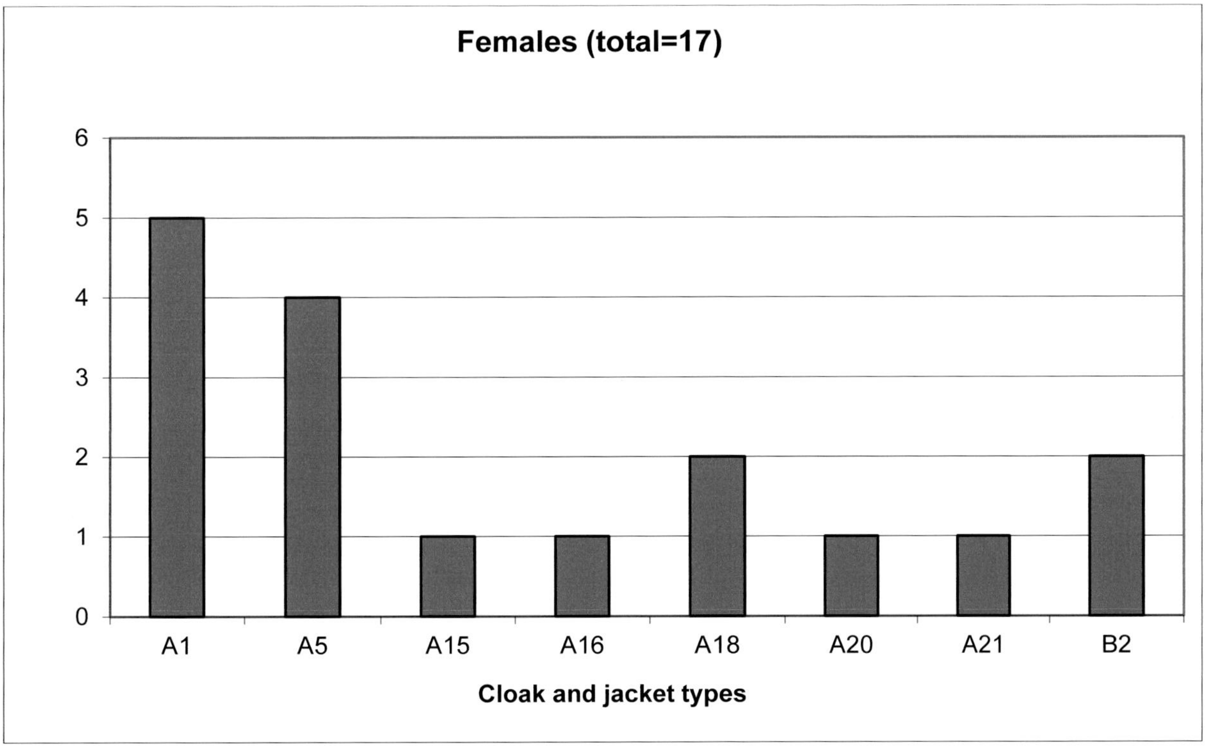

Fig. 5.6 Distribution of female cloak and jacket types in the jewellery.

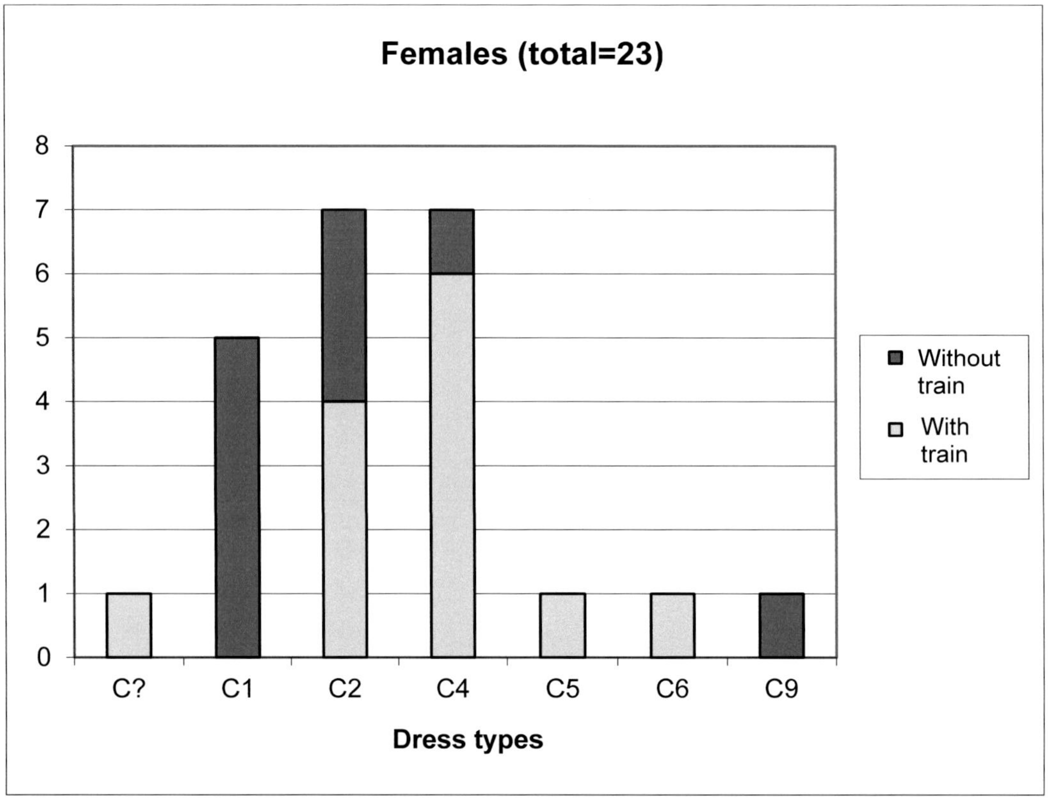

Fig. 5.7 Distribution of female dress types in the jewellery.

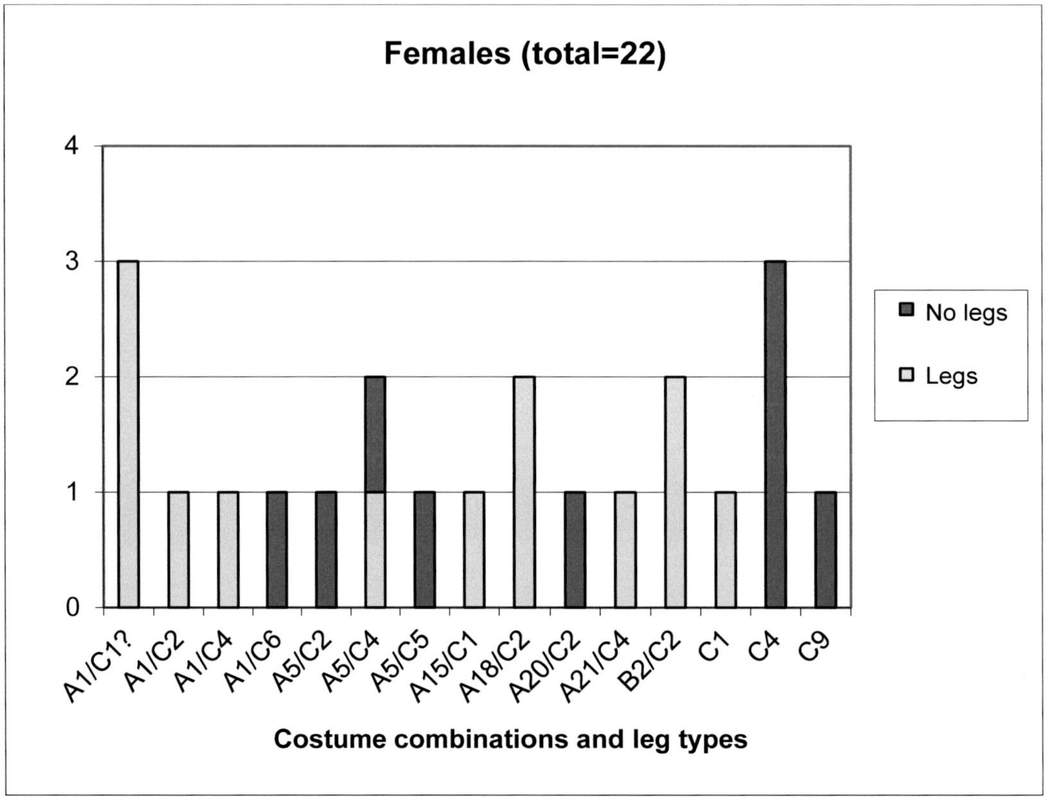

Fig. 5.8 Distribution of female costume combinations and leg types in the jewellery.

Fig. 5.9). This type is seen among the gold-foil figures only in the patterned (L9) version. Other trouser types are the knee-length breeches (L4), or the three-quarter length wide breeches (L7). Usually knee-length breeches are seldom found among costume representations, and are only seen among the gold-foil figures in three cases (Sorte Muld 162, Sorte Muld ris, and Uppåkra 16). The brooch from Lousgård (6, Fig. 5.1) with the depiction of a man in knee breeches is attributed to the Late Germanic Iron Age 2A period, which corresponds to the dating of the gold-foil figures. The wide breeches are recorded for only one male figure attributed to the Viking Age (29, Fig. 5.4). However, they are also frequently represented in the tapestry category. The remaining three male figures dated to the Viking Age wear plain tunics (E7) and are thus not especially characteristic. In turn, the blouse and trouser ensembles tend to occur in the earlier period corresponding to the Early Germanic Iron Age. Similar to the tunics, the blouses, too, have long sleeves. A single male representation in a short tunic is interpreted as a wolf warrior (28, Fig.5.4). Wolf warriors or males with animal heads are relatively rare, but occur also in the helmet plate and tapestry categories. Finally, in male costume, patterned blouse and trouser combinations and tunics tend to be portrayed on objects that are attributed to the Early Germanic Iron Age, while the plain tunic is primarily encountered in the finds dated to the Late Germanic Iron and Viking Ages. Moreover, it appears that the patterning of the blouse and trouser ensembles often constitute an integral part of the object ornamentation (with most found on the relief brooches), and it is uncertain whether this patterning should be interpreted as actual costume patterning.

Conclusion

In general, the female costumes depicted on the jewellery are homogeneous. They consist of a dress with a train that, in most cases, is combined with an item of outerwear. The cloak is primarily of the type that is closed in front and is seen evenly distributed between plain and patterned types, of which the vertically-striped is the most common. The jacket resembles those seen on the gold-foil figures, but is longer, with more evident sleeves. Furthermore, a cloak type special to this object category and the tapestries is recorded. As in the case with the gold-foil figures, the most common female costume ensemble constitutes a cloak and a dress, but the combinations demonstrate no special tendencies. One figure may possibly wear an apron. Most female figures are attributed to the Viking Age, while a single one is dated to the 6th–8th centuries AD. The specimen from Hagebyhöga (13, Fig. 5.2) with

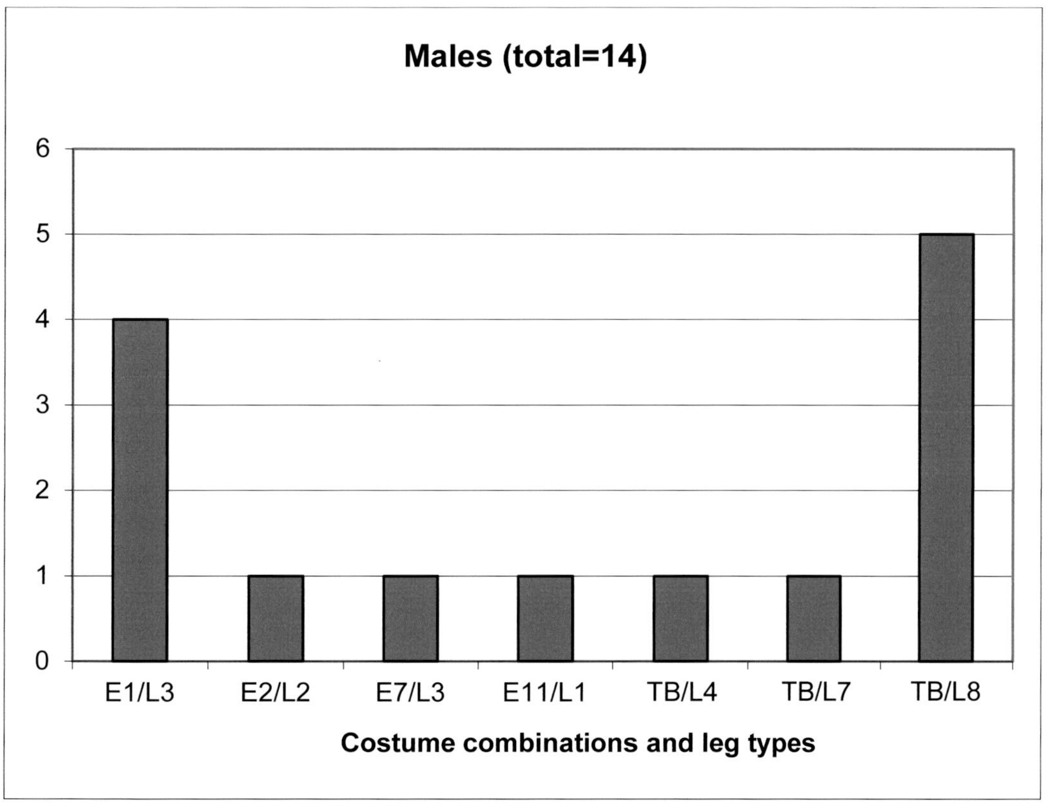

Fig. 5.9 Distribution of male costume combinations and leg types in the jewellery.

Table 5.3 Costume combinations in the jewellery material selected for this study.

Costume combination	*Males on single figures*	*Males on double figures*	*Females on single figures*	*Females on double figures*	*Total*
Tunic & trousers	5	2			7
Blouse & trousers	7				7
Cloak & dress			11	3	14
Cloak, dress & apron			1		1
Jacket & dress				2	2
Dress			1	5	6
Costumes in total	12	2	13	10	37

its depiction of a disc-on-bow fibula has a clear link to the Late Germanic Iron Age, which renders it probable that, in its find context, it is of very old provenance. In many ways, the female costume types observed in this object category tally well with the general tendencies recorded for Costume Group 3 of the gold-foil figures, *i.e.* the singly represented females. Concurrently, the depictions without visible feet probably represent a general characteristic in common with Costume Groups 1 and 2 of the gold-foil figures. Males are only depicted in tunic or blouse and trouser combinations. The figures in the latter category wear a costume with evident sleeves, and the blouse is either cropped at the waist or tucked into the trousers. A striking detail about the blouse is the clearly added sleeves which are also known from several figures in the bracteate category. Males on the double-figure representations wear long-sleeved tunics and, at times, trousers, which is a general trait among the gold-foil figures. As the male jewellery representations are distributed differently to the female representations in chronological terms, it is difficult to ascertain any clear tendencies. However, it is evident that the blouse and trouser ensembles are most common among the finds from the early period, while the plain tunic predominates in the later period. None of the males wear cloaks unlike in some of the other object categories. Moreover, it is remarkable that there are no kaftan-clad men in this group either, despite males being encountered depicted with both weapons and helmets, and occurring on their own without females.

6

Tapestries

In Scandinavia, at least nine tapestries of various sizes and states of preservation dating from the Viking Age to the Scandinavian Middle Ages have been recovered (Table 6.1). Of these, four are dated to the Viking Age. Those from Oseberg, Rolvsöy and Grønhaug in Norway are grave finds, dated by their find context (Christensen *et al.*, 1992, Opedal 1998; Christensen and Nockert 2006; Vedeler 2013), while the tapestry from Överhogdal in Sweden, found in a church, is radiocarbon dated (Nockert and Franzén 1992; Nockert and Possnert 2002, 69–75). The other tapestries are primarily dated to the Scandinavian Middle Ages, and as the finds originate from an unspecified ecclesiastical context, their dating is primarily based on stylistic features.

Although the tapestries from Skog in Sweden and Baldishol and Höylandet in Norway are undoubtedly of medieval/Christian origin, opinion is divided on the extent to which several of the narratives on the tapestries may be ascribed to a pre-Christian or Christian context (Nockert and Franzén 1992, Göransson 1999, 113–117, 119–120). Another method of dating the tapestries is by looking at the figures' posture where an S-shaped body posture indicates a Viking Age ideal, while a straighter body posture is indicative of a medieval ideal (Göransson 1998). This ideal body posture indicates that the tapestry from Överhogdal is related to the somewhat later tapestry from Skog, while that from Rennebu in Norway resembles the Viking Age tapestries (Nockert and Franzén 1992; Göransson 1999, 117, 123). Moreover, the production of the different tapestries links them across time. For example, those from Oseberg and Rolvsöy are crafted in an intricate brocading technique with contours outlined in soumak thread. No less than 21 different patterns are employed on the various figures on the Oseberg fragments (Krafft 1955; Christensen and Nockert 2006). The fragments from Överhogdal Ia-III and Skog, however, are produced in the less complex soumak technique covering the overall surface. Both the Överhogdal VI and Rennebu specimens are made as double weaves, although only the latter displays stylised figures, while the former is entirely pattern woven. In terms of format, the fragments from Oseberg, Rolvsöy, Överhogdal, Rennebu and Skog that are woven in long narrow strips resemble one another, while those from Höylandet, Rön and Baldishol are much larger in size. The latter examples are especially noteworthy as they are either embroidered or woven using the tapestry technique. Known for its integral role in late Antique ornamentation and costume, these techniques are only first attested in Scandinavia by these finds (Geijer 1994, 285). It is likely that the introduction of these two techniques is connected to the larger formats.

Due to the many stylistic and technical similarities among the tapestries from Oseberg, Rolvsøy, Överhogdal, Rennebu and Skog, they form the core of this costume analysis. The fragment from Grønhaug is not included because the only recognisable figures on the tapestry are birds (Bender Jørgensen 1986, 257; Opedal 1998, 71–73), while that from Rön is too fragmented to be used for analytical purposes. Although the number of tapestries under scrutiny is few, each contains a large number of figures. In most cases, they feature clearly clothed figures. However, the detail in the clothing and its usefulness to

Table 6.1 Scandinavian tapestries found in Sweden and Norway dated to the Viking Age and Middle Ages.

Location	*Country*	*Dating*
Oseberg	Norway	AD 834
Rolvsöy	Norway	*c.* AD 900
Överhogdal	Sweden	AD 800–1100
Grønhaug	Norway	AD 880–950
Rennebu	Norway	Middle Ages, possibly Viking Age
Höylandet	Norway	First half of the 12th century AD
Rön	Norway	Middle Ages
Baldishol	Norway	*c.* AD 1180
Skog	Sweden	13th century AD

the costume analysis vary. The selected material is briefly described in the following.

Oseberg: The ship burial from Oseberg in Norway excavated in 1904 is dendrochronologically dated to AD 834 (Christensen *et al.*, 1992, 278). Among its many unique objects were found several fragments of a tapestry, which with its richness of detail is highly useful to textile studies. The burial contained two adult females, of whom the younger is interpreted as a high-ranking person (perhaps the legendary queen, Åsa Haraldsdottir of Agder) and the elder as her slave (Christensen *et al.*, 1992, 267–269). Altogether, an incomparably fine collection of different textile tools deposited in various places in the grave chamber, as well as several different textile fragments belonging to the women's clothing were found in the grave (Arwill-Nordbladh 1998).

In total, 25–30 small and large fragments of the tapestry are preserved. However, as many fragments still adhere together in complex layers, it is difficult to obtain a clear overview of the narrative elements in the tapestry. In 1955, Sofie Krafft published watercolour drawings of some of the fragments, here numbered from 1–12. A further three fragments, drawn by Mary Storm, were published in 1940 (Hougen 1940). Since the textile fragments are in a very fragile condition today, these drawings constitute the most reliable documentation of this find and form the basis of the costume analysis.[1]

The textile fragments from the tapestry (termed *revler* in Norwegian, *i.e.* strips) vary from 16 cm to 23 cm in height, while its original length is not known. This almost miniature format is intriguing, but most likely, it indicates that these strips were originally placed on the walls above a row of benches in a large room or hall (Hougen 1940, 91–92). Indeed, the same dimensions recur in the other selected tapestries. The warp threads of the weavings, as well as the pattern threads, are made of wool, but the weaving originally had a double weft system, a pattern weft in wool which is still preserved and a binding weft in plant fibres that has now disappeared. This is also known as brocading. The brocading itself occurs in 21 different patterns, which were used to fill out the figures and features. Furthermore, the contours of each figure were marked with a thread in a contrasting colour, enabling the shape to stand out more clearly (Christensen *et al.*, 1992, 186). When unearthed, the tapestry still had clear colours, but these faded rapidly, and today it is primarily only the red colours that are still visible (Christensen *et al.*, 1992, 187).

Fragments 1 and 2, which probably belonged together, are the two largest pieces (Figs 6.1 and 6.2). They are interpreted as the depiction of a large train of wagons comprising, among others, a horse rider, two horse-drawn wagons, a horse-drawn carriage with two women sitting in it, as well as several men and women on foot (Hougen 1940, 113–114). These two fragments were later put together in a reconstruction which contains even more figures than were seen in Krafft's watercolours (Hougen 1940, 113; Christensen and Nockert 2006, 353–367). Likewise, Krafft's fragment 2 (Fig. 6.2) displays an extension of the picture to the right, with horses and figures not seen in Hougen's reconstruction (Krafft 1955, 31). Therefore, these two versions do not completely correspond in terms of the content, and are here combined. It has been suggested that this strip may depict the legendary king, Gudrød the Hunter, who was perhaps the husband of the Oseberg queen (Ingstad 1992), thus, one of the women in the carriage might be the Oseberg queen herself. The procession has a peaceful and elevated atmosphere and seems to be heading towards the front-facing figure, interpreted as a shaman. Similarly, it was suggested that the carpets of the wagon conceal the statues of gods (Christensen *et al.*, 1992, 234–236).

Fragment 3 depicts a battle scene with a horseman, several Valkyries, warriors, berserkers (bear- and wolf-skin wearing warriors), four lines of spears shaped into a pointed roof and a wagon containing a man (Fig. 6.3). This is interpreted as a representation of the Battle of Bråvalla, where an old and blind Harald Wartooth leads the battle between the tribes of Danes and Swedes from his wagon (Hougen 1940, 114–116).

Fragment 4 comprises four different strips pressed together, on which several men and women are seen alongside other figures of unknown gender (Fig. 6.4). One of these fragments depicts, among others, seven men with upraised arms, the so-called adorers. In another place, quite a large part of a horse is visible, and among others, a frontally depicted female figure. This figure is interpreted as the goddess Freyja (Christensen *et al.*, 1992, 243).

Fragment 5 shows the bow of a ship resembling the Oseberg ship sailing in, in front of a bridge with raised spears (Fig. 6.5).

Fragment 6 shows Odin's tree in the sacred grove, with several men hanging in the tree and other men and women standing around it (Fig. 6.6). The front part of a horse, too, can be glimpsed. This image is interpreted as a sacrifice to the goddess Freyja (Christensen *et al.*, 1992, 245). The person in front at the top left is considered a woman (Christensen *et al.*, 1992, 243).

Fragment 7 depicts two opposite rows of women with upraised arms. In the bottom row, the upper part of a figure with a bird's head is seen (Fig. 6.7).

Fragment 8a displays a tree with birds as well as parts of a man and a woman (Fig. 6.8). Fragment 8b was perceived by Krafft as a dragon, but if the drawing is turned 180°, it is most likely the front part of a horse (Krafft 1955, 36). To the left of this stands a man (Fig. 6.9).

On fragment 9, two spear-carriers stand outside two houses with dragons on the roofs. One of them might be a woman (Christensen *et al.*, 1992, 246) (Fig. 6.10).

On fragment 10, the back end of a horse with its rider and a figure in a feathered pelt in front of a house with dragon heads are seen (Fig. 6.11), perhaps along with yet another figure glimpsed above the roof. Ingstad considers

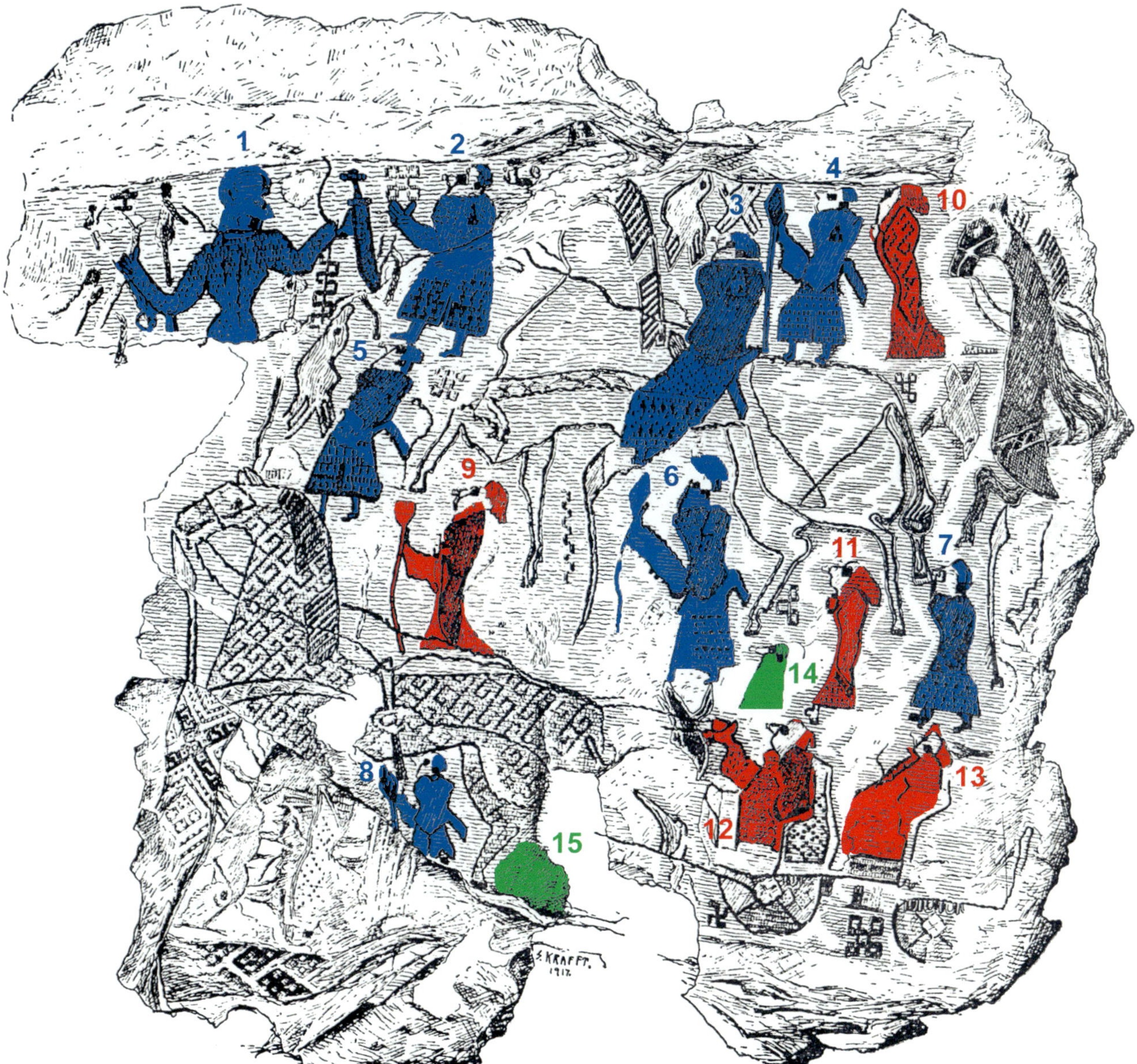

Fig. 6.1 Drawing of a tapestry fragment from the Oseberg grave in Norway, fragment 1 with the figure numbers and gender signature used in the costume analysis: female (red), male (blue) and neutral (green) (drawing: Pernille Foss; after Krafft 1955).

this as depicting the king and Freyja in her feathered pelt in front of a sacred building (Christensen *et al.*, 1992, 246).

Fragment 11 depicts a horse-drawn wagon/sledge with dragon heads (Fig. 6.12). The feature visible in the middle of this wagon/sledge cannot be identified. To the left of the wagon/sledge is a woman and to the right is a man.

Fragment 12 is composed of two different strips with different narrative directions (Fig. 6.13). One depicts six men and two figures of unknown gender in two rows, while the other shows a row of three men and a figure of unknown gender.

On fragment 13, a horse and rider, two men, two women and two figures of unknown gender are visible (Fig. 6.14).

On fragment 14, a flock of six riders, including, at least, two walking females and a figure of unknown gender are depicted (Fig. 6.15).

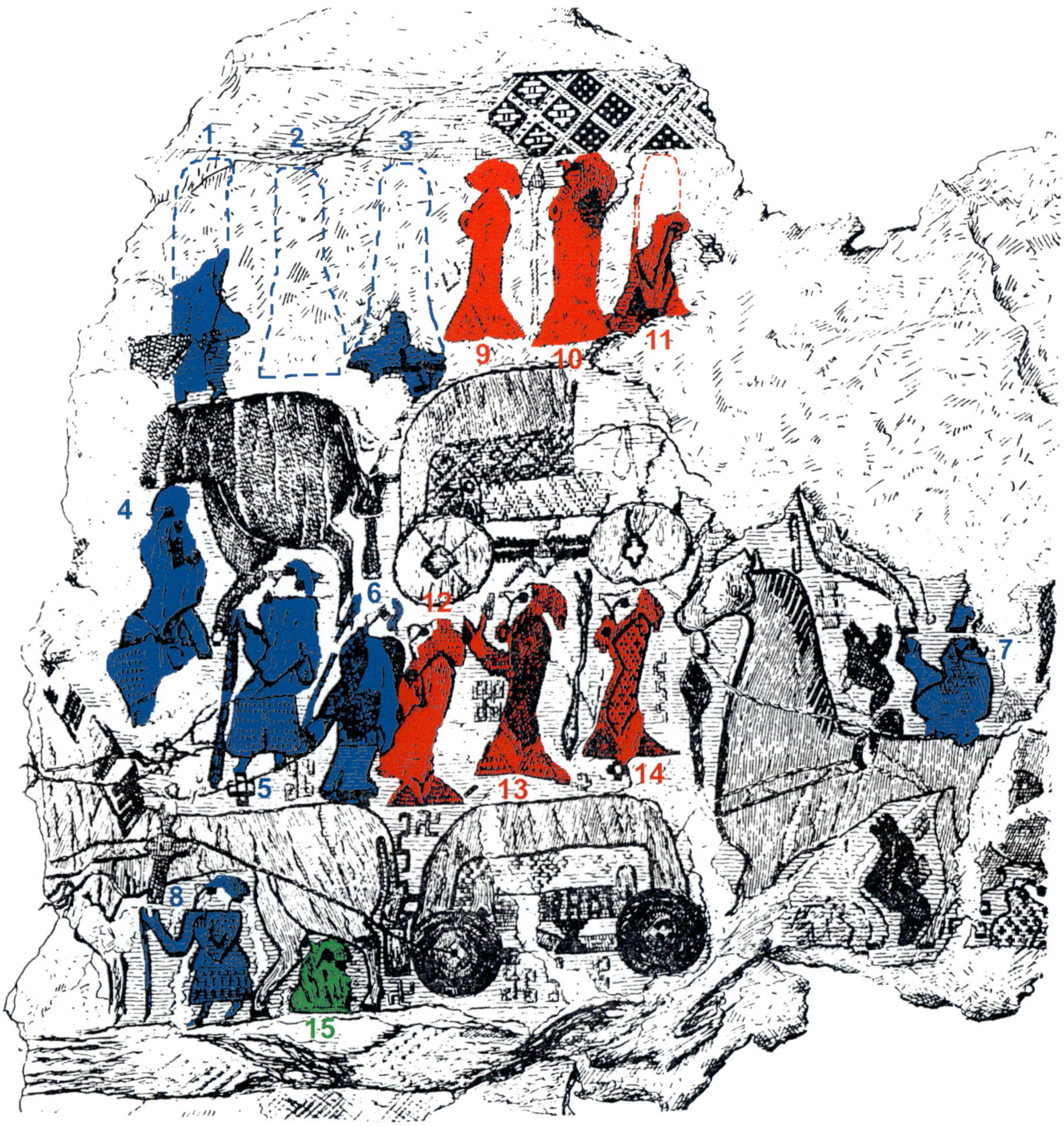

Fig. 6.2 Drawing of a tapestry fragment from the Oseberg grave in Norway, fragment 2 with the figure numbers and gender signature used in the costume analysis: female (red), male (blue) and neutral (green) (drawing: Pernille Foss; after Krafft 1955).

Fragment 15 shows the front part of two horses, probably with riders (Fig. 6.16).

Altogether, 132 figures are recorded from the sketches of fragments 1–15 (Table 6.2). This number of figures has to be considered a minimum. Among the 132 figures are 64 males and 41 females. Moreover, two bird figures and two bear figures, as well as 23 other figures whose gender cannot be determined on the basis of the criteria used for the costume analysis, are included. Most of the figures are depicted in profile, with a clear movement of the narrative sequence from right to left. Although the narrative interpretations of the motifs of the individual fragments vary from one to another, a homogenous depiction strategy may be seen. The male group consists of at least 11 riders, eight hanging figures, seven adorers, and, largest of all, a group of standing figures. The men are often depicted with short or no hair. Weapons are carried by 27 men, with an equal number carrying spears and swords,

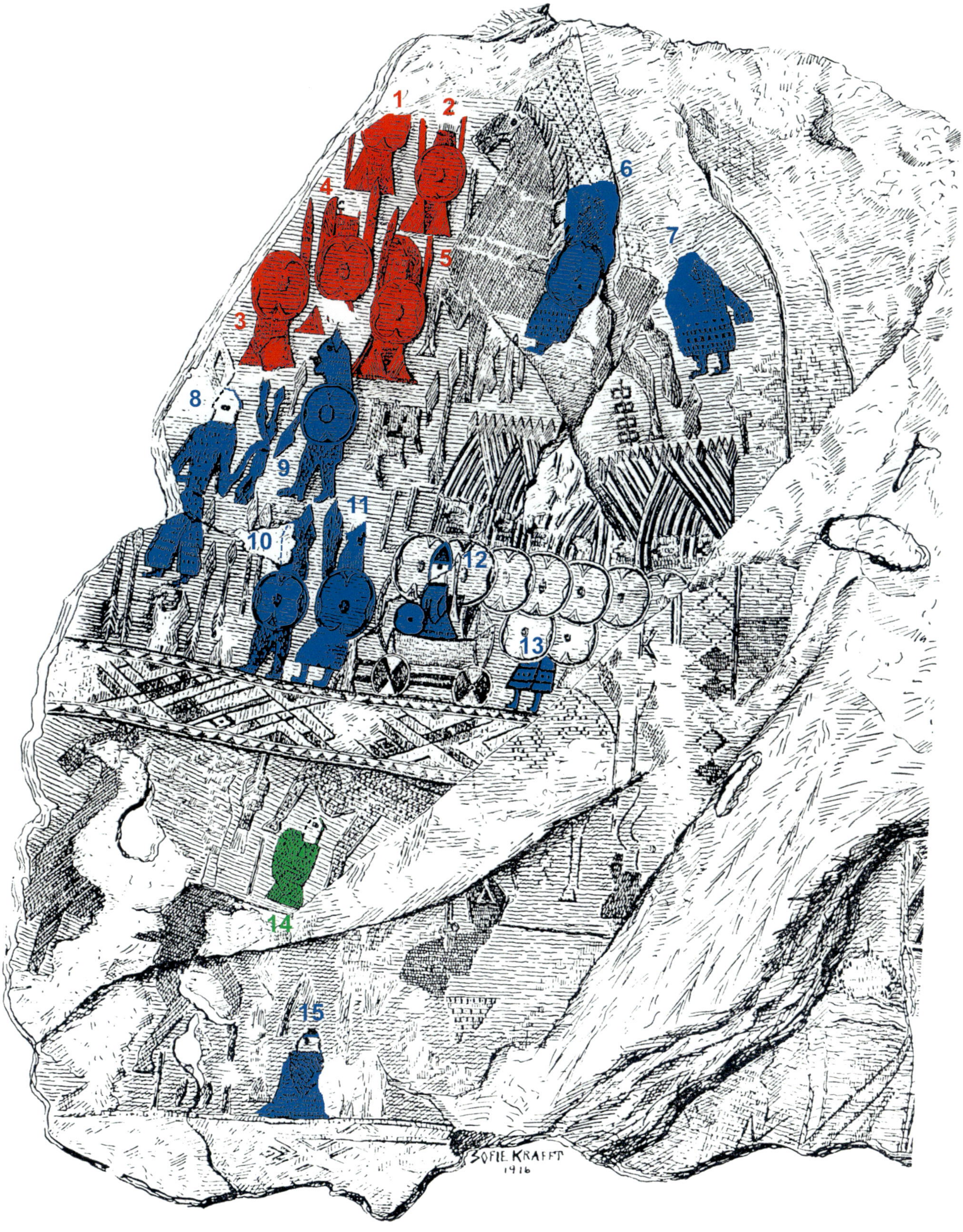

Fig. 6.3 Drawing of a tapestry fragment from the Oseberg grave in Norway, fragment 3 with the figure numbers and gender signature used in the costume analysis: female (red), male (blue) and neutral (green) (drawing: Pernille Foss; after Krafft 1955).

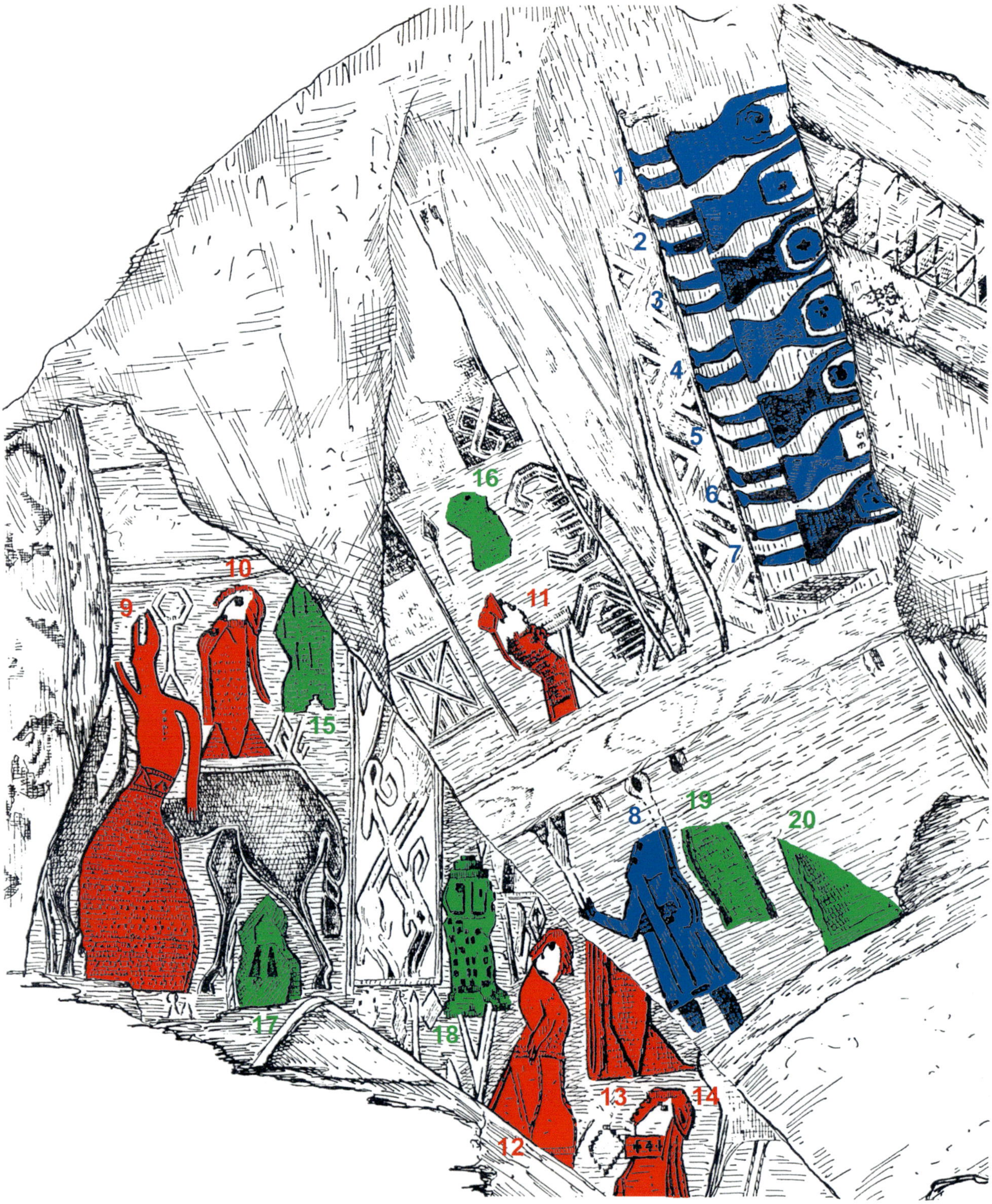

Fig. 6.4 Drawing of a tapestry fragment from the Oseberg grave in Norway, fragment 4 with the figure numbers and gender signature used in the costume analysis: female (red), male (blue) and neutral (green) (drawing: Pernille Foss; after Krafft 1955).

although these objects are sometimes also seen together. Substantially fewer bear a shield in addition; a trait that precisely defines the female warriors in fragment 3. These women presumably also carry a spear or sword. Axes occur in, at least, six cases. Females are nearly always displayed standing, although two figures are depicted

Fig. 6.5 Drawing of a tapestry fragment from the Oseberg grave in Norway, fragment 5 with the figure numbers and gender signature used in the costume analysis: female (red) and neutral (green) (drawing: Pernille Foss; after Krafft 1955).

seated in a sledge or wagon. Their hair is sometimes swept up, although hair knots cannot be discerned. Another possibility is that the figure wears a cap, or, that the hair ends in a long ponytail. It is not possible to see whether the figures wear any jewellery (Krafft 1955). However, in the drawing published in 1940 (Hougen 1940, colour plate) a number of women are depicted with a little circle on the chest, which may be interpreted as a dress fastener. The greatest difference between the two genders are thus not seen in relation to their hairstyle or attributes, but the extent to which the figure wears a long costume with no feet visible or a shorter costume that permits the legs to be seen.

Figures of indeterminate gender primarily consist of fragmented figures, including half-size figures. The costume that they wear often envelops the entire body. These figures are recorded as wearing cloaks, even though the actual appearance and function of this costume are uncertain.

Rolvsöy: A chamber grave was discovered in 1864, during fieldwork, at Rostad-Haugen on the island of Rolvsöy in Norway, but it was never fully examined. Three years later, several fragments of a silk ribbon and,

Fig. 6.6 Drawing of a tapestry fragment from the Oseberg grave in Norway, fragment 6 with the figure numbers and gender signature used in the costume analysis: female (red), male (blue) and neutral (green) (drawing: Pernille Foss; after Krafft 1955).

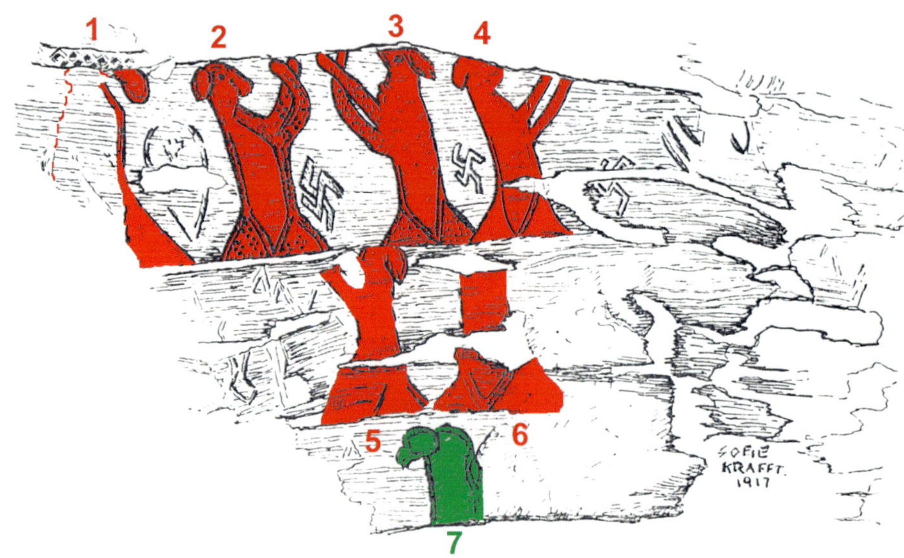

Fig. 6.7 Drawing of a tapestry fragment from the Oseberg grave in Norway, fragment 7 with the figure numbers and gender signature used in the costume analysis: female (red) and neutral (green) (drawing: Pernille Foss; after Krafft 1955).

not least, a part of a tapestry were found (Brøgger 1921, 20–27). The grave is dated to *c*. AD 900 on the basis of the preserved objects. Today, the tapestry from Rolvsöy measures 20 × 14 cm, but nonetheless, it contains many costume details (Krafft 1955, 39). The fragment consists of a strip with a border at the bottom (Fig. 6.17). It is made in

Fig. 6.8 Drawing of a tapestry fragment from the Oseberg grave in Norway, fragment 8a with the figure numbers and gender signature used in the costume analysis: female (red) and male (blue) (drawing: Pernille Foss; after Krafft 1955).

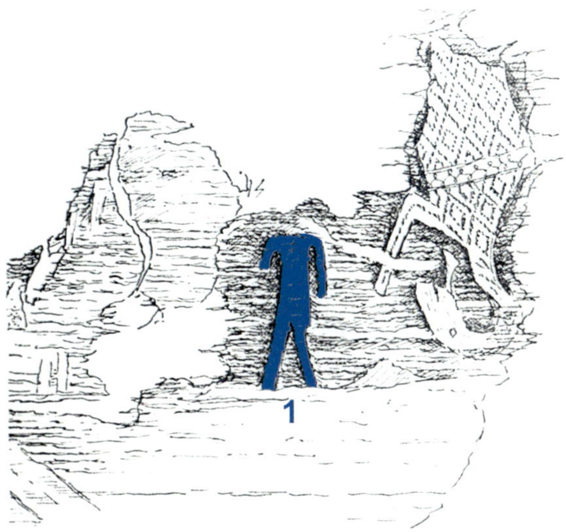

Fig. 6.9 Drawing of a tapestry fragment from the Oseberg grave in Norway, fragment 8b with the figure numbers and gender signature used in the costume analysis: male (blue) (drawing: Pernille Foss; after Krafft 1955).

the soumak technique with pattern brocading (Göransson 1999, 118; Christensen and Nockert 2006). The colours are still well preserved and consist of red, yellow, blue and green hues on a white background (Fig. 6.18). On the left side, part of a ship may be seen, and to the right two rows of figures one above the other. The top row contains three figures, a woman and two men, whose heads are not fully preserved. It is possible that there is another, not fully preserved, female figure further to the right. The lower row contains three men and a woman. All the figures are depicted in profile, facing the stern or bow of the ship (Brøgger 1920–21, 28;

Fig. 6.10 Drawing of a tapestry fragment from the Oseberg grave in Norway, fragment 9 with the figure numbers and gender signature used in the costume analysis: male (blue) (Drawing: Pernille Foss; after Krafft 1955).

Krafft 1955, 39; Christensen *et al.*, 1992, 253). The two women standing in front of the men wear an ensemble of a cloak and dress with a train, with no feet visible (Table 6.3). The choice of colours renders it probable that the woman in the top row is wearing a dress in two layers. The cloak has a pointed end and is most likely of the type that is open at the front. The top row of women wear a red cloak together with a blue dress, the other a blue cloak with a red and blue dress. Their heads are not preserved, but the hair of both women ends in a long ponytail. The five men all wear long cloaks that end in pointed corners on each side of the figure. From the outline, it is likely that the cloak is open at the left side. The male cloaks have various colours: two are red, one yellow, one blue and one green. Colours also indicate that the men are clad in long, close-fitting trousers, three males are wearing green trousers, while two are wearing yellow ones. The males have their heads covered by the cloaks, perhaps as a sign of grief.

This fragment is indubitably the remains of a much larger item whose iconographical contents cannot be fully determined (Christensen *et al.*, 1992). Only seven figures (two females and five males) are seen on the preserved fragment, of which the three topmost figures are incomplete, while the other figures are very clear and detailed (Fig. 6.17). The well-preserved colours that give life and originality to the tapestry permit the individual costume items to be clearly differentiated from one another. General features such as leg or costume length are identical to what is seen on the Oseberg tapestry. None of the figures have their arms displayed, nor do they bear any attributes.

Överhogdal: Five narrow textile strips were unearthed in 1910 during a renovation of Överhogdal Church in Härjedalen in Sweden, and most of the remaining pieces were located the following year. Previously, it was believed that the tapestries originated in the Scandinavian Middle Ages, but the radiocarbon dating of each item confirms a production period of between AD 800 and 1100, which attests

Fig. 6.11 Drawing of a tapestry fragment from the Oseberg grave in Norway, fragment 10 with the figure numbers and gender signature used in the costume analysis: male (blue) and neutral (green) (drawing: Pernille Foss; after Krafft 1955).

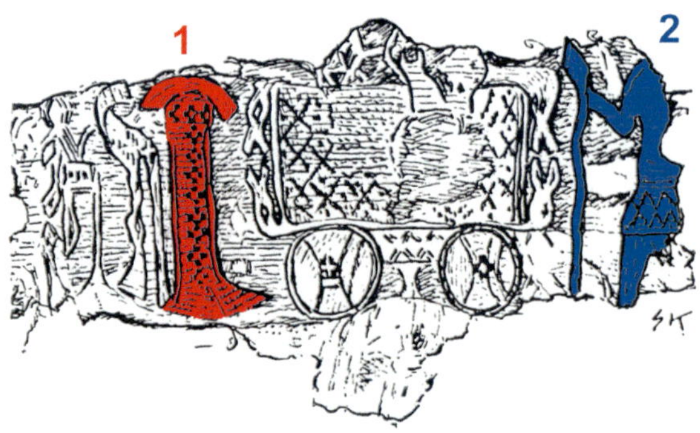

Fig. 6.12 Drawing of a tapestry fragment from the Oseberg grave in Norway, fragment 11 with the figure numbers and gender signature used in the costume analysis: female (red) and male (blue) (drawing: Pernille Foss; after Krafft 1955).

to their connection to the Viking Age. Although Överhogdal Church was not built until 1746, it contains a chapel that dates back to 1466, and thus, the tapestries were most likely a long forgotten gift once bestowed to the church (Nockert and Possnert 2002, 69–71; Oscarsson 1994, 4–6, 16). The weavings are made in the soumak technique in red, blue and yellow wool threads on an undyed or natural coloured linen background. The colours are still well preserved, and colour analyses have demonstrated that the colours derive from madder, woad and dyer's weed (Oscarsson 1994, 8–9). Ever since the tapestries came to light, there has been disagreement on whether they reflect pagan myths or scenes from Christianity, and evidence supporting both views can be found in the many details on the weavings. As the weavings are highly stylised in regard to the depiction of the human figures, this makes it hard to determine the gender in a similar manner to the other object categories in this study, and limits information on the costume. Based on their body posture, the figures clad in a long costume are interpreted as women and those in knee-length costume are considered to be men (Göransson 1999, 108).

Strip IA measures 184 cm in length, 35 cm in height and the image area is encircled above and below by a narrow border (Fig. 6.19). It is probably not preserved in its original length. At first glance, the viewer is presented with a confusing mix of people, animals (most are four legged), at least two houses, a ship and many other remarkable elements. At the centre of the image stands a large tree, and the strip is, more or less, consistently divided into three horizontal rows. All animals are depicted in profile, and the majority move from right to left. Of the human figures identifiable, 63 are women and 45 men. The positioning of their legs mainly suggests a left-facing profile. In addition to this, is a rider with arms upstretched in worship and a second figure, both of undeterminable gender.

Strip IB measures 195 cm in length and 35 cm in height (Fig. 6.19). The image area is framed above and below by a narrow border. It is probably not preserved in its original length. At first glance, the motifs do not appear as compressed as strip IA, but the tapestry contains the same mix of humans and four-legged animals divided into three horizontal rows. With the exception of one figure, they all move from right to left. Of the human figures identifiable, 45 are women, who, with the exception of a seated figure, are all standing. To these can be added two riders and two riding worshippers.

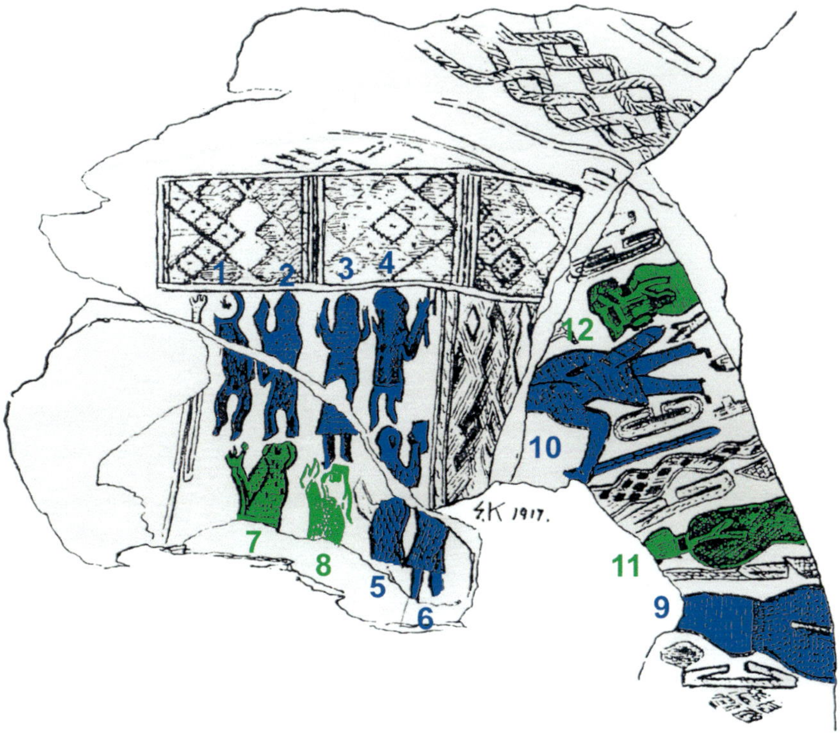

Fig. 6.13 Drawing of a tapestry fragment from the Oseberg grave in Norway, fragment 12 with the figure numbers and gender signature used in the costume analysis: male (blue) and neutral (green) (drawing: Pernille Foss; after Krafft 1955).

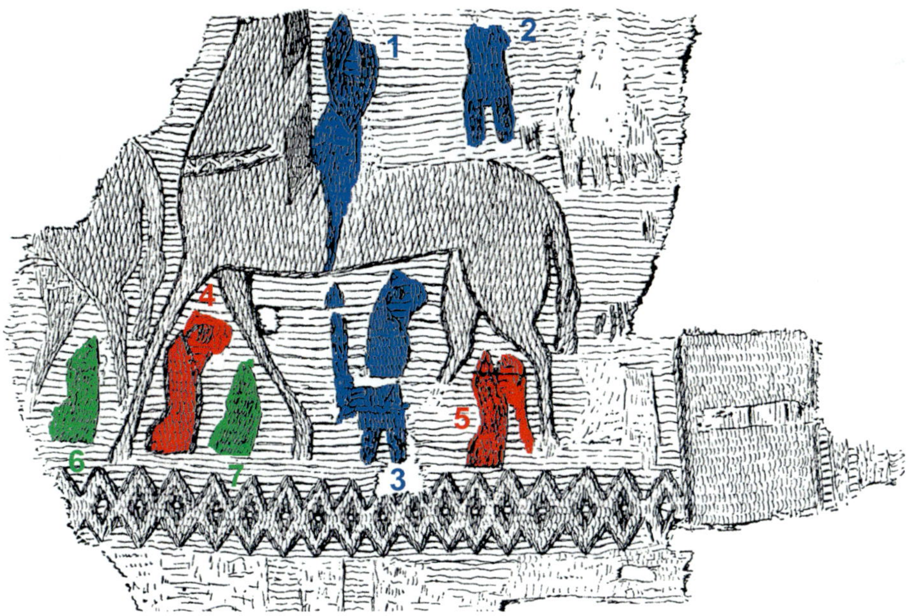

Fig. 6.14 Drawing of a tapestry fragment from the Oseberg grave in Norway, fragment 13 with the figure numbers and gender signature used in the costume analysis: female (red), male (blue) and neutral (green) (drawing: Pernille Foss; after Hougen 1940).

Strip II measures 184 cm in length and 37 cm in height (Fig. 6.20). The image area is framed on all four sides by a narrow border and is presumably preserved in its original length. At first glance, it resembles strip IA, but appears slightly more organised in three equal rows. It contains the same mixture of humans and four-legged animals, and with a tree at the centre. The overall direction of movement is from right to left. Of the identifiable human figures, 57 are

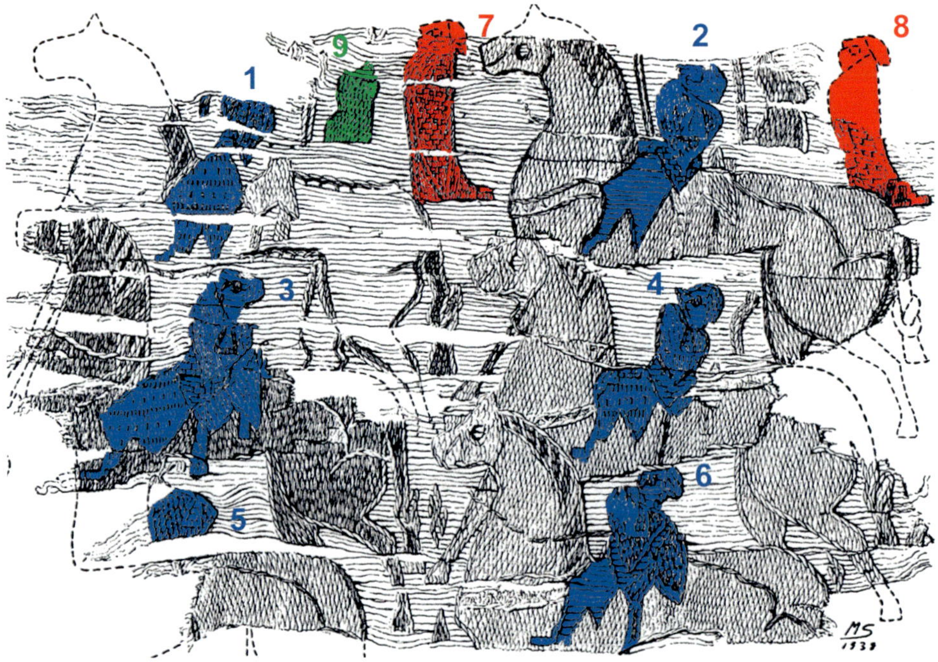

Fig. 6.15 Drawing of a tapestry fragment from the Oseberg grave in Norway, fragment 14 with the figure numbers and gender signature used in the costume analysis: female (red), male (blue) and neutral (green) (drawing: Pernille Foss; after Hougen 1940).

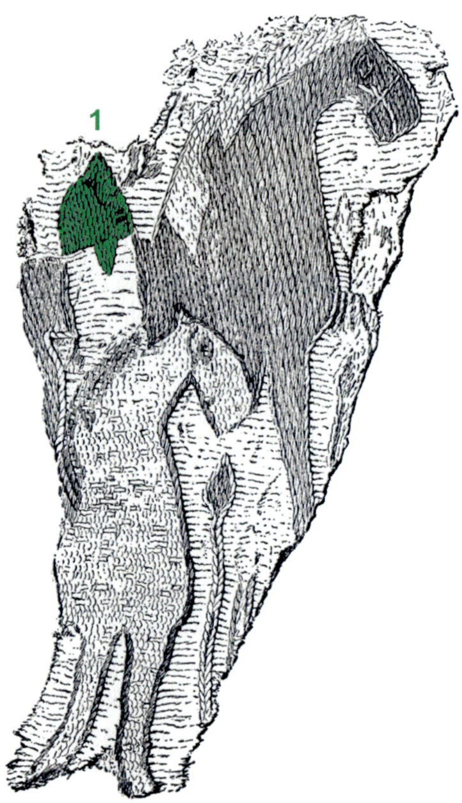

Fig. 6.16 Drawing of a tapestry fragment from the Oseberg grave in Norway, fragment 15 with the figure numbers and gender signature used in the costume analysis: neutral (green) (Drawing: Pernille Foss; after Hougen 1940).

women and 43 are men, 16 are riders and two figures are of undeterminable gender.

Strip III: This strip measures 67 cm in length and 24 cm in height (Fig. 6.21). On three sides, the image area is framed by a narrow border, thus indicating that it is not preserved in its original length. In contrast to the other strips, this strip contains only two rows of figures. Of the identifiable human figures, 22 are women, two are men, two are riders and six figures are placed in a ship. The gender of the latter cannot be determined.

Strip IV measures 176 cm in length and 30 cm in height. It deviates from the others by being only pattern-woven, and is thus not included in the costume analysis.

On the tapestry strips IA–III from Överhogdal, the figures are clearly more stereotyped and simply depicted than in the two previous tapestries (Nockert and Franzén 1992). This not only renders the determination of gender more uncertain, but also signifies that not many costume details can be identified. The figures as well as the animals on all four strips primarily move from right to left, and from the placing of the feet, it appears that the majority of the figures are portrayed in profile. Most figures have no arms, and only in a few cases does a figure bear any attributes. The heads are primarily depicted in outline, with no hint of hair. A total of 309 figures are recorded on the four strips.

Based on the present methods for gender determination, figures in a long costume are determined as female, and those with legs clearly seen and a knee-length costume as male (Göransson 1999, 108). On strips IA, IB, II and III no less than 187 females and 113 males are visible, while the

Table 6.2 Costume analysis of the Oseberg tapestry from Norway.

Fragment/ figure	*Costume not identified*	*Gender*	*Outerwear*	*Dress*	*Train*	*Tunic*	*Leg type*	*Belt*	*Head seen from R=right L=left F=font*	*R=rider S=standing T=sitting H=hanging A=adorer*	*H=hair knot/hat S=short L=long hair*	*Number of arms*	*Number of legs*	*Helmet with horns*	*S=shield P=spear W=sword T=staff A=axe*
1.1		M				TB	L7?		F	S	-	2	2	X	W
1.2		M	A19			TB	L7		L	S	S	1	2		W
1.3		M	A19			TB	L7		L	R	S	1	1		W?
1.4		M	A19			TB	L7		L	S	S	1	2		PW
1.5		M	A19			TB	L7		L	S	S	0	2		W?
1.6		M	A19			TB	L7		**S**	S	S	1	2		PW
1.7		M	A19			TB	L7		L	S	S	0	2		
1.8		M	A19			TB	L7		L	S	S	1	2		PW
1.9		F	A20	C?	X		L6		L	S	H	1	0		T
1.10		F	A20	C?	X		L6		L	S	H	0	0		
1.11		F	A20?	C?	X		L6		L	S	H	0	0		
1.12		F	A3	C?	X		L6		L	T	H	1	0		A
1.13		F	A3	C?	X		L6		L	T	H	0	0		
1.14		?	X?				L6		L	T	?	0	0		
1.15		?	X?				L6		L	T	?	0	0		
2.1		M	A19			TB	L7		L	S	S	1	2		PW
2.2		M	A19			TB	L7		L	S	S	0	2		W
2.3		M	A19			TB	L7		L	S	S	1	2		T
2.4		M	A19			TB	L7		L	S	S	1	2		PW
2.5		M	A19			TB	L7		L	S	S	1	2		PW
2.6		M	A19			TB	L7		L	S	S	1	2		T
2.7		M?	A19?			?	?		L	S?	S	1	?		
2.8		M				TB	L7		L	S	S	1	2		P
2.9		F	A20	C?	X		L6		L	S	H	1	0		
2.10		F	A20	C?	X		L6		L	S	H	0	0		
2.11		F	A20	C?	X		L6		L	S	H	0	0		
2.12		F	A20	C?	X		L6		L	S	H	0	0		
2.13		F	A20	C?	X		L6		L	S	H	1	0		
2.14		F	A20	C?	X		L6		L	S	H	0	0		
2.15		?	X?				L6		L	T	H	0	0		
3.1		F	A20	C?	X		L6		L?	S	?	0	0		SP?
3.2		F	A20	C?	X		L6		L	S	L	0	0		SW?
3.3		F	A20	C?	X		L6		L	S	L	0	0		SW?
3.4		F	A20	C?	X		L6		L	S	?	0	0		SW?
3.5		F	A20	C?	X		L6		L	S	L	0	0		SW?

(Continued)

Table 6.2 Costume analysis of the Oseberg tapestry from Norway. (Continued)

Fragment/ figure	*Costume not identified*	*Gender*	*Outerwear*	*Dress*	*Train*	*Tunic*	*Leg type*	*Belt*	*Head seen from R=right L=left F=font*	*R=rider S=standing T=sitting H=hanging A=adorer*	*H=hair knot/hat S=short L=long hair*	*Number of arms*	*Number of legs*	*Helmet with horns*	*S=shield P=spear W=sword T=staff A=axe*
3.6		M	?			TB?	L7		L	R	?	1?	1		S
3.7		M	A19			TB	L7		L	S	?	0?	2		
3.8		M				TB	L7		F?	S	S?	2	2		2P?
3.9		M/bear				TB	L3		L	S	-	0	2		SP
3.10		M/bear				TB	L3		L	S	-	0	2		SP
3.11		M	?			TB?	L7		L	S	S	0	2		SW?
3.12	X	M?	A19?				?		L	T		0	0	X	S
3.13		M	?			TB?	L7		L	S	?	0	2		S
3.14		F?	A20	C?			L6		L	S?	?	0	0		
3.15		M?	A19?			TB?	L7?		L	S?	?	0	0		
4.1		M				E1	L1		F?	A	-	2	2		
4.2		M				E1	L1		F?	A	-	2	2		
4.3		M				E1	L1		F?	A	-	2	2		
4.4		M				E1	L1		F?	A	-	2	2		
4.5		M				E1	L1		F?	A	-	2	2		
4.6		M				E1	L1		F?	A	-	2	2		
4.7		M				E1	L1		F?	A	-	2	2		
4.8		M				E1	L1?		L	S	-	1	2		
4.9		F		SB	%		L6	X	?	S	-	2	0		
4.10		F	A20	C?	X		L6		L	S	L	0	0		
4.11		F	A1?	C?	X		L6		R	S	L	0	0		
4.12		F		SB	X		L6	X	L	S	S?	1	0		
4.13		F	A20	C?	X		L6		?	S	?	0	0		
4.14	X	F	?	?			?		L	S?	L	?	?		
4.15	X	?	X				L6		?	S?	?	0	0		
4.16	X	?	X				L6		L?	T	S	0	0		
4.17	X	?	X				L6		?	T?	?	0	0		
4.18	X	?	X				L6		?	?	?	0	0		
4.19	X	?	X				L6		?	S?	?	0	0		
4.20	X	?	X				L6		?	S?	?	0	0		
5.1		F	A20	C?	X		L6		R	S	H	1?	0		
5.2		F	A20	C?	X		L6		L	S	L	0	0		
5.3	X	?	X				L6		L	T?	S	0	0		
5.4	X	?	X				L6		L	T?	S	0	0		
6.1		M				E1	L3		F	R	?	2	2		

(Continued)

Table 6.2 Costume analysis of the Oseberg tapestry from Norway. (Continued)

Fragment/ figure	*Costume not identified*	*Gender*	*Outerwear*	*Dress*	*Train*	*Tunic*	*Leg type*	*Belt*	*Head seen from R=right L=left F=font*	*R=rider S=standing T=sitting H=hanging A=adorer*	*H=hair knot/hat S=short L=long hair*	*Number of arms*	*Number of legs*	*Helmet with horns*	*S=shield P=spear W=sword T=staff A=axe*
6.2		M				E1	L3		F	R	S	2	2		
6.3		M				E1	L3		F	R	S	2?	2		
6.4		M				E1	L3		F	R	?	2	2		
6.5		M				E1	L3		F	S?	?	2	2		
6.6		M				E1	L3		F	R	?	?	2		
6.7		M				E1	L3	X	F	R	?	2	2		
6.8		M				E1	L3		F	R	?	2	2		
6.9		M				E1	L3		F	R	?	2	2		
6.10		M				E1	L3		F	R	S	0	2		
6.11	X	F?	?	?			L6?		L	S	L	?	0		
6.12		F	A20	C?	X		L6		L	S	L	2?	0		?
6.13		F	?	SB?	X		L6		L	S	L	1?	0		?
6.14		F	A20	C?	X		L6		L	S	?	1?	0		?
6.15		F	A20	C?	X		L6		L	S	?	0	0		
6.16		M?				TB?	L7?		L	S	S	2	?		W
6.17		?	X				L6		L	S	S	0	0		
7.1	X	F	A20?	C?	X		L6		L	S	H	0	0		
7.2		F	A20	C?	X		L6		R	S	H	2	0		
7.3		F	A20	C?	X		L6		L	S	H	2	0		
7.4		F	A20	C?	X		L6		R	S	H	2	0		
7.5	X	F	?	C?	X		L6		L	S	H	2?	0		
7.6		F	A20	C?	X		L6		?	S	?	?	0		
7.7		Bird	X				?		L	S	-	0	?		
8a.1		M	?			TB?	L7		L	S	?	?	2		
8a.2		F	A20	C?	X		L6		?	S	?	0	0		
8b.1		M				TB	L8?		F	S	?	2	2		
9.1		M				TB	L7	X	L	S	?	1	2		P
9.2		M				TB	L7	X	L	S	?	1	2		P
10.1		M	?			TB?	L7		L	R	?	0	1		S
10.2		Bird	A				L3?		L	S	-	0	2		
10.3	X	?	X				?		L	?	L?	0	?		
11b.1		F	?	C?	X		L6		L	S	?	0	0		
11b.2		M				TB	L8?		L	S	?	1	2		P
12.1		M				TB	L8		L	S	?	1?	2		
12.2		M				TB	L8		L	S	?	1	2		

(Continued)

Table 6.2 Costume analysis of the Oseberg tapestry from Norway. (Continued)

Fragment/ figure	*Costume not identified*	*Gender*	*Outerwear*	*Dress*	*Train*	*Tunic*	*Leg type*	*Belt*	*Head seen from R=right L=left F=font*	*R=rider S=standing T=sitting H=hanging A=adorer*	*H=hair knot/hat S=short L=long hair*	*Number of arms*	*Number of legs*	*Helmet with horns*	*S=shield P=spear W=sword T=staff A=axe*
12.3		M				E1	L3		F?	A?	?	2	2		
12.4		M				TB	L8		F?	A?	?	2	2		
12.5		M				E1	L3?		F?	A?	?	2	?		
12.6		M				E1	L3		?	S?	?	?	2		
12.7	X	?	X				?		L	?	S	1	?		
12.8	X	?	X				?		L	?	L?	1	?		
12.9		M				TB	L7	X	L?	S	?	0	2		
12.10		M				E1	L1		F	S	?	2	2		PW
12.11	X	?					?		L	S	S	0	?		
12.12	X	?					?		L	?	H?	0	?		
13.1		M	A19			TB?	L7?		L	R?	?	0	?		
13.2		M				E1	L3?		L?	S	?	0	2		
13.3		M				E1	L3?		L	S	S	1	2		W?
13.4		?	X		?		L6		L	S	H	0	0		
13.5		F	X		?		L6		L	S	L	0	0		
13.6		?	X		?		L6		?	S?	?	0	0		
13.7		?	X		?		L6		?	S?	?	0	0		
14.1		M	A19			TB	L7		L	R	?	0	1		A
14.2		M	A19			TB?	L7		L	R	S	0	1		A
14.3		M	A19			TB?	L7		L	R	S	0	1		A
14.4		M	A19			TB?	L7		L	R	S	0	1		A
14.5	X	?					?		L	R?	?	?	?		?
14.6		M	A19			TB?	L7		L	R	S	0	1		A
14.7		F	A20	C?	X		L6		L	S	H	0	0		
14.8	X	F	A20?	C?	X		L6		L	S	?	0	0		
14.9		?	X				L6		?	T?	?	0	0		
15.1	X	?					?		R	R?	?	?	?		

The numbering in the list refers to numbers marked in the corresponding images in Figs 6.1–6.16.

gender of nine figures cannot be determined (Table 6.4). The vast majority of the 187 female figures stand upright and only one is seated on a chair. None of the females on strips IA–II have arms, or are connected to any attributes, while strip III includes several females with arms. Among others, one stands inside a house with an unidentified object in her hand. Furthermore, it is primarily the females that are observed inside houses. On strip III, yet another house with a depiction of two figures and a loom closely resembling that of the Rennebu tapestry may be observed. The figures on either side of the loom are depicted differently. These have earlier been identified as females, but with the present methods of gender determination, the figure to the right, with legs clearly portrayed, could also be a male. Most

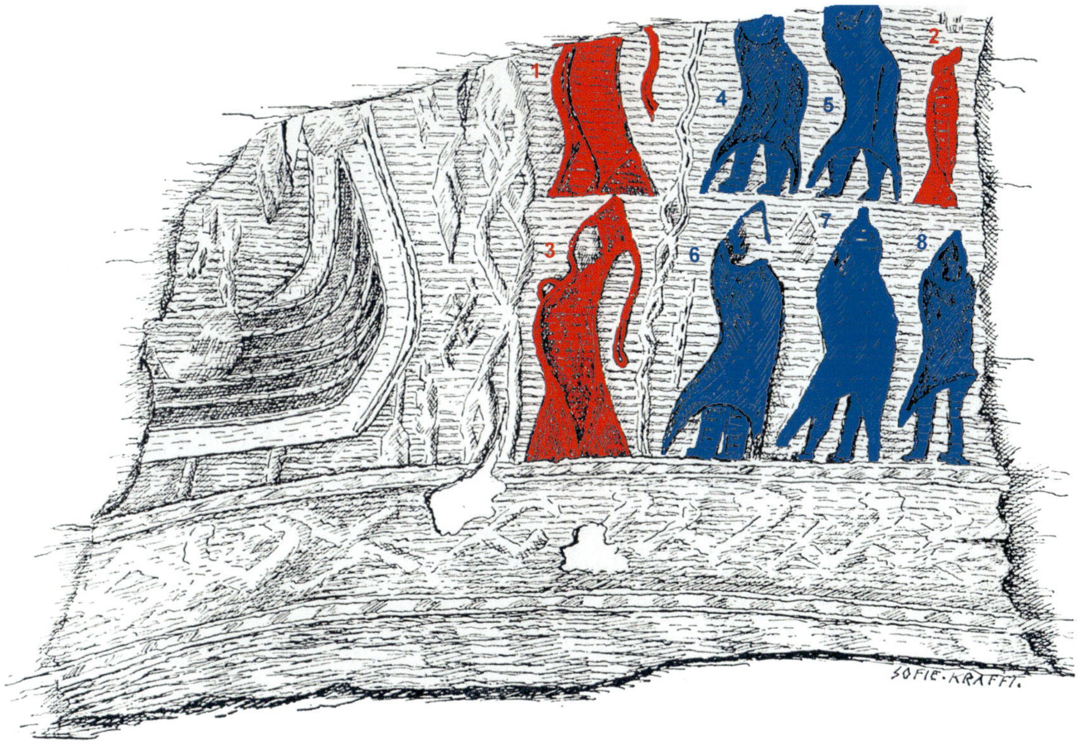

Fig. 6.17 Drawing of a tapestry fragment from Rolvsöy in Norway with the figure numbers and gender signature used in the costume analysis: female (red) and male (blue) (drawing: Pernille Foss; after Christensen et al., 1992).

female figures are quite small, and, in many cases, smaller than the males, although strip III also includes some larger females. Males are primarily depicted standing upright, although two figures lie fallen and 23 are riders. The mounted figures are assumed to be male, although legs are only visible in the case of two riders.

Three of the riders may be termed riding worshippers, as they are depicted from the front with both arms upstretched. Few of the remaining riders have arms. In general, depictions of figures with arms holding attributes rarely occur, and are primarily seen on fragment II, where one figure carries an axe and three carry swords. On fragment III, two men with axes are observed. A figure of indeterminable gender, who presumably sits in a horse-drawn sledge, holds a long object, possibly an axe.

Rennebu: A tapestry depicting human figures made in a similar technique to that of strip IV from Överhogdal was found in Rennebu Church in Norway. It measures 44 cm in length and 13 cm in height with borders along the top and bottom edges (Fig. 6.22). Part of the tapestry is missing. The pattern is made in blue and red wool yarn on a white linen background and the figures appear relatively stylized due to the weaving technique. To the left, the tapestry depicts two right-facing female figures, with arms and legs in an almost right-turned profile, inside a building, holding a loom between them, and to the right, a right-facing male figure holding a long-handled axe in front of him (Nockert and Franzén 1992, 66–69). The piece is dated to the Scandinavian Middle Ages, although, it is believed that the body posture and the motifs link it to the Viking Age (Göransson 1999, 122–123).

The two women and the man are recorded based on the same identification principle as with the Överhogdal tapestry (Nockert and Franzén 1992). The women are clad in ankle-length costumes, one of which may be in two parts and the other with a cloak, while the man wears a blouse with trousers or knee breeches (Table 6.5). Regrettably, further costume details are difficult to discern.

Skog: This tapestry, woven in the same Soumak technique as the Överhogdal tapestry (Nockert and Franzén 1992, 51–53 Oscarsson 1994, 7) was found in 1912 in Skog Church in Sweden (Fig. 6.23). Radiocarbon dating of the linen weave indicates that it was made in the period AD 1245–1290, while dating of a wool thread used for the pattern has given a time frame of between AD 1275 and 1395 (Nockert and Possnert 2002, 75–77). The first date is believed to be the most reliable (Göransson 1999, 115).

Over the years, the content of the tapestry has been the subject of debate, and it has been alternatively interpreted as depicting a Christian milieu and a heathen one. On this tapestry, 21 females and 43 males respectively are recorded. The central motif on the tapestry is a church with a belfry. Turning towards the building on either side of it are humans and animals. The largest figures are the

Fig. 6.18 Drawing of the tapestry from Rolvsöy in Norway in its original colours (drawing: Tone Strenger 1986; photo: Eirik Irgens Johnsen; Creative Commons licence CC BY-NC-ND 3.0 https://creativecommons.org/licenses/by-nc-nd/3.0/).

Table 6.3 Costume analysis of the Rolvsöy tapestry from Norway.

Fragment/Figure	*Costume not identified*	*Gender*	*Outerwear*	*Dress*	*Train*	*Tunic*	*Leg type*	*L=body turned to the left*	*S=standing*	*H=hood L=long hair*	*Number of arms*	*Number of legs*
1.1		F	A21	C9	X		L6	L	S	L	0	0
1.2		M	A11			-	L1	L	S	H	0	2
1.3		M	A11			-	L1	L	S	H	0	2
1.4		F	A21	C9	X		L6	L	S	L	0	0
1.5		M	A11			-	L1	L	S	H	0	2
1.6		M	A11			-	L1	L	S	H	0	2
1.7		M	A11			-	L1	L	S	H	0	2

The numbering in the list refers to numbers marked in the corresponding images in Fig. 6.17.

three males on the left-hand side. The figures in and around the buildings are middle-sized, while the rest of the male figures, positioned in various places among the animals, are even smaller and nearly the same size as the females. The male figures have both legs visible, and are portrayed either from the front with arms raised in adoration or worship, or in profile with one or no arms visible. The vast majority of the males stand upright, with five figures standing next to horses. Whether or not they should be interpreted as riders is uncertain. The female figures are the smallest figures depicted. Similar to the fragments from Överhogdal, they are reduced to mere outlines, without costume details, faces or limbs (Table 6.6). The various sizes serve to emphasise the importance of the figures and especially place the female figures in a peripheral role in relation to the more active male figures.

In conclusion, more than 515 figures are identified in the tapestries selected for this study, with males outweighing

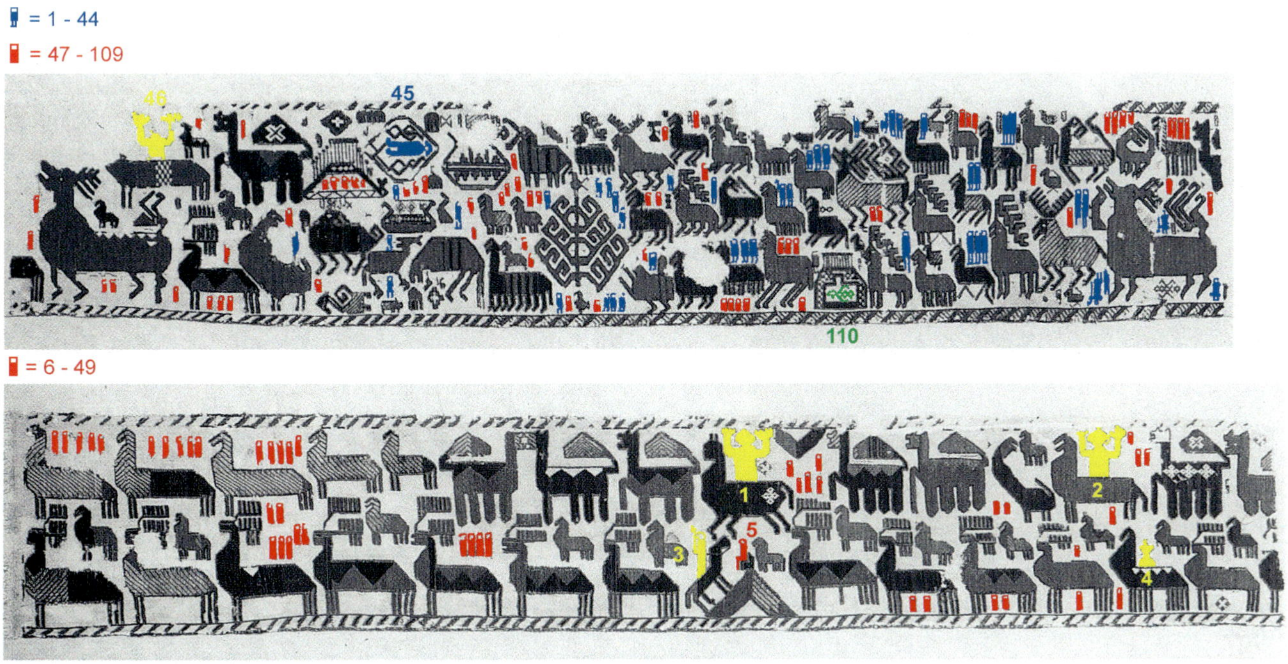

Fig. 6.19 Picture of the tapestry from Overhögdal in Sweden, strip IA and IB with the figure numbers and gender signature used in the costume analysis: female (red), male (blue), rider (yellow) and neutral (green) (drawing: Pernille Foss; after Nockert and Franzén 1992).

Fig. 6.20 Picture of the tapestry from Overhögdal in Sweden, strip II with the figure numbers and gender signature used in the costume analysis: female (red), male (blue) and rider (yellow) (drawing: Pernille Foss; after Nockert and Franzén 1992).

females by a slight majority (Table 6.7). In contrast to the other objects categories, the tapestries are more varied in their appearance, which makes it difficult to use the exact same recording method. No differences in the use of colours or patterns in regard to gender or specific costume items could be detected. The greatest possible contrast and variation were employed when the tapestries were designed and woven. On the Överhogdal strip IA, a diagonal-striped pattern is used for some animals and humans, as well as for a mounted adorer on strip IB. However, the pattern does not seem to have been used in any consistent way and most figures have no patterning. On the tapestry fragments from Oseberg, the brocading of patterns is an integral element of the figures, but regrettably, the sketches do not permit the determination of whether specific patterns were used on specific figures. There are many indications, however, that the wide breeches worn by the men have a specific dotted pattern. Generally, the colours and patterning of the individual figures depend on the technique and colours employed on the tapestry, and are thus not a parameter that is directly comparable to the patterning of, *e.g.* the costumes on the gold-foil figures.

Despite the tapestry fragments from Oseberg being the most degraded among the selected tapestries, and the Rolvsöy tapestry being the smallest, it is primarily these two finds that provide the clearest costume details. The costumes on the tapestries from Överhogdal and Rennebu are much more difficult to identify, as the figures are mostly only shown in outline. The later tapestry from Skog assumes a midway position, with very simple female and more detailed male figures.

Fig. 6.21 Picture of the tapestry from Overhögdal in Sweden, strip III with the figure numbers and gender signature used in the costume analysis: female (red), male (blue), rider (yellow) and neutral (green) (drawing: Pernille Foss; after Nockert and Franzén 1992).

Fig. 6.22 Picture of the tapestry fragment from Rennebu in Norway with the figure numbers and gender signature used in the analysis: female (red) and male (blue) (drawing: Pernille Foss; after Nockert and Franzén 1992).

Fig. 6.23 Picture of the tapestry fragment from Skog in Sweden with the figure numbers and gender signature used in the costume analysis: female (red) and male (blue) (drawing: Pernille Foss; after Nockert and Franzén 1992).

Female outerwear and innerwear

Female outerwear is attested on the Oseberg and Rolvsöy tapestries (Fig. 6.24). The former includes 30 figures and the latter two. It consists primarily of a floor-length cloak of Outerwear Group 1c (OG1c) with a pointed end at the centre of the figure. As the patterning and colours

Table 6.4 Costume analysis of the Överhogdal tapestry from Sweden.

Fragment/figure	*Costume not identified*	*Gender*	*Female costume*	*Train*	*Tunic*	*Leg type*	*Body turned to L=left R=right F=front*	*R=rider L=lying T=standing A=sitting D=driving S=sailing*	*Hair*	*Number of arms*	*Number of legs*	*A=axe S=sword*
IA.1-44		M			E?	L3	L	T	–	0	2	
IA.45		M			E?	L3	F	L	–	0	2	
IA.46		M?			B?	?	F	R	–	2	?	
IA.47–109		F	X	-		L6	?	T	–	0	0	
IA.110	X	?					F	L	–	2?	0	
IB.1-2	X	M?			B?	?	F	R	–	2	0	
IB.3		M?			B?	?	R	R	–	2	0	A
IB.4		M?			B?	?	L?	R	–	0	0	
IB.5		F	X			L6	L	A	–	0	0	
IB.6–49		F	X			L6	?	T	–	0	0	
II.1–14		M?			B?	?	L	R	–	0-1	0	
II.15		M			TB	L8	L	R	–	1	1	
II.16		M			TB	L8	L	R	–	0	1	
II.17		M			TB?	L7?	L	T	–	0	2	A
II.18–24		M			TB?	L7?	L	T	–	0	2	
II.25–26		M			TB?	L7?	F?	T	–	0	0	
II.27–29		M			E?	L3	L	T	–	1	2	S
II.30–59		M			E?	L3	L	T	–	0	2	
II.60		F	X	X		L6	L	T	–	0	0	
II.61–116		F	X	–		L6	L	T	–	0	0	
II.117		?	?		?	L6	L	D	–	0	0	
II.118		?	?		?	L6	L	D	–	1	0	A?
III.1–2		M?			B?	?	L	R	–	2	0	
III.3–4		M			TB	L7?	L	T	–	1	2	A
III.5–7		F	X	X		L6	L	T	–	0-1	0	
III.8–9		F	X	X		L6	L	T	–	1	0	
III.10		F	X	X		L6	R	T	–	0	0	
III.11		F?	X	X		L5	L	T	–	0	2?	
III.12		F	X	–		L6	L	T	–	1	0	
III.13–24		F	X	–		L6	L	T	–	0	0	
III.25–26	X	F	X	?		L6	L	T	–	0	0	
III.27–32		?	?		?	?	?	S	–	0	0	

The numbering in the list refers to numbers marked in the corresponding images in Figs 6.19–6.21.

of the outerwear recorded on the Oseberg tapestry cannot be determined with any certainty, they are all recorded as plain, while the cloaks from the Rolvsöy tapestry are recorded as dotted. In comparison, two female figures from the jewellery category dated to the Viking Age, too, are recorded as wearing OG1c cloaks. A single cloak of the OG1a type is recorded for the Oseberg tapestry, but as the figure is indistinct, the identification is uncertain. The two figures seen driving on fragment 1 are recorded as wearing an OG2 cloak type A3, but as they are seated, the changed

Table 6.5 Costume analysis of the Rennebu tapestry from Norway.

Fragment/Figure	Costume not identified	Gender	Outerwear	Dress	Train	Tunic	Leg type	Head seen from R=right L=left F=ffront	R=rider S=standing L=lying/sitting	Hair	Number of arms	Number of legs	A=axe L=loom
1.1		F		BS			L5	R	S	–	2	2	L
1.2		F	?	C?	X		L5	R	S	–	2	2	L
1.3		M				TB	L4	R	S	–	1	2	A

The numbering in the list refers to numbers marked in the corresponding images in Fig. 6.22.

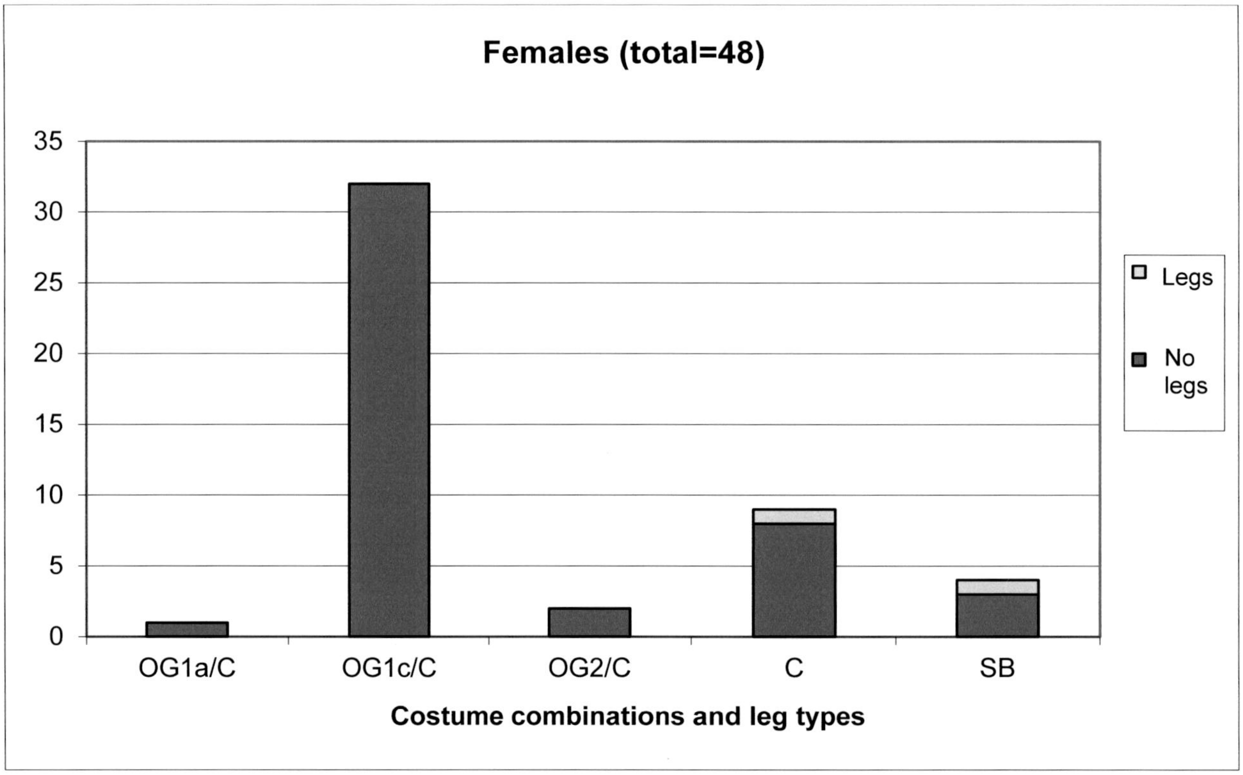

Fig. 6.24 Distribution of female costume combinations and leg types in the tapestries.

posture may be the main reason for the different look of the garment in comparison to the other figures. On the simpler figural representations from Överhogdal and Skog, the costume is only shown in a square outline, with no extra details. As the feet are not visible, they are considered to be females, but as it is impossible to ascertain whether the costume comprises an outer or inner garment, they were excluded from the overall analysis.

As the outerwear on the figures in the Oseberg and Rolvsöy tapestries conceals most of the body, it is only possible to see the lowest part of the innerwear. A characteristic feature of this innerwear is that it is floor-length, and, in most cases, furnished with a train (Fig. 6.24). It is most likely a dress, although this cannot be determined with certainty. Just as with the outerwear, all innerwear on the Oseberg strips is recorded as plain (C), while the innerwear on the Rolvsöy fragment is recorded as dotted (C9). Fragments 4 and 6 from Oseberg include three representations of women in a skirt and blouse ensemble without any outerwear. One figure (Fig. 6.4, figure 4.9) is remarkably clear due to its size which is double that of the other female representations. Another figure (Fig. 6.4, figure 4.12) resembles this woman in stature, but the costume identification is uncertain due to

Table 6.6 Costume analysis of the Skog tapestry from Sweden.

Fragment/Figure	*Costume not identified*	*Gender*	*Female costume*	*Train*	*Tunic*	*Leg type*	*Belt*	*Head seen from R=right L=left F=front*	*R=rider S=standing L=lying/sitting*	*Hair*	*Number of arms*	*Number of legs*	*Crown*	*S=shield P=spear W=sword T=staff A=axe B=bell*
1.1		M			TB	L7		L	S	?	0	2		
1.2		M			E?	L3	X	F	S	–	2	2	X	A
1.3		M			E?	L3	X	F	S	–	2	2	X	A?
1.4		M			TB	L8		L	S	–	1	2		T
1.5		M			E?	L3	X	F	S	–	2	2	X	TW?
1.6		M			E?	L3		L	S	?	1?	2		
1.7-8		M?			E?	?		L	S?	–	0	?		
1.9		M			E?	L3		F	S	–	2	2		
1.10–11		M			E?	L3		L	S	–	0	2		
1.12		M			E?	L7		F	S	–	2	2		
1.13		M			TB	L7		L	S	–	2	2		
1.14		M			TB	L7		L	S	–	1	2		
1.15		M			TB	L7		L	S	–	2	2		
1.16		M			E?	L5		F	S	–	2	2		
1.17–19		M			TB	L7		F	S	–	2	2		
1.20		M			TB	L7		F	S	–	1	2		
1.21		M			E?	L5		F	S	–	2	2		
1.22		M			E?	L5		F	S	–	1	2		T
1.23		M			E?	L5		L	S	–	1	2		
1.24		M			TB	L7		R	S	–	2	2		B
1.25		M			TB	L7		R	S	–	1	2		B
1.26		M			TB	L7		L	S	–	1	2		B
1.27		M			E?	L5		L	S	–	1	2		
1.28–32		M			E?	L5		L	S	–	0	2		
1.33–37		M			E?	L3		L	R?	–	1	2		A?
1.38–42		M			E?	L5		L	S	–	1	2		
1.43		M			E?	L5		L	R?	?	?	?		A?
1.44–64		F	X	X		L6		?	S?	–	0	0		

The numbering in the list refers to numbers marked in the corresponding images in Fig. 6.23.

its poor state of preservation. The costume of the third figure (Fig. 6.6, figure 6.13) is best described as a skirt and blouse combination, but could also be another kind of a cloak and dress ensemble. This figure has a very clearly marked train, which is not seen in the other two representations. One of the two women on the Rennebu tapestry is possibly depicted in a dress, and the other in a skirt and blouse ensemble. Owing to the composition of the image, a more precise identification is impossible. These two female representations are also the only female figures in this object category depicted with feet. Furthermore, one figure on strip II and seven on strip III of the Överhogdal

Table 6.7 Summary of the recorded figures on the tapestries selected for this study.

Location	*Males*	*Females*	*Unidentified*	*Total*
Oseberg	64	41	27	132
Rolvsöy	5	2		7
Överhogdal	113	187	9	309
Rennebu	1	2		3
Skog	43	21		64
Total	226	253	36	515

tapestry wear a very simple costume that corresponds to the descriptions of the outerwear, but with a train clearly attached. As trains are a feature that often appears together with innerwear, this is likely to be a dress.

Male outerwear and innerwear

Among males, the presence of outerwear is documented on the Oseberg and Rolvsöy tapestries. The former includes 23 figures and the latter five figures. The type of cloak seen on the Rolvsöy tapestry consists of the typical male cloak with pointed corners on either side of the figure, which corresponds to Outerwear Group 3 (OG3). A special feature of the male cloak on the Rolvsöy tapestry is that it either has a hood or is large enough to be pulled over the head as a hood. This is a unique trait and is not encountered in any other representations. On the Oseberg tapestry, a shorter version of the teardrop-shaped cloak corresponding to OG1c is recorded. This type is depicted with a pointed corner in the centre front, and on this basis, it is likely that it fastens on the right shoulder. As far as may be ascertained, no male figures in outerwear appear on the tapestries from Rennebu, Överhogdal or Skog.

Altogether 175 items of male innerwear consisting primarily of tunics or blouse and trouser ensembles are recorded. The tunics of the male figures on the Oseberg tapestry are long-sleeved and either thigh- or knee-length, while those from Skog are knee-length or longer, and widen out considerably at the lower end. A striking feature among the figures from Skog is the clearly defined waist displayed on the tunics and the blouse and trouser ensembles, which makes it difficult to differentiate between the two costume types. As the belt or narrowed waist is seldom encountered together with tunics in the other object categories, this is a new way of depicting and wearing the tunic. The vast majority of the figures from Överhogdal wear a simple costume, similar in style to the female ones, but the straight border at the lower end is indicative of a tunic. A further group of figures in a blouse and trouser ensemble is included here. Several of the male figures on the Oseberg tapestry also wear a blouse and trouser combination. The blouses are waist length and presumably long sleeved. Six different types of leg wear are recorded. Of these, the most widely represented are a plain lower leg or narrow-legged trousers, followed by wide breeches on the tapestries from Oseberg, Överhogdal and Skog. Wide breeches are otherwise rare and only encountered in the jewellery category.

Male costume combinations

The cloaks recorded on the tapestry fragments from Oseberg are always combined with wide breeches, while on the Rolvsöy tapestry, they are worn together with narrow-legged trousers (Fig. 6.25). It is likely that the figures on the former also wear a blouse. In several instances on fragments 1 and 2 from Oseberg, the arms of the men are portrayed using another colour and pattern than that seen beneath the outerwear. Whether this is a result of the inner garment being sleeveless cannot be determined from the available documentation. As far as it may be ascertained, the same dotted pattern is nearly always chosen for leg wear. On the tapestries from Oseberg, Överhogdal and Skog, the tunic is always combined with a plain lower leg or narrow-legged trousers. The legs are very short on the tapestry from Skog, as the costume is relatively long. On the same tapestries, blouses are combined with wide breeches, or less often with narrow-legged trousers. As the two aforementioned examples from Överhogdal are mounted figures, their costume is difficult to determine with certainty. On the Rennebu tapestry, the male figures wear a blouse with knee-length trousers, which is only encountered in a few cases in the other object categories. Other types of figures difficult to identify are, for example the bird people on the Oseberg tapestry. It is possible that they wear a cloak, but as the gender of the figures cannot be determined, they are not recorded. The same applies to the two bear people on fragment 3 from Oseberg.

Conclusion

Altogether, 469 costume combinations from the 515 figures were recorded on the five selected tapestries. Of these, 204 are recorded as male and 265 as females. Due to uncertainty as to whether the costume consists of outer or innerwear, 217 female costumes were not included. Altogether 252 costume combinations, of which 204 are male and 48 female costumes, form the basis of the costume analysis, which can be further condensed into three different male and female costume combinations (Table 6.8). Two different combinations of cloaks and trousers belong to the male costume. The ensemble of a short cloak with wide breeches is typical for the figures on the Oseberg tapestry, and unique in comparison with the other object categories. The cloak and trouser ensemble on the Rolvsöy

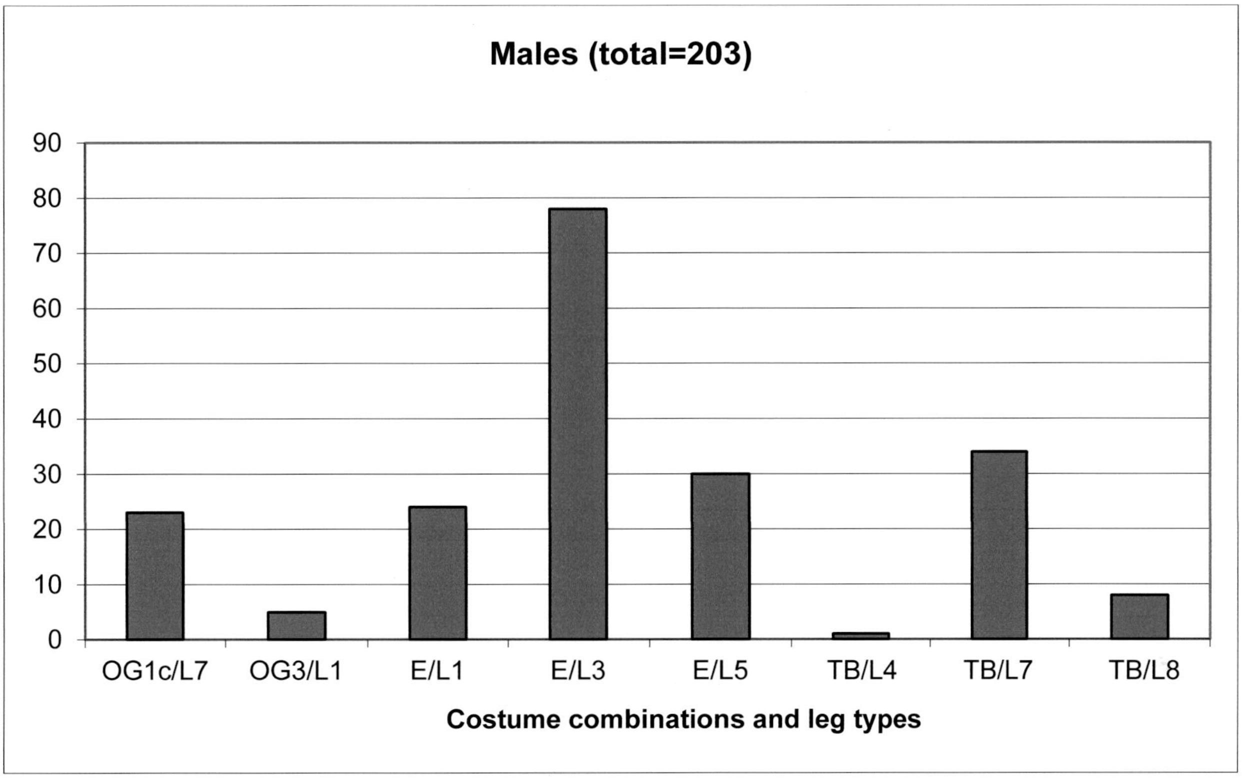

Fig. 6.25 Distribution of male costume combinations and leg types in the tapestries.

tapestry resembles the costumes of the gold-foil figures to a great extent, except for their heads being covered with either the cloak or a hood. Male representations without outerwear consist partly of tunic and trouser, or blouse and trouser ensembles. The tunics from the Oseberg tapestry greatly resemble those on the other object categories, *i.e.* a relatively narrow costume that is thigh- or knee-length. They are often combined with narrow-legged trousers, while the blouse and trouser ensembles are often combined with wide breeches. On the Skog tapestry, the male figures are attired in a long tunic that is also fuller at the lower end of the costume. Thus, the legs appear very short here. The tunic and trousers constitute the male costume ensemble most commonly seen. It is often knee-length or longer, and the shortest are encountered amongst the Oseberg figures. Thus, the material displays a tendency for the length of the tunic to increase with time. Furthermore, on the Oseberg fragments, a connection seems to exist between the warrior status and the cloak and wide breeches combination, while men without weapons, as, for instance the hanged figures or the adorers are clad in tunic and trousers. When taking into consideration the many different figures incorporated in the narratives on the tapestries, including warriors, it is surprising that no males in kaftan are depicted here, not even among the mounted figures. The riders on the Oseberg tapestry wear the same costume as the warriors on foot, and on the Överhogdal tapestry, the costumes of the riders are not specific in any way. Thus, no males in kaftan or other types of jackets are encountered among the figures on the tapestries, making kaftans exclusive to the object categories linked to the Late Germanic Iron Age, or specific geographical regions, such as Sweden in the case of the helmet plates and the south-eastern Scandinavian area in the case of the gold-foil figures.

Female costume comprises cloaks and dress ensembles along with a few skirt and blouse combinations similar to those observed particularly among the Norwegian gold-foil figures. The shield-bearing females on fragment 3 are clad in the same costume as most of the other females, namely a dress with a train combined with a cloak. The identification of the female costume on the Rennebu tapestry is not clear-cut, and is merely a suggestion. The simple female representations from the Överhogdal and Skog tapestries, as well as the tiny figures from the Oseberg tapestry are not included in the enumeration as the costume cannot be determined with any accuracy. However, a few women on the Överhogdal tapestry do wear a garment with a train, which makes it likely that it represents a dress. These figures are also depicted in a larger format than the other women, more in line with the men. No doubt, the selected tapestries follow the same iconographical rules for male and female costume as recorded in the other object

Table 6.8 The costume combinations on the tapestries selected for this study.

Costume combination	*Males*					*Females*					*Total*
	OB	*RO*	*OH*	*RB*	*SK*	*OB*	*RO*	*OH*	*RB*	*SK*	
Cloak and trousers	23	5									28
Tunic and trousers	24		78		30						132
Blouse and trousers	17		14	1	12						43
Female costume						(17*)		(179*)		(21*)	–
Cloak and dress						33	2				35
Dress								8	1		9
Skirt and blouse						3			1		4
Costumes in total	64	5	92	1	42	36	2	8	2	0	252

Oseberg (OB), Rolvsöy (RO), Överhogdal (OH); Rennebu (RB); Skog (SK). * As these figures are only depicted in contour form, the costume elements cannot be determined more closely, and therefore, they are not included in the summing up.

categories. The tapestries attributed to the first part of the Viking Age (Oseberg and Rolvsöy) include most details, and it seems that the simplification of the figures, most markedly in the female costume, increases over time. In contrast, the shape of the male costumes from the Scandinavian Middle Ages (Rennebu and Skog, and also Överhogdal) varies over time. The tapestries thus have the greatest similarities with the jewellery category attributed to the Viking Age, which strengthens the argument that the iconographical material in this category displays contemporary costumes.

Note

1 In 2006 a comprehensive book on the Oseberg textile finds (Christensen and Nockert 2006) was published, but too late for its excellent descriptions of the tapestries to be included in the present study.

7

Clothing and costume

From an archaeological perspective, the Scandinavian Late Iron Age is a period when large-scale social and technological changes took place. A development that left traces and had a profound impact on the production of textiles, textile technology and clothing typology. In the past decades, we have gained new knowledge of Late Iron Age textile production through the textiles themselves (Hald 1980; Bender Jørgensen 1986; Andersson Strand *et al.*, 2010), and also the organisation of textile crafts through sources, such as tools (Andersson 1999a–c; 2003a; 2003b), settlement structures (Ulriksen 1998; Møller Hansen *et al.*, 1999) and the study of fibre materials and resources (Robinson 1993; 1994; Robinson *et al.*, 1995; Henriksen and Harild 2005). Although the introduction of the sail would have had a decisive influence on the organisation of the textile craft, a more specialised textile production can only be first documented during the Viking Age (Andersson Strand 2005; 2011; Zagal Mach Wolf 2013).

Scandinavia has a long tradition for textile research founded on its large and unique collection of textiles and skin materials from primarily burial finds. Despite most archaeological textiles and skin objects often being highly fragmented and degraded due to the poor preservation conditions, thorough analyses of the many thousands of fragments have made it possible to map and describe the development of textile technology, especially in Scandinavia and northern Europe from the Bronze Age to the Viking Age (Hald 1980; Bender Jørgensen 1986; 1992a; Nockert 1991; Gleba and Mannering 2012). Studying the technical aspects of textiles enables us to see that, material produced in the Early Germanic Iron Age and Viking Age is easily distinguishable from each other, even at first glance, while the material produced in the Late Germanic Age and the Viking Age is very similar. Moreover, the choices made in terms of colours, patterns and decorations show a clear difference between the textiles produced in the Early Germanic Iron Age, the Late Germanic Iron Age and the Viking Age (Tables 7.1 and 7.2).

Table 7.1 Different Scandinavian textile types and their characteristics as defined by Bender Jørgensen (1986).

Textile type	*Weave*	*Spin direction*	*Thread relation*	*Fibre material*
Haraldskær	2/2 twill	z/z	Balanced	Wool
Gerlev-Dråby	Tabby	z/z	Balanced	Flax
Nørre Sandegård	Tabby	z/z	Unbalanced	Wool
Birka	2/2 diamond twill	z/z		Wool
Hessens/ Elisenhof	2/2 twill or 2/2 broken/ diamond twill	z/s	Unbalanced	Wool
Gudmingegård	Tabby	z/spin pattern	Unbalanced	Wool
Ribbed twill	1/2, 2/1 twill	z/z	Unbalanced	Wool

Although the number of Scandinavian textile finds is unprecedented in comparison to most parts of Europe, it is difficult to work with prehistoric clothing, as the finds are rarely complete enough to give a full overall picture of what the clothing looked like. In Scandinavia, archaeological clothing history begins with the extraordinary Danish oak-coffin burials from the Early Bronze Age around 1700 BC (Broholm and Hald 1939; 1940). These are the first clothing items from Scandinavian prehistory made of loom-woven material, and the finds demonstrate from the outset a well-established familiarity with both the weaving technology and the construction of patterns. Some of these items of clothing are clearly modelled on patterns used for skin clothing. With the transition to the Iron Age, around 500 BC, it is possible to see a shift towards a clothing tradition where many items of clothing were shaped and finished on the loom. Exactly when and how rapidly this change took place is difficult to determine, as the textile material from the Late Bronze Age is scarce and highly fragmented and does not contain any complete clothing items.

In contrast to the Bronze Age tradition of cutting and sewing, the excellently preserved textiles, including garments, from the Pre-Roman Iron Age are based on the use of various rectangular textiles to wrap or drape around the body. A unique feature of this period is that, it is almost impossible to differentiate between male and female costumes. In most other periods, it is possible to see differences in the design of male and female clothing in the archaeological record, and, in this regard, the clothing tradition in the Pre-Roman Iron Age deviates from that of all other prehistoric periods (Hald 1980; Mannering 2011).

Following an unusual period dominated by unisex wrap-around textiles, and sewn, but still unisex skin garments, costume design again takes a turn towards the production of more gender specific and tailored costumes in the Roman Iron Age around the Common Era. Especially in the North German bogs and weapon deposits, but also in Danish burials, well-preserved and almost complete garments dating to the Roman Iron Age have been recovered. The most recent find of a complete tunic in Norway is an excellent example of Roman Iron Age clothing, although its context as a stray find makes its link to a specific gender less certain (Vedeler and Bender Jørgensen 2013). Altogether, these finds play a significant role in the understanding of the Scandinavian prehistoric clothing tradition as a whole and have been used as a basis for the construction of a Germanic/north European clothing identity (Schlabow 1976; Hald 1980; Möller-Wiering 2006).

Our knowledge of clothing in the Early Germanic Iron Age primarily derives from a few large cemeteries, such as Sejlflod and Hjemsted, although the textile material is scarce and not very well preserved (Bender Jørgensen 1986; Ethelberg 1986; 1990; Demant 1996; Ræder Knudsen and Mannering 2007). From Sweden and Norway, a number of graves with well-preserved, but not complete, garment finds have provided invaluable information on male clothing (Bender Jørgensen 1986; Nockert 1991).

More textiles are dated to the Late Germanic Iron Age, although, no completely intact costumes have been found as yet. In this period, information on the clothing is based on the use and position of dress fasteners, primarily deriving from rich female graves from Bornholm in Denmark (Bender Jørgensen 1986; Jørgensen and Jørgensen 1997; Mannering 1997). A tunic from Bernuthsfeld in Germany, radiocarbon dated to AD 680, is certainly comparable to the Scandinavian finds and possibly represents a male garment (Schlabow 1976; van der Sanden 1996, 127–128).

Textile material from Viking Age burials occurs quite frequently in Scandinavia, but information on clothing is primarily based on finds from the Birka cemetery in Sweden (Geijer 1938; Hägg 1971; 1974; 1986; Larsson 2007) and the harbour and settlement complex in Haithabu in Germany (Hägg 1984a; 1991a). The textile material from Birka has been presented in several scholarly publications, although, a single publication combining male and female costumes has yet to come. To this must be added the very rich male and female burials from Mammen, Fyrkat, Hvilehøj and Hørning in Denmark and Oseberg in Norway that have yielded promising information on the clothing of the elite (Ingstad 1988; Iversen 1991; Christensen *et al.*, 1992; Hedeager Krag and Ræder Knudsen 1999; Hedeager Krag 2005; Christensen and Nockert 2006; Pentz *et al.*, 2009).

More recently, Old Norse clothing from South Greenland and the bog finds from Moselund and Kragelund in Denmark, Skjoldehamn in Norway and Bocksten in Sweden, together with some general works, have shed new light on Viking Age and medieval clothing in Scandinavia (Gjessing 1938; Nockert 1997; Østergård 2004; Ewing 2006; Vedeler 2007a; Wiking-Faria 2008; Frandsen *et al.*, 2010; Glæsel 2010).

Altogether, the above-mentioned finds provide, in each their own way, a glimpse of the development of male and female clothing from the Bronze Age to the Scandinavian Middle Ages. However, seen from a source critical perspective, this information is difficult and problematic to use for extrapolating general characteristics, as most items of clothing are unique in terms of preservation and context, rendering it problematic to see them as representing typical items from these periods. Nevertheless, they are all we have. Using a source-critical approach, they still provide us with an excellent basis for building up an overall picture of the Scandinavian clothing tradition in these times.

Table 7.2 Different Scandinavian textile types as defined by Bender Jørgensen (1986) and their chronological distribution.

	Haraldskær	*Gerlev-Dråby*	*Nørre Sandegård*	*Birka*	*Hessens/ Elisenhof*	*Gudminge-gård*	*Ribbed twill*
Early Germanic Iron Age	█						█
Late Germanic Iron Age		█	█	█	█	█	█
Viking Age		█	█	█	█	█	

Thus, bearing the caveat in mind that most costume finds are dispersed in terms of time, geography and gender, and that preservation conditions entail that primarily clothing items pertaining to the elite are preserved, while less is known about everyday wear or work wear, the following constitutes a comparative discussion of the archaeological textiles and clothing finds from primarily Scandinavia and the costumes observed in the iconography. Most of the clothing finds selected for this study were those with the most details originating from periods prior to the selected iconographical material, and were included in the discussion in order to provide a broader perspective on the development of clothing in the Late Iron Age, as well as to characterise the clothing tradition that the iconographic costumes belong to.

In other words, the aim of this comparison is thus to show how the selected iconographical costume material relates to the available archaeological clothing finds, and to determine the extent to which the iconographical costumes are typical for their time or present archaic or unknown elements. The examples are thus chosen with the aim of giving the analysis a new perspective and not intended as a complete presentation of the entire gamut of Scandinavian archaeological clothing material in all its nuances.

Outerwear

Cloaks

The cloak is a costume item linked to male and female gold-foil figures, male and female figures in the tapestry category, and female figures in the jewellery category. Three different outerwear groups were identified, which are differently distributed among male and female costumes (see Fig. 1.10).

Among the archaeological textile finds, the cloak seems to be a universal costume item found in various shapes and materials throughout prehistory. Cloaks can be made from animal skin or woven materials, each affecting the shape of the finished product. Regrettably, the archaeological clothing finds are neither sufficiently preserved nor representative to establish a complete chronological development of skin and woven cloaks throughout prehistory, and in the following, only a few examples are highlighted.

Animal skins and fur cloaks are primarily known from the Danish Early Iron Age bog finds, where the preservation conditions are optimal for proteinaceous materials. More than 30 skin capes have been recovered from Danish bogs (Hald 1980; Mannering *et al.*, 2010) (Fig. 7.1). In Late Iron Age burials there are frequent recordings of traces of animal

Fig. 7.1 Two examples of Early Iron Age sheepskin cloaks found in Danish bogs. The top one was found in True and has a symmetrical pattern. The lower cloak belonged to the Huldremose Woman and has an asymmetrical pattern. The schematic patterns of the two skin cloak types are shown to the right (scale = 10 cm) (National Museum of Denmark).

skin and fur, as in the case of the graves from Mammen and Hvilehøj (Hald 1980, 102–120; Iversen 1991), but due to the poor preservation conditions, it is rarely possible to determine if they derive from clothing or the lining of the grave. It is a well-known phenomenon in Danish Early Bronze Age oak-coffin burials to line the coffins with animal hides, and the frequent but poorly preserved fragments of fur in graves indicate that this was a custom that continued in the Iron Age. Thus, unless traces of, for instance sewing are recorded, the presence of skin or fur cannot be deduced automatically as belonging to clothing.

In general, the Early Iron Age skin cloaks are constructed according to highly intricate patterns including cutting and sewing. The extant examples are all relatively short, with a height of 70–90 cm, thus only covering the upper body. In several instances, the find contexts demonstrate that several cloaks were worn one on top of each other with an inner cloak with the fur side turned against the body and an outer cloak with the fur side turned outwards (Mannering 2011; Hald 1980, 316–317). These cloaks were primarily fastened on the shoulders or in the front with straps and toggles and are thus not associated with the use of dress fasteners or fibulae. Furthermore, it is known that, in the Early Iron Age male and females wore the same type of skin cloak, and no gender difference can be detected in regard to this item of clothing. Skin and fur fragments from the Late Iron Age have been found in many graves as well as in the two afore mentioned high-status graves from Mammen and Hvilehøj. The fur recovered in these graves was identified as marmot, and due to its exotic status, at least, in these two cases, the fur can be interpreted as remains of a fur cloak. In a reconstruction made in 1990 of the so-called Mammen costume, the fur was used to line a woven cloak (Munksgaard 1989), but this is not the only possibility, and the fur could equally well have been used for a complete clothing item (Fig. 7.2). As most of the preserved skin and fur cloaks mentioned above originate from a period prior to the selected iconographical depictions dealt with here, and as Late Iron Age cloak types cannot be specified more closely, it is difficult to compare these clothing items with the depictions. However, there are clear similarities between the contours of the skin cloaks with their curved lines and Outerwear Group 1. For instance, especially among the female gold-foil figures, short cloaks, which may be compared to the Early Iron Age skin cloaks, are frequently portrayed. An argument against the depicted cloaks being made of skin is the costume patterning that most likely refers to specific textile types. The irregular pointed patterns seen on Cloak Type A15 (see Fig. 1.10) have been linked together with representations of fur or feathers. This pattern is a characteristic feature of gold-foil figures belonging to Costume Group 3 and singly represented female figures, and several of these depictions have been interpreted as the Nordic goddess Freyja in her feathered pelt (Arrhenius 1962; Axboe 1986; Simek 2002, 110–111). However, the same pattern is also used on many male tunics, making the identification as fur less certain. Although, in general, it cannot be excluded that many of the depicted items of outerwear represent skin cloaks.

Cloaks made of textiles are known throughout the entire Scandinavian prehistoric period from many different contexts, especially bog finds, burials and weapon deposits dated to the Early Iron Age (Fig. 7.3). The vast majority of these finds are largely intact, while similar finds dated to the Late Iron Age, are highly fragmented. In the latter period, fragments of woven cloaks are primarily found in graves, but are also known from Haithabu harbour.

Most well preserved Iron Age cloaks have a rectangular shape, and vary from simple lengths of cloth taken straight from the loom to more sophisticated forms woven to shape (Lindqvist *et al.*, 1925; Broholm and Hald 1940; Hald 1980;

Fig. 7.2 The reconstructed male cloak from the Mammen chamber grave in Denmark dated to AD 971. The embroidered wool fabric is lined with marmot fur (National Museum of Denmark).

Hägg 1971; Schlabow 1976). This variation is not indicative of any chronological sequence of development. The cloak recovered from the Gerum bog in Sweden (Fig. 7.4), dated to the Pre-Roman Iron Age (360–100 BC) is, with its oval shape, a type that harkens back to the Bronze Age (Lindqvist *et al.*, 1925; Nockert and Possnert 2002, 62–63) (Fig. 7.5).

Remains of cloaks in male and female graves dated to the Roman Iron Age suggest that rectangular shapes were the most commonly used (Wild and Bender Jørgensen 1988, 65–67; Nockert 1991; Schlabow 1976) (Fig.7.6). An excellently preserved rectangular cloak was found in the Skærsø bog in Denmark, and although this item was

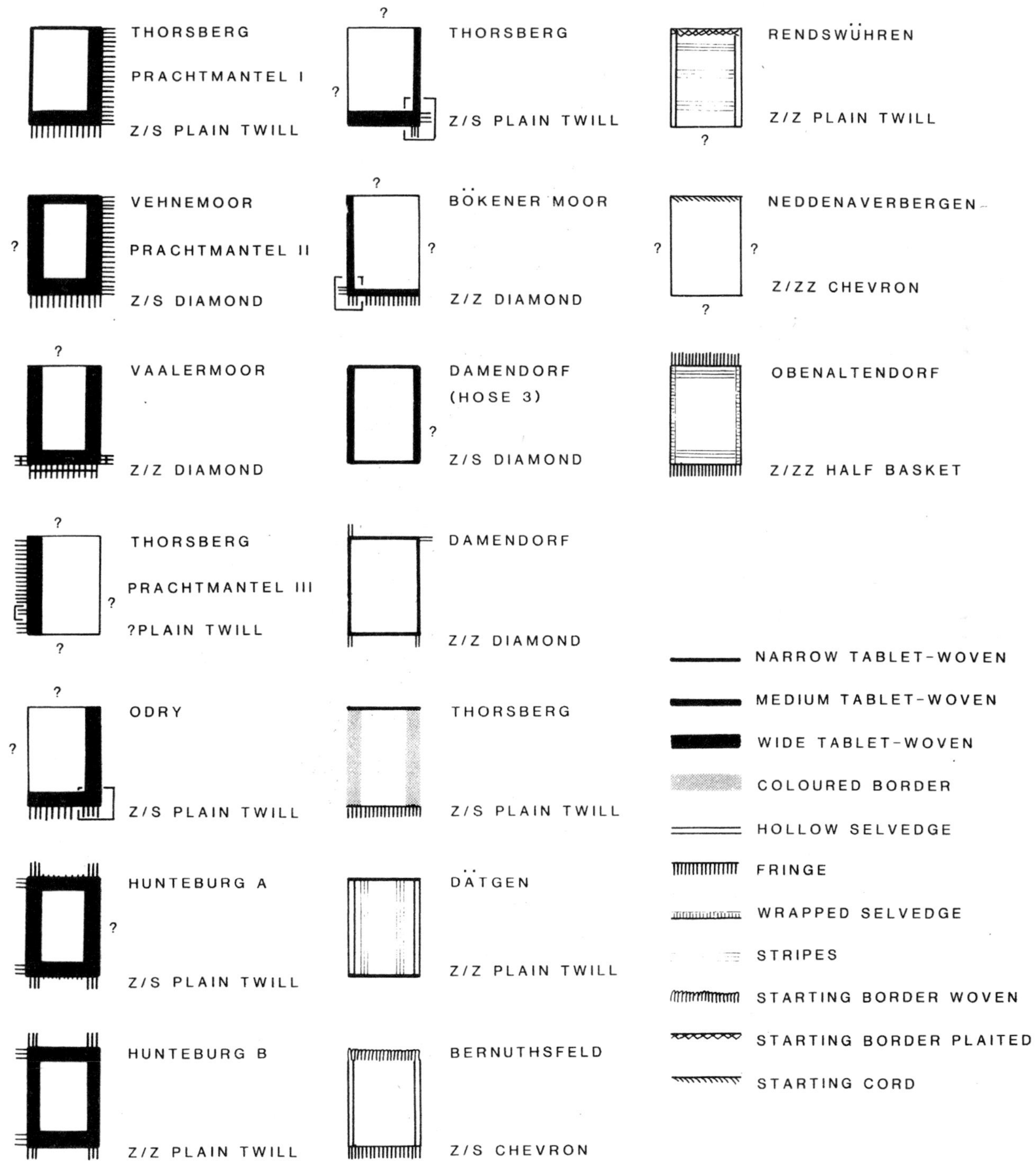

Fig. 7.3 Schematic drawings of different textile cloaks dated to the Roman Iron Age found in Germany (drawing: Courtesy of John Peter Wild).

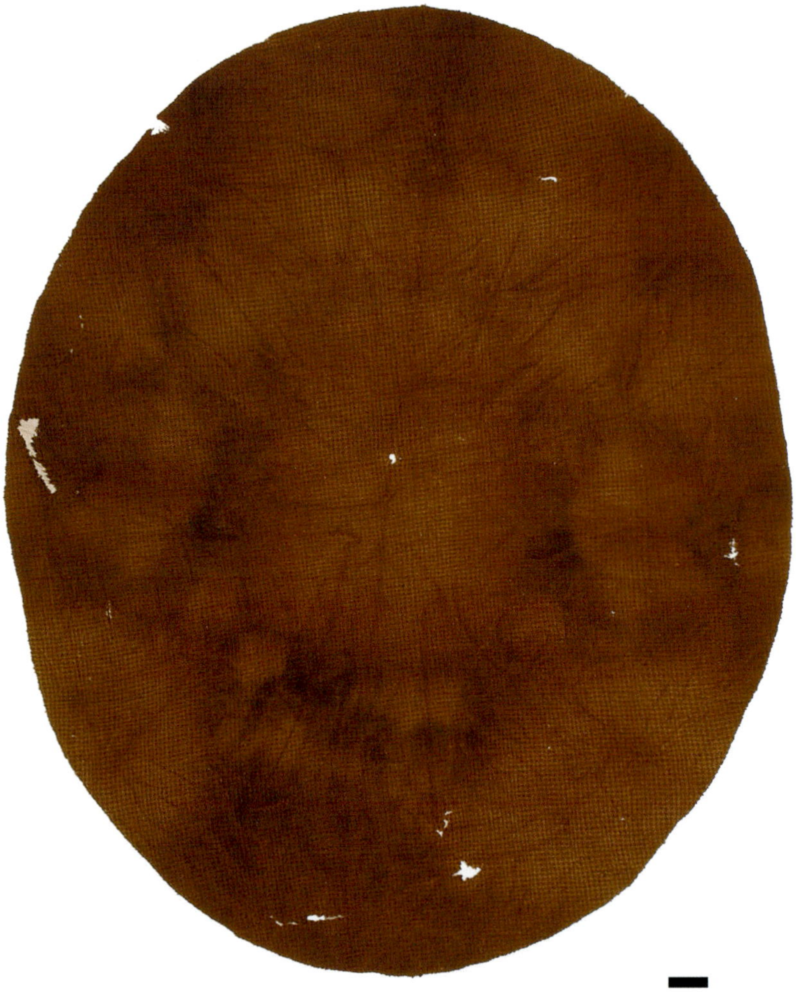

Fig. 7.4 The Gerum cloak from Sweden. The oval wool cloak is woven in 2/2 twill and measures 250 × 199 cm. The find is radiocarbon dated to 360–100 BC (scale = 10 cm) (Swedish History Museum; photo: G. Hildebrand).

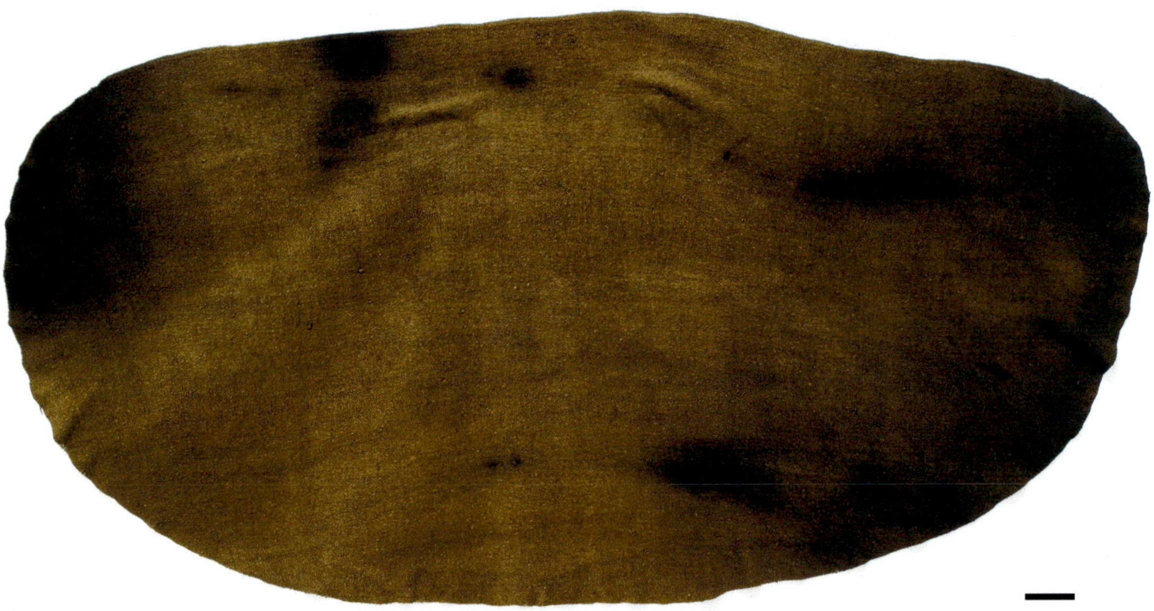

Fig. 7.5 The wool tabby cloak found in the Early Bronze Age oak coffin from Muldbjerg in Denmark. The find is dated to 1365 BC (scale = 10 cm) (National Museum of Denmark).

Fig. 7.6 The wool cloak from Thorsberg in Germany. The greenish parts are what remain of the cloak. The brown part is a modern support fabric. The find is dated to the 3rd century AD (scale = 60 cm) (National Museum of Denmark).

Fig. 7.7 The cloak from Skærsø in Denmark. The wool cloak is woven in 2/2 diamond twill and measures 206 × 150 cm. It is radiocarbon dated to 350 BC–AD 90. The middle part of the cloak is bleached due to its exposure to sunlight during exhibition (scale = 10 cm) (Museet på Koldinghus).

radiocarbon dated to the Pre-Roman Iron Age, in regard to weaving technology and shape, it greatly resembles the Late Roman Iron Age cloaks (Fig. 7.7). A specific Roman Iron Age male item is the so-called magnificent cloaks known for their wide, tablet-woven borders, but whether this was a type that continued in use in the Late Iron Age, is difficult to determine (Wild and Bender Jørgensen 1988, 83; Nockert 1991; Ræder Knudsen 1998; 2013). No intact cloaks from the Late Iron Age have as yet been found, and the existence of cloaks is primarily documented from the location of fibulae in graves. The finds from Evebø/Eide in Norway and Högom in Sweden (Magnus 1982; Bender Jørgensen 1991a; Nockert 1991) indicate that high-status males used large rectangular cloaks of a type evocative of the large cloaks from the Roman Iron Age (Wild and Bender Jørgensen 1988) (Figs 7.8 and 7.9). If worn on the one shoulder this kind of cloak would resemble Outerwear Group 3 with its pointed corners (see Fig. 1.10). Although evidence suggests that females, too, wore rectangular cloaks, the location of the fibulae in graves indicates that, most likely women wore their cloaks fastened on the chest, which corresponds to the appearance of Outerwear Group 1. Thus, there seem to be congruence between the observed use of fibulae and the outerwear groups recorded in the iconography.

In conclusion, the depictions attest to a gender differentiated mode of wear corresponding to the location of the fibulae in the burials. However, neither the images nor the archaeological textile finds demonstrate any discernible difference in the overall shape between male and female cloaks. Male cloaks seen on the depictions have primarily plain or patterned surfaces that can be related to weave patterns or give the impression of a flowing costume. The same applies to female cloaks. Similar to the costume patterning, the fibulae, too, are depicted in exaggerated dimensions. This most likely indicates that these patterns were a key iconographical feature, and that they reflect specific, known textile types. No evidence in archaeological textile finds has hitherto been able to substantiate a clear correspondence between the use of cloaks and specific textile types. Yet, on the basis of the Viking Age material from Haithabu, it has been argued that specific items of clothing were produced in specific textile types (Hägg 1984a, 215–216). Until now, it has not been possible to ascertain to what extent this applies to the entire Late Iron Age period.

Fig. 7.8 Reconstruction drawing of the Evebø/Eide male clothing items found in Norway showing the cloak, tunic and trousers. The grave is dated to the Early Germanic Iron Age c. *AD 475 (drawing: Courtesy of Søssa Magnus, Oslo).*

Jackets and kaftans

The jacket is a costume item that in the male depictions is associated with gold-foil figures and helmet plates, and in the female depictions is associated with gold-foil figures and a single object from the jewellery category. There are three different basic types depending on whether they are worn by men or women. The kaftan occurs most frequently, and is worn exclusively by males. The other jackets are female types, with the exception of a type common to both (see Fig. 1.10).

In Scandinavia, the jacket cannot be documented in the archaeological clothing finds prior to the Viking Age, but among the textile finds from Birka dated to the 9th and 10th centuries AD, male and female jackets are documented. The male jacket is defined as a calf-length tailored item with sleeves and an open front[1] (Hägg 1974, 92–94). It is often made of a fine quality wool or linen textile. Moreover, it

Fig. 7.9 Reconstruction of the Högum male clothing items found in Sweden consisting of a cloak, tunic and trousers. The grave is dated to the Early Germanic Iron Age c. AD 500 (Museum of Cultural History, University of Oslo; photo: Eirik Irgens Johnsen; licence: CC BY-NC-ND 3.0: https://creativecommons.org/licenses/by-nc-nd/3.0/).

can have silk borders with passementerie decorations made of silver and gold (Hägg 1974, 99). Remains of this type of jacket were recovered in 18 male burials from Birka. Of these jackets, five were buttoned in front using small rows of bronze buttons; three had leather belts with mounts; while the rest had fabric belts, often made of silk with richly worked end pieces. Trimmings of marten, beaver or squirrel fur were also a possibility (Hägg 1984b, 207–209). Most likely, the fibula placed on the right shoulder is associated with this item of clothing, and, in certain cases, a fibula could also be placed in the lower front corner, probably where the front pieces overlapped (Hägg 1984b, 207–209). The drawings of these graves reveal that, the rows of buttons belonging to the jacket lay in a vertical line across the chest (Arbman 1940; 1943). It is, however, not possible to ascertain if this costume item should be characterised as a jacket or a kaftan according to the types seen in the iconography.

In the female costume from Birka, the jacket[2] constitutes an item of clothing roughly of similar shape to the male costume, but, in these cases, the female jacket was fastened with a fibula passed through two fabric eyelets, undoubtedly indicating a jacket with a symmetrical closing corresponding to Outerwear Group 5 (OG5) (Arbman 1940; Geijer 1938; Hägg 1983, 334). The female jacket was embellished with strips of silk and tablet-woven bands, although generally, it was not quite as elaborate as the male types and was shorter in length (Hägg 1974,

Fig. 7.10 Reconstruction drawings of the various Viking Age female clothing items found in the Birka graves in Sweden. The drawings are based on Inga Hägg's interpretations (drawing: Flemming Bau).

Fig. 7.11 Reconstruction drawings of the various Viking Age male clothing items found in Haithabu in Germany (redrawn by Sidsel Frisch; after Hägg 1984; 1991a; and Inga Hägg's reconstruction drawings in Elsner 1989).

94–95, 112) (Fig. 7.10d). In the textile material from Haithabu, many fragments originating from a jacket-like lined item of clothing with a slanted, overlapping front and contrasting trimmings along the edges were identified from both the harbour and settlement areas (Fig. 7.11b). The textiles used for this garment are, in most cases, a dense wool twill with a slightly felted surface, while the trimmings have an almost fur-like roughened surface (Hägg 1984a, 73–75, 183–185; 1991a, 48–50). The garment was presumably hip-length and worn with a belt (Hägg 1984a, 86, 188).

The Scandinavian archaeological textile material from the Viking Age can thus attest to the existence of several types of jackets for both males and females. Of these jackets, at least one resembles the iconographical kaftans. The difference in the iconographical jacket types portrayed in Outerwear Group 5 (jacket with straight hemline, solely worn by women) and Group 6 (jacket with slanted, overlapping front edges, only worn by men) is, however, difficult to document in the textile material from Birka due to the poor preservation, and the archaeological textile finds seem to indicate the presence of an item of clothing similar in shape for both males and females. In certain cases, it looks as if the upper part of the male jacket had a straight opening, while the lower part was more slanted.

While the depictions indicate the presence of a jacket, and especially a kaftan earlier than the archaeological textile finds themselves demonstrate, the latter in turn attests that this special clothing item was definitely in use, perhaps with an altered significance in the Viking Age. This shift of focus from an iconographical to an archaeological clothing item, with the jacket and kaftan disappearing in the iconography, is possibly due to a change in its use and connotation. It has often been suggested that, Scandinavia was influenced by Byzantine fashions through the Frankish/ Carolingian court. The special Byzantine kaftan, the *scaramangion*, was produced in the imperial workshops in Constantinople and was used as a diplomatic gift. It is assumed that it was through these connections that the garment became part of western European costume in the 9th–10th centuries AD (Kondakov 1924, 11–14; Cumont 1926; Hägg 1974, 99–101). Although Inga Hägg compares the Viking Age male jacket with the Asiatic kaftan, she however emphasises that this influence primarily lay in the embellishment and not in the cut of the jacket (Hägg 1984b, 211). The exotic and foreign elements in this jacket should thus be primarily perceived as symbolic, or as signifiers of rank added to a known Scandinavian clothing type through the influence of Byzantine fashion. Indeed, the frequent occurrence of elements, such as borders of silk and passementerie in gold and silver, serves as an argument for the jacket itself being locally produced (Hägg 1984b, 210). Thus, the crucial question is when the jacket or kaftan appears in the Scandinavian archaeological clothing material, and from where it originates. A study of the development of the sleeved jacket in the context of eastern European and Near Eastern depictions demonstrated that the jacket has roots in the 1st half of the 1st millennium BC (Knauer 1985, 607–609). Thereafter, it appears regularly in depictions in, for example Persian, Parthian, Scythian, Sassanian, and later Byzantine art, and is considered a typical equestrian garment among the so-called horse-riding nomads (Knauer 1978, 21–23; Vogelsand Eastwood 2000). Elfriede Knauer further differentiates between a kaftan and a sleeved jacket (*kandys*) and defines the former as a tight-fitting, jacket-like item of clothing that can be fastened in various ways at the front. It can be either short- or long-sleeved, thigh- or ankle-length, and is often provided with a belt (Knauer 1985, 608). The sleeved jacket resembles the kaftan to a great extent, but is worn over this without a belt. Another significant feature is that, it is often draped on the shoulders, the sleeves rarely being used, and often closed off at the ends (Knauer 1985, 610). Both types can be crafted of skin or textile and trimmed with fur. While the kaftan is primarily a male costume, the sleeved jacket also appears to be a female garment. In the 4th century BC, Greek women adopted the Persian sleeved jacket (*kandys*), albeit with the difference that the sleeves were open at the ends (Knauer 1985, 610). The sleeved jacket is occasionally depicted in both Greek and Roman female iconography (Fig. 7.12), and these examples demonstrate that the sleeved jacket influenced and constituted western European female fashion at a very early stage. Thus, evidence suggests that the jacket corresponding to Outerwear Group 5 (OG5) recorded in a few instances on female gold-foil figures most likely has its origins in a real garment, but how it entered the Scandinavian clothing tradition is not straightforward.

In the Late Roman period, Roman cavalry was dominated by riders of either Greek-Hellenistic (*cataphracti*) or Persian-oriental (*clibannarii*) origin. Apart from special weaponry and helmets, the equestrian equipment also comprised thick protective outerwear made of textiles of the kaftan type (Engström 1997, 249–250). Military organisation in the Scandinavian area was transformed in the 5th and 6th centuries AD into a society of warrior elite, where basic uniform weaponry, a muster of housecarls or personal bodyguards or troops, as well as an increase in the use of cavalry played a significant role. On the basis of the burial material, it is often emphasised that the transformations occurred as a result of influence from, and with direct links to, the Merovingian warrior elite (Nørgård Jørgensen 1991a, 195–197; Watt 2003, 40–41). However, it has been suggested that the weaponry in the burials from Vendel and Valsgärde in Sweden reflects

Fig. 7.12 Copy of a Greek relief of Medea and the Peliades dated to 410/20 BC now in the Antikensammlung Berlin, in Germany. In the course of this study, it was discovered that the figure to the left is wearing a sleeved jacket. This feature had not been observed earlier (photo: Marcus Cyron; http://commons.wikimedia.org/wiki/File:Relief_of_Medea_and_the_Peliades_Antikensammlung_Berlin.jpg).

more eastern connections (Engström 1997, 249–250). With this link to eastern cavalry, there is possibly also a more direct explanation for the occurrence of the kaftan in the Scandinavian iconography. Thus, evidence suggests that along with cavalry, armament and military organisation, the corresponding clothing traditions, too, were adopted, explaining why depictions of kaftan-clad men reflect such a male-dominated sphere with clear military connotations. It is thus possible that the kaftan in this period is literally used as a symbol of the new warrior elite, or as a kind of ceremonial costume (Watt 2003, 42).

The only possible North European find of a kaftan dated earlier than the Viking Age was recovered in a 7th century AD male burial from Castledyke South in England. Among the textile fragments recovered from this grave was a tablet-woven band placed vertically across the body. The position indicates that it belonged to a kaftan-like garment, which was open at the front (Walton Rogers 1998, 279). As this is the hitherto earliest find of a possible kaftan in the North European textile record, this opens for future possibilities for testifying its existence in the Scandinavian textile material possibly already in the Germanic Iron Age.

It has been suggested that among the textile remains from the Valsgärde grave 5 dated to the Late Germanic Iron Age, a patterned linen weave with floating threads could be the remains of a kaftan border. Nevertheless, as the deceased, at the time of the burial, most likely was not truly clad in the textiles found in the grave, but only covered by a cloak, while the remainder of the textiles were spread over the entire grave, this interpretation appears uncertain (Arrhenius 1980; 1985; Malmius 1997; 2003, 55–57). The textile type that this costume item is linked to has also been found in several other graves in Valsgärde, as well as in two female burials from Bornholm[3] (Bender Jørgensen 1992a, 261–262; Mannering 1997, 125–126). In the Alamanni/Frankish area, in contrast, this textile type is found in male burials, although it cannot be linked to a specific garment and therefore, this mixed distribution cannot, at present, validate the proposed interpretation (Hundt 1987).

In conclusion, it is possible to interpret the kaftan portrayed in the depictions from the 6th and 7th centuries AD on the gold-foil figures and helmet plates as an expression of the warrior elite that developed in Scandinavia either through western European connections or directly through connections further to the east. At the same time, this explains why the kaftan does not occur on the slightly earlier bracteates, as they are modelled on classical Roman traditions with the emperor as the dominant figure. That the kaftan as a signifier of rank so quickly loses its significance in the depictions, presumably already by the end of the 7th century AD, is perhaps linked to a fundamental transformation in the military organisation caused by the development of the sail and naval organisation (Nørgård Jørgensen 1999a, 201–202). However, there is evidence indicating that the kaftan, after its adoption as a Scandinavian male warrior garment, was slowly transformed into a more universal clothing item, worn by men and women, at least, in the higher social strata.

Archaeological parallels for jackets in Outerwear Group 4 (B1 and B2) worn by men and women are more difficult to account for. This type can be considered a hybrid between a cloak and a jacket, and whether or not this costume item had sleeves is uncertain. A possible parallel can be seen among the items of clothing from

Fig. 7.13 The semi-circular cloak of Queen Bathilde from France. The find is dated to AD 680 (Museo Municipal de Alfred Bonno).

Queen Bathilde's grave in Chelles, France, dated to AD 680, where a semi-circular cloak fringed on its curved edge was found (Laporte and Boyer 1991, 26–27) (Fig. 7.13). The fringe is a characteristic feature of type B1, and is a well-known phenomenon in the Scandinavian archaeological textile material found in many burials from the Late Iron Age (Hald 1980, 99–100; Bender Jørgensen 1986, 210–228; Mannering 1997, 129–130). In archaeological textiles, fringes are often associated with blankets or blanket-like cloaks (see, *e.g.* Schlabow 1976; Munksgård and Østergaard 1988). However, the French find indicates that fringes can also be combined with an item of outerwear cut or woven to shape, and it is possible that this costume item has a semi-circular shape like those also known from the much earlier grave finds from Verrucio in Italy (von Eles 2002, Stauffer 2012; Ræder Knudsen 2012). At present, it is not possible to determine if OG4 reflects a realistic physical shape or if it is a special and limited iconographical costume item. This type is primarily limited to a few gold-foil figures from Sweden, and contrary to most other depictions, it is identical for both genders, which is unusual in a Scandinavian Late Iron Age context.

Fig. 7.14 Reconstruction drawing of the Danish Early Bronze Age tubular dress and a suggestion of how this clothing may have been worn. The reconstruction is based on the wool tabby textile found in the female oak coffin from Skrydstrup in Denmark which measures 395 × 145 cm. The find is dated to c. 1300 BC (National Museum of Denmark; Drawing: Marie-Louise Andersson).

Innerwear

Dress

The dress is the commonest type of female innerwear depicted, and in the period under consideration, its shape stays, more or less, unchanged. The costume is often quite long, ankle or floor-length, covering the legs and sometimes also the feet, but is otherwise difficult to describe in detail. Presumably, the dress had long sleeves and it seems that it only comprises a single layer of textile, often ending in a train. The dress may also be combined with an additional front piece like a kind of apron, but its actual construction is difficult to document based on the depictions.

In Scandinavia, the dress can be detected among archaeological clothing finds as early as the Early Bronze Age (Broholm and Hald 1940) (Fig. 7.14), but the best known dress is probably the tubular garment from Huldremose in Denmark dated to the Pre-Roman Iron Age (350–30 BC) (Hald 1980; Mannering *et al.*, 2010) (Fig. 7.15). This garment has been considered as having been worn like the ancient Greek peplos (Fig. 7.12), fastened at the shoulders with a pair of fibulae, placed one on each shoulder. This fibulae constellation occurs frequently in burial material in Scandinavia from the Roman and Early Germanic Iron Ages (Jørgensen and Jørgensen 1997, 59; Bennett 1987, 105–107). Moreover, it has been argued that this tubular garment was the dominant female dress type in Scandinavia until the beginning of the Late Germanic Iron Age where the

Fig. 7.15 The tubular dress from Huldremose in Denmark. The wool textile is woven in 2/2 twill and measures 256 × 173 cm. The find is radiocarbon dated to 350–30 BC (National Museum of Denmark).

position of the fibulae recorded in the graves moves from the shoulders to the area of the chest (Hägg 1968).

The most characteristic feature of the tubular dress is its cylinder-shape that required some kind of fastening at the shoulders, thereby creating armholes, but leaving the shoulders and arms bare. Its length could be adjusted by a fold in the top edge of the dress, or by a belt, thereby rendering this a very flexible type. A similar, but not quite as large, dress was recovered in the female burial from Hammerum in Denmark, which is dated to the 3rd century AD (Mannering and Ræder Knudsen 2013) (Fig. 7.16). In the burial finds dated to the Early Germanic Iron Age, the presence of buttons, hooks and eyes at the wrists further indicates that, in this period, the tubular dress could consist of two layers of textile, and that, it was most likely combined with a long-sleeved undergarment (Bennet 1987, 105–106; Es and Ypey 1977, 124–125; Ethelberg 1986, 44–46). It has been suggested that the tubular dress was linked to a definite Scandinavian textile type, the Huldremose type that disappears at the end of the Early Roman Iron Age (Bender Jørgensen 1986, 139–140). However, it appears that this fundamentally Scandinavian dress type continued to be in use right up to the beginning of the Late Germanic Iron Age, in spite of the textile technology changing over time.

Generally, the Early Germanic Iron Age dress tradition contains old and new elements, and new ways of combining different layers and textiles, although its appearance in its entirety is still subject to discussion. The most radical transformation of the female dress occurs at the onset of the Late Germanic Iron Age where the use of the fibulae that held the dress in place, moved from the shoulders to the chest. This position then remained unchanged until the middle or end of the Viking Age where the dress fibula disappears from the grave finds (Jørgensen and Jørgensen 1997, 58–59; Hedeager Madsen 1990). No completely intact dress from the Late Iron Age has yet been recovered in Scandinavia. However, the textile material from the Late Germanic Iron Age cemetery of Nørre Sandegård Vest on Bornholm suggests that, at this point, the dress comprised an inner dress made of linen and an outer dress made of wool (Mannering 1997, 138–139). As the types of textiles and the position of jewellery in the graves from this period highly resemble that of the Viking Age, it is most likely that, Viking Age female dress had the same overall appearance as in the Late Germanic Iron Age. The reconstruction of Viking Age female clothing is primarily based on a few large finds, such as the textiles from Birka and Haithabu (Geijer 1938; Hägg 1974; 1984a; 1991a) as well as a handful of lavishly furnished female burials, such as Hvilehøj, Hørning and Oseberg (Hald 1980,100, 111–113; Bender Jørgensen 1986, 221, 227; Hedeager Krag and Ræder Knudsen 1999; Christensen *et al.*, 1992) (Figs 7.10, 7.17 and 7.18).

In the textile material from Birka in Sweden, textile fragments preserved on the backs of oval brooches have remains of straps and intricate sequences of different layers of textiles. A pleated linen inner dress and a wool strap dress constitute the most common clothing items. A reappraisal of the costume construction took place during investigations in the 1930s (Fig. 7.17) and 1970s (Fig. 7.10), yet however without getting closer to a definite interpretation (Geijer 1938; Hägg 1974, 52–54, 109–110). Thus, many unanswered questions still remain concerning the detailed construction of the dress, and several diverse suggestions for a reconstruction exist in scholarly works as well as in museums and re-enactment groups that all take their point of departure from the same sources (*e.g.* Almgren 1967, 199; Bau 1981). Furthermore, it has become apparent that the finds from Birka should not be considered representative for all of Scandinavia. For instance, the seemingly more mundane textile finds from Haithabu harbour demonstrate a completely different type of female dress, tailored and relatively tight-fitting, wide at the bottom and with long sleeves, that did not require the use of fibulae (Fig. 7.18). This type, too, comprises a double-layered dress constructed of various characteristic textile types (Hägg 1984a, 171–173; Elsner 1989, 47). The more tight-fitting dress shape has close similarities to the medieval Scandinavian clothing finds, and it is possible that this type of dress is precisely the one placed in the many Viking Age female graves which contain no dress fibulae (Hedeager Madsen 1990). Probably the same dress type is seen in the female burials from Oseberg and Hørning. Thus, the Scandinavian Late Iron Age female dress comprises a much larger variation than the few, albeit large, find complexes with their limited preservation conditions demonstrate.

In conclusion, the dress types portrayed on the depictions are considerably more stylised than those documented in archaeological sources, and it is difficult to ascertain whether or not their construction is identical and follows the same development as in the archaeological textile finds. First and foremost, the depictions portray a single-layered costume. Moreover, the depicted fibulae are always linked to the outerwear, while fibulae are never seen holding a dress together as the archaeological finds suggest. Thus, the cut of the dress seen on the depictions mostly resemble the Viking Age dress documented in the material from Hvilehøj, Hørning, Oseberg and Haithabu, while the typical Late Iron Age dress type fastened with a pair of fibula placed on the chest cannot be identified. In the case of the gold-foil figures, there are important differences between the depictions and the contemporaneous archaeological clothing. However, this is most likely due to a simplification of the depictions, where emphasis was placed on displaying textile patterns and borders in exaggerated forms similar to the case of the jewellery. This suggests that, clothing and textile design were important constituent elements signifying rank, status and social connections just as in the case of the dress fasteners.

Fig. 7.16 The tubular dress from Hammerum in Denmark during excavation with the outline and different characteristics highlighted. The find is dated to the 3rd century AD (National Museum of Denmark/Museum Midtjylland; graphics: Lise Ræder Knudsen).

Fig. 7.17 Reconstruction drawings of various Viking Age female clothing items found in the Birka graves in Sweden. The drawings are based on Agnes Geijer's interpretations (after Almgren 1967, 200; drawing: Åke Gustavsson).

Fig. 7.18 Reconstruction drawings of various Viking Age female clothing items found in Haithabu in Germany (redrawn by Sidsel Frisch; after Hägg 1984; 1991a; and Inga Hägg's reconstruction drawings in Elsner 1989).

Skirts and blouses

The skirt and blouse ensemble is another type of innerwear depicted, although not as widespread as the dress. It only occurs in a few examples on bracteates, gold-foil figures and tapestries, but these depictions are dated to both the Germanic Iron Age and the Viking Age. The shape of the skirt closely resembles the dress and can also end in a train. The accompanying blouse presumably has long sleeves. The skirt and blouse ensemble is documented in the archaeological textile finds from the Early Bronze Age to the Roman Iron Age. Several examples of finely tailored blouses are known from the oak-coffin burials of Skrydstrup, Borum Eshøj C and Egtved in Denmark (Broholm and Hald 1940; Hald 1980, 347). In the case of the Egtved Girl, the blouse was combined with a short corded skirt (Fig. 7.19), and in the case of the Skrydstrup Woman, the blouse was combined with a large tubular garment that could be draped around the body like a dress or a skirt (Broholm and Hald 1940). In the bog find from Damendorf in Germany, a short skirt sewn together from several pieces was recovered together with the remains of a young woman (Schlabow 1976, 95-96; Sanden 1996, 129, 167 and 193). The find is radiocarbon dated to the 9th century BC, which makes it comparable to other finds of short, knee-length skirts which were an important item of female clothing in the Bronze Age. The find of small bronze tubes with the remains of cords inside them further shows that corded skirts were widespread in Bronze Age female clothing (Broholm and Hald 1940, 150–154).

Fig. 7.19 Blouse, belt and corded skirt found in the Early Bronze Age female oak coffin from Egtved in Denmark. The find is dated to 1370 BC (scale = 10 cm) (National Museum of Denmark).

Fig. 7.20 The outer skin cape, scarf and skirt found together with the bog body of the Huldremose Woman in Denmark in 1879. The find is radiocarbon dated to 350–41 BC (National Museum of Denmark).

Several skirts dated to the Early Iron Age have been recovered from the Danish bogs. The most well known is that from Huldremose which was found in 1879 together with the body of a woman (Fig. 7.20). The clothing consisted of two skin cloaks, a woven checked skirt with a leather string gathering the waist and a rectangular scarf that was wrapped around the neck (Hald 1980, 47-49; Mannering 2010). A recent examination of this find has revealed that the woman also wore an undergarment made of linen beneath the skirt, although the shape of this is not known (Frei *et al.*, 2009; Mannering 2010). In the bog find from Krogens Mølle in Denmark, too, an almost intact skirt was found (Hald 1980, 15–19) (Fig. 7.21). From the 1st century AD, the much more fragmented textile remains recovered in 1969 from the female burial from Lønne Hede in Denmark indicate the presence of a skirt (Fig. 7.22). A pair of fibulae placed on the shoulders, normally associated with a tubular dress, was found in this grave and interpreted as a blouse-like upper part (Fig. 7.23) (Munksgård and Østergård 1988, 54; Demant 2007). Based on the Huldremose and Hammerum finds, the woman could have also worn a close-fitting dress beneath the skirt.

Compared to the short skirts from the Bronze Age, the above-mentioned Early Iron Age examples are all long, covering most of the leg, which correspond to the impression of the lengths shown on the depictions. However, there are no archaeological textile finds documenting with

Fig. 7.21 The wool skirt from Krogens Mølle in Denmark as it looks today (top) and as it may have looked when it was made (below). The skirt is woven in 2/2 twill in a white or greyish wool, and measures 80 × 350 cm. The three blue stripes are placed slightly off centre. The find is radiocarbon dated to 399–181 BC (scale = 10 cm) (National Museum of Denmark).

Fig. 7.22 Some of the textiles and jewellery found in the Lønne Hede female grave excavated in 1969. The grave is dated to the 1st century AD (National Museum of Denmark).

Fig. 7.23 Reconstruction of the Lønne Hede 1969 clothing made and modelled by Anna Nørgård (after Bonniers Store Håndarbejdsleksikon; photo: Mikael Müller).

certainty that the skirt and blouse combination was used in the Late Iron Age. Yet, it cannot be excluded that in graves where no dress fibulae are found, the deceased could have been buried in a skirt and blouse ensemble. In a burial dated to the 7th century AD in Flurlingen in Switzerland, the remains of a 3–4 year-old girl were found buried in a costume consisting of a skirt and blouse (Windler *et al.*, 1995). Similarly, in a 6th century AD female burial from Waging am See, Germany, the clothing was interpreted as a skirt (Bartel and Knöchlein 1993, 425–428).

In conclusion, the skirt and blouse ensemble is intrinsic to Bronze and Early Iron Age archaeological clothing, while the depictions demonstrate that it also belonged to the Late Iron Age clothing tradition. On the few bracteates representing women, these figures are attired in a skirt, although it is impossible to determine what was worn on the upper body, while the gold-foil figures from Slöinge (see Fig. 2.17, Slöinge 5, 7 and 9) display a skirt worn together with a long-sleeved, tight-fitting blouse. With the exception of a few grave finds, the archaeological clothing finds have only little to add, in contrast to the iconography which shows a wider chronological use than the Scandinavian archaeological textile finds can document. The train-like ending of the skirt and the dress is a phenomenon recorded for many of the female depictions, which has not yet been securely documented in the archaeological textile finds. This is probably primarily due to the textile remains from this part of the body being rarely preserved. Yet, as the train is such a prominent feature on the depictions, it is highly likely that, it was also a true element in the Late Iron Age clothing tradition.

Tunics

The tunic constitutes a key item of male costume on the depictions throughout the entire period under consideration. It is often thigh- or knee-length, relatively tight-fitting and is either short- or long-sleeved. It is never combined with a kaftan, but rather with a cloak. The tunic is patterned in various ways, presumably reflecting the different textile types or volume of fabric, however, it generally has the same shape except in the chronologically youngest depictions selected for the study. The tunics seen on the depictions dated to the Scandinavian Middle Ages are longer and wider at the bottom than the earlier ones, as demonstrated by the tapestry from Skog in Sweden.

The presence of tunics in the archaeological textile material is documented by several finds from the Iron Age. The earliest and best preserved examples of tunics originate from the German bogs and weapon deposits, and are dated predominantly to the Roman Iron Age (Sanden 1995, 151; van der Plicht 2004). The tunics display great variation in their construction, yet a common denominator is that they are made of straight fabric lengths without much cutting (Hald 1980, 335–337; Schlabow 1976, 69–71; Wild

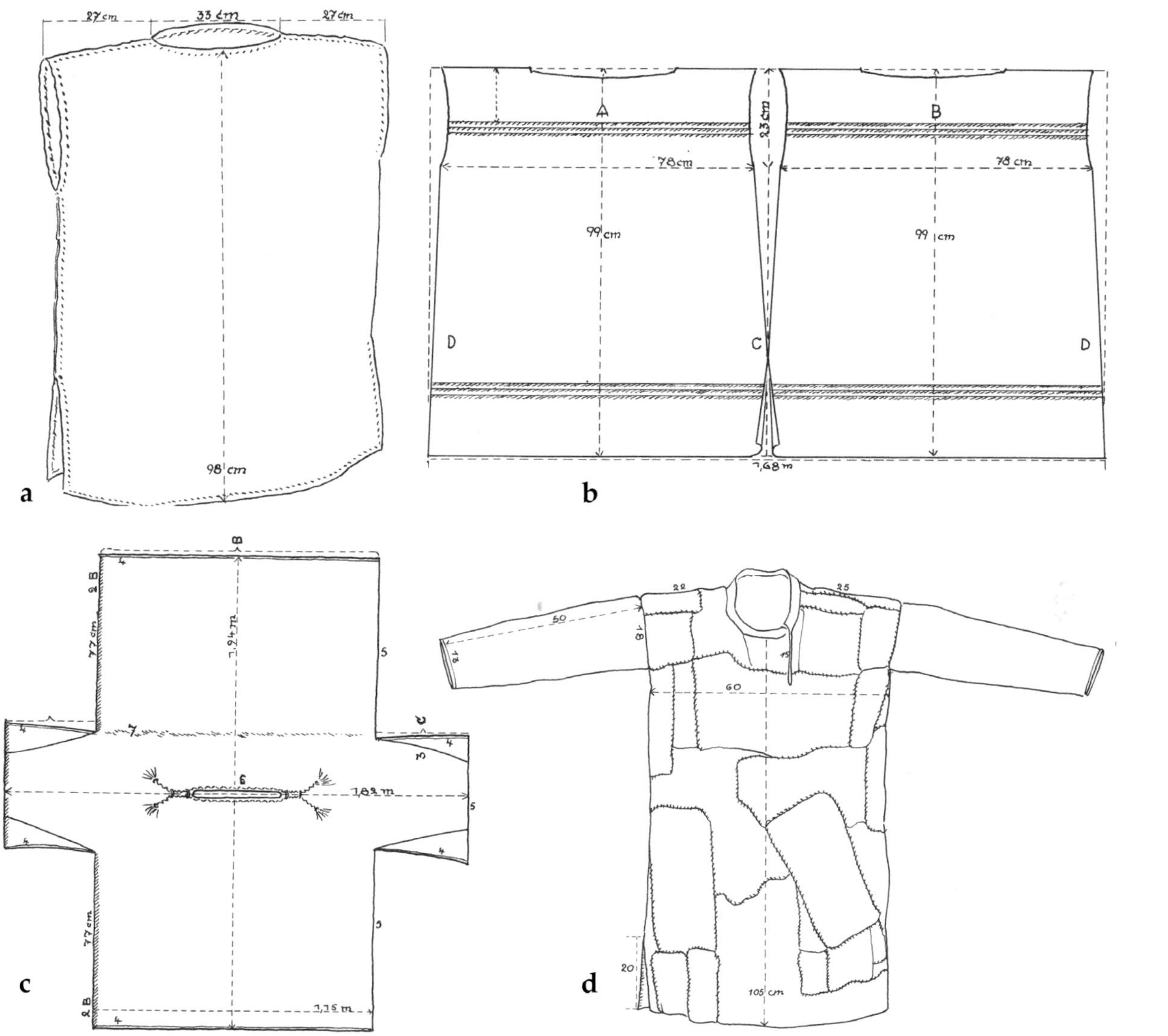

Fig. 7.24 Reconstruction drawings of various Iron Age tunics found in Germany: a) Marx-Etzel, b) Obenaltendorf, c) Reepsholt and d) Bernuthsfeld (after Schlabow 1976).

1985, 369–372). Tunics basically comprise two types: a loose sleeveless tunic, such as those from Marx-Etzel and Obenaltendorf (Fig. 7.24a and b), and a more tight-fitting sleeved tunic, such as that from Reepsholt (Fig. 7.24c) or Thorsbjerg, all in Germany (Fig. 7.25). The former has been considered a Gallic type with clear links to Roman clothing, while the latter has been regarded a Germanic type (Munksgård 1974). The Thorsbjerg garments have thus long been perceived as representative of the Scandinavian male clothing tradition, and, in archaeological textile research, used as a prototype for many reconstructions of Scandinavian or Germanic male garments from the Roman to the Viking Ages (Munksgård 1989, 1991; Bender Jørgensen 1991a; Høigaard Hofseth 1998). However, it has been suggested that the Thorsbjerg garments ought not to be perceived entirely as part of an original Scandinavian clothing tradition, but rather as a representative of a Continental Germanic auxiliary (warrior) or equestrian clothing tradition (Bender Jørgensen 1986, 154–155; Hägg 1984a, 219; Möller-Wiering 2011). Recently, a well-preserved tunic was found in Lendbreen in Norway dated to the 3rd century AD demonstrating that the tunic was part of the Scandinavian clothing tradition (Vedeler and Bender Jørgensen 2013) (Fig. 7.26).

No intact tunics dated to the Early Germanic Iron Age have been found in Scandinavia, but the burial finds, nevertheless, offer some hints as to their appearance. The best points of departure are the three male burials from Högom in Sweden and Evebø/Eide and Snartemo in Norway respectively which contained textile fragments interpreted as tunics (Hougen 1935; Magnus 1982; Høigaard Hofseth 1998; Bender Jørgensen 1991a; 2003) (Figs 7.8 and 7.9).

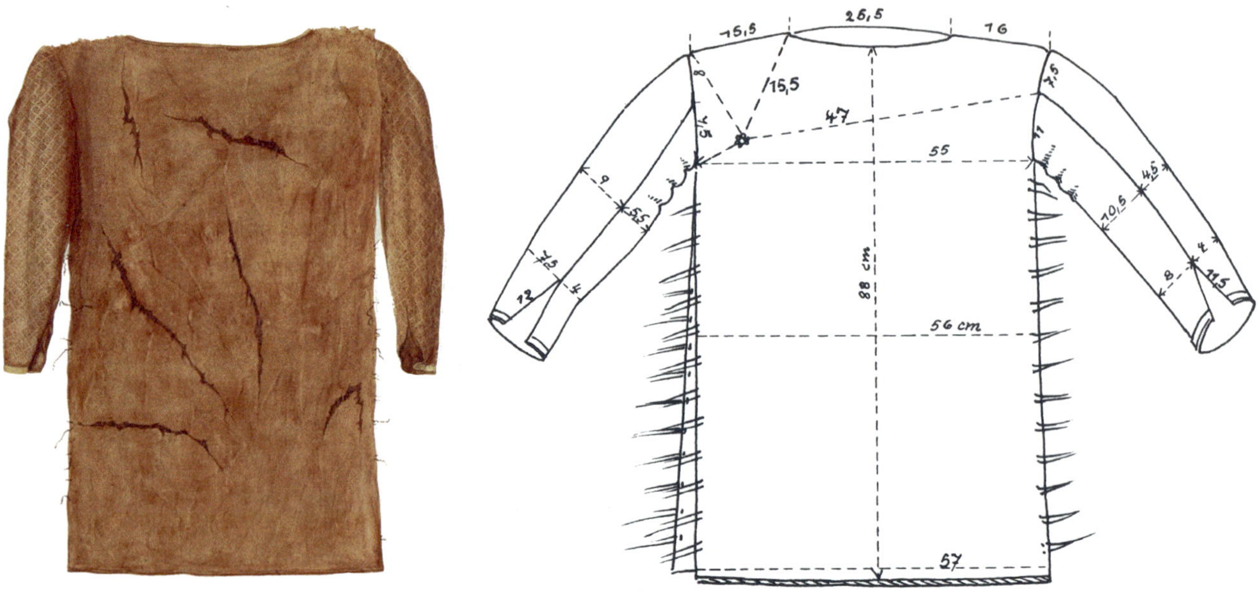

Fig. 7.25 Reconstruction drawings of the Thorsberg tunic found in Germany. The find is dated to the 3rd century AD (after Engelhardt 1863 and Schlabow 1976).

Fig. 7.26 The tunic from Lendbreen in Norway. The find is radiocarbon dated to AD 230–390 (Museum of Cultural History, University of Oslo; photo: Courtesy of Marianne Vedeler).

The most intact of these examples is the tunic from Högom. It is tight-fitting with long sleeves, resembling, to some extent, the Thorsbjerg tunic, although it has a different and more intricate cut (Nockert 1991, 130). Similarly, details such as the cuffs on the sleeves, and the frequent use of wide, patterned, tablet-woven bands constitute specific Early Germanic Iron Age clothing features. To this can be added two long sleeves, most likely belonging to a tunic, recovered in a male burial from Døsen in Norway (Nørlund 1941, 31; Bender Jørgensen 1986, 254; Nockert 1991, 141). This burial is dated to a period between the second half of the 5th century AD to the beginning of the 6th century AD.

As textile finds from male burials in the Late Germanic Iron Age are so rare, it is difficult to deduce any information on the appearance of the tunic in this period. It is, however, assumed that the most common clothing item was a tunic, and the tunic from Bernuthsfeld in Germany recently radiocarbon dated to AD 680–774, *i.e.* the end of the Late Germanic Iron Age, shows a similar construction to the earlier finds (Sanden 1995, 141–142; 1996, 127–128) (Fig. 7.24d). This tunic constitutes of a long-sleeved, knee-length garment, sewn together of no less than 43 pieces (Schlabow 1976, 72–73; Farke 198, 2001). Similarly to the Lendbreen tunic, the extensive patching on the garment illustrates its value that made the repeated repair worthwhile. The same tendency of recycling and making second-hand textiles and garments is also attested in the Roman clothing from Mons Claudianus in Egypt (Bender Jørgensen 2000; Mannering 2000) and was probably more common than the archaeological textile finds testify.

In the Viking Age, the finds from Haithabu harbour and the Birka graves provide the foundation for interpreting textile remains as tunics (Hägg 1971, 149; 1984a, 175–176; 1986, 67; Hald 1980,102–104) (Fig. 7.11). In the male grave from Mammen, some of the textile fragments were identified as belonging to the tunic and used to reconstruct a tunic based on a drawing of Canute the Great in the manuscript *Liber vitae*, dated to AD 1031 (Munksgård 1989, Bender Jørgensen 1992b, 32–34; Østergård 1991, 123–126). Another important find worth mentioning is the almost intact linen tunic from Viborg in Denmark dated to the 11th century AD (Fentz 1987, 23) (Figs 7.27 and 7.28c). Altogether these Viking Age tunic finds indicate that, in this period, male innerwear could consist of several layers and be made

Fig. 7.27 The linen tabby tunic found in Viborg as it is preserved today. The tunic is dated to c. AD 1050 (Viborg Museum; photo: Courtesy of Margit Petersen).

of different materials. The clothing found in Skjoldehamn in Norway, dated to AD 995–1029, is another example of this two-piece inner garment (Gjessing 1938; Holck 1988; Nockert and Possnert 2002, 59–60; Vedeler 2007a)[4] (Fig. 7.28d). The use of two tunics is further corroborated by Old Norse language usage, as in the difference between *skyrta* and *kyrtill*, the latter being of wool and never worn directly next to the skin (Falk 1919, 141–146). On this basis, it is likely that the garment from Viborg should be interpreted as an inner tunic (Fentz 1987, 33–34). Furthermore, it is likely that the development leading to the use of two tunics, one of linen and one of wool, began already in the Late Germanic Iron Age, where linen textiles prevail in the burial finds (Bender Jørgensen 1986, 164–165; Mannering 1997, 126–127). The greatest transformation in the appearance of the tunic presumably occurs in the Late Viking Age where the tunic developed into a more extensively cut and voluminous garment of sizeable length. Examples of this are the clothing items from the Old Norse burials in Herjolfsnæs in South Greenland, the Swedish find from Bocksten, and the Danish bog finds from Kragelund and Moselund, all dated to the Scandinavian Middle Ages (Nørlund 1935; Nockert 1985, 73–76; Hald 1980, 39, 60–61, 340–344; Østergård 2003) (Fig. 7.28a and b). The tunics from Kragelund (Fig. 7.29) and Moselund (Fig. 7.30) are radiocarbon dated to

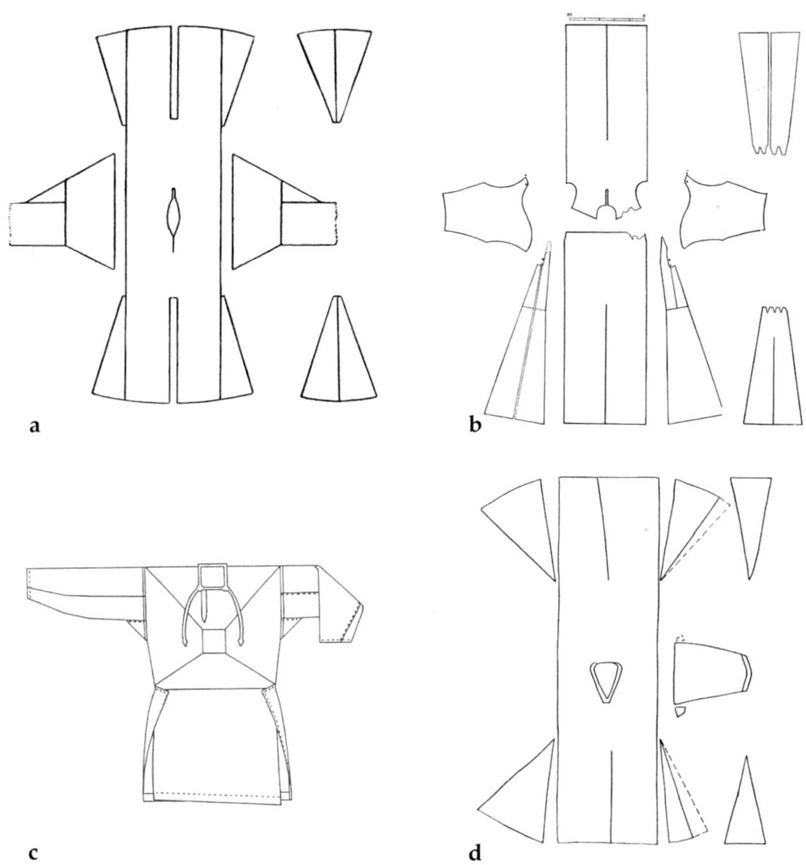

Fig. 7.28 Reconstruction drawings of various tunics found in Denmark and Norway: a) Kragelund, b) Moselund, c) Viborg and d) Skjoldehamn (after Hald 1980; Gjessing 1938).

Fig. 7.29 The wool tunic from Kragelund in Denmark. The find is radiocarbon dated to AD 1045–1155 (National Museum of Denmark).

Fig. 7.30 The wool tunic from Moselund in Denmark. The find is radiocarbon dated to AD 1050–1155 (National Museum of Denmark).

AD 1045–1155 and AD 1050–1155 respectively (Østergård 2003, 124–127, 232). Most of these garments are more than knee-length, and based on this feature alone, impossible to categorise as male or female clothing items.

A blouse is also recorded among the male depictions. In its construction, this resembles the tunic to a great extent, but it ends at the waist and is always combined with trousers, and perhaps a belt. The blouse can be either short- or long-sleeved and is often tight-fitting. The blouse and trouser ensemble is found in all object categories, although it does not occur especially frequently, except among the tapestries where it is primarily combined with wide breeches. A distinct feature of some of the male blouses, also seen in the bracteate category, is a round-cut sleeve, which to some extent corresponds with the design of the tunic from Högum (Nockert 1991, 32; Bender Jørgensen 2003, 69–70). No clothing items can be directly compared with the description of the blouse in the iconography, and most likely the blouse should be interpreted as a tunic tucked into the trousers. The combination of blouse and trousers in male costume iconography could also be a way of depicting an inner tunic.

In conclusion, during the period from the Roman Iron Age to the Scandinavian Middle Ages, the tunic in the archaeological textile finds generally tends to become more flowing and intricately cut, and to comprise several layers. Similarly, the tunics from the Roman and Germanic Iron Ages, as well as the Viking Age, are relatively short and never more than knee-length, while the tunic in the Scandinavian Middle Ages develops into a long, loose-fitting garment, difficult to differentiate from the corresponding female dress (Nørlund 1941, 31; Østergård 2003). Furthermore, if the tunic has sleeves, they are most often long. The development of the tunic portrayed on the depictions corresponds in many ways to the development described above. However, whether or not the tunic is single or multi-layered and always has long sleeves cannot be determined from the depictions. A small number of male figures on the helmet plates are attired in short-sleeved, dotted tunics, and presumably these short sleeves can be used as an argument for the depictions constituting chain-mail rather than tunics made of textiles.

Trousers

Legwear constitutes an important item of male innerwear on the depictions that can be combined with a tunic or a blouse. The depictions demonstrate that several different types of trousers feature in different periods. Likewise, the archaeological clothing finds demonstrate that trousers were subject to great variation, and that many different types were worn in the same period. Thus, trousers are an item of clothing that unmistakably constituted an integral part of Scandinavian Iron Age male clothing, at least from the first centuries AD and onwards, but precisely when it was integrated within the Scandinavian clothing tradition is still not fully evident.

Roman written sources attest that, the peoples living north of the Roman Empire were clad in long trousers (*bracae*) improper to wear in the city of Rome (Croom 2000, 55–57). Roman men did wear trousers beneath their tunics, but these were short and not intended to be seen (Cleland *et al.*, 2005, 22, 24, 68). Especially, old men wore trousers in winter, and trousers were also adopted by soldiers in the army for practical purposes. In the German bogs, several types of trousers have been found, for example in Thorsbjerg, Damendorf, Dätgen, Obenaltendorf and Marx-Etzel (Fig. 7.31). These are all radiocarbon dated to the Roman Iron Age (Sanden 1995, 146, 151; 1996,

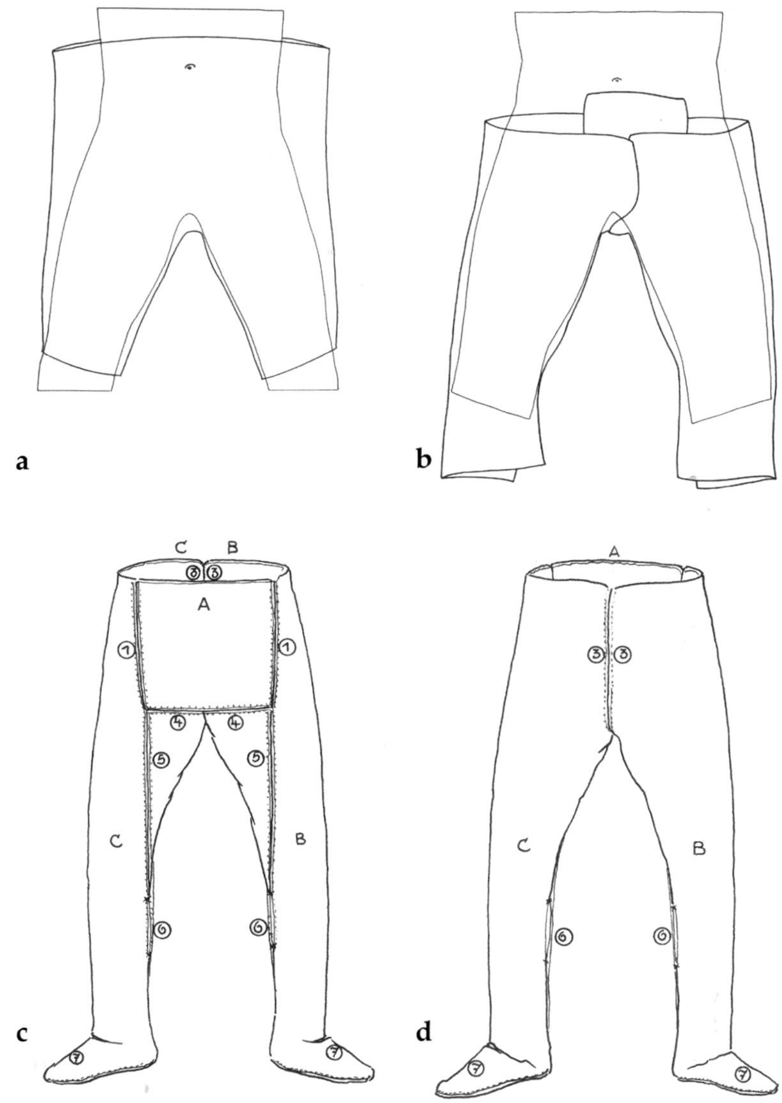

Fig. 7.31 Reconstruction drawings of various Iron Age trousers found in Germany: a) Marx-Etzel, b) Dätgen, c) Thorsberg (back) and d) Thorsberg (front) (after Schlabow 1976).

192–193; van der Plicht 2004). The Thorsbjerg trousers have feet attached, one of which is still preserved (Fig. 7.32). Reconstructions of this type of trousers have demonstrated that they are not especially suitable for walking, while they are eminent for riding. This is one of the arguments in favour of interpreting the Thorsbjerg garment as equestrian attire. However, a passage from *Njal's Saga* (chapter 134) suggests the opposite, namely that trousers with feet attached are unsuitable for riding (Falk 1919, 119; Nockert 1991, 127–128). The trousers from Damendorf, too, are long legged, although without sewn-on feet (Schlabow 1976, abb. 173 and 178). The shaped trouser legs were presumably constructed for easing the leg through the relatively narrow trouser leg. In contrast, the trousers from Obenaltendorf, Dätgen and Marx-Etzel are knee-length (Schlabow 1976, 76ff) (Fig. 7.31a and b), and evidence suggests that this is the kind worn by Roman soldiers (Croom 2000, 55-56). This, however, does not exclude that they were a universal type used across cultural barriers.

Another way of covering the lower legs was to use leg wrappings in the shape of long bands, that could both be used on their own or together with long trousers (Schlabow 1976, 88–90). There are examples of these from, for instance Obenaltendorf (Sanden 1996, 127) (Fig.7.33d). The rectangular leg bindings from the bogs of Søgård (Fig. 7.34) and Thorup (Fig. 7.35) in Denmark were used in a similar way to cover the lower legs (Hald 1980, 34–37, 42–43). These rectangular leg wrappers are dated to the Pre-Roman Iron Age, *i.e.* a period where no trousers have been documented among the Scandinavian clothing finds (Mannering *et al.*,

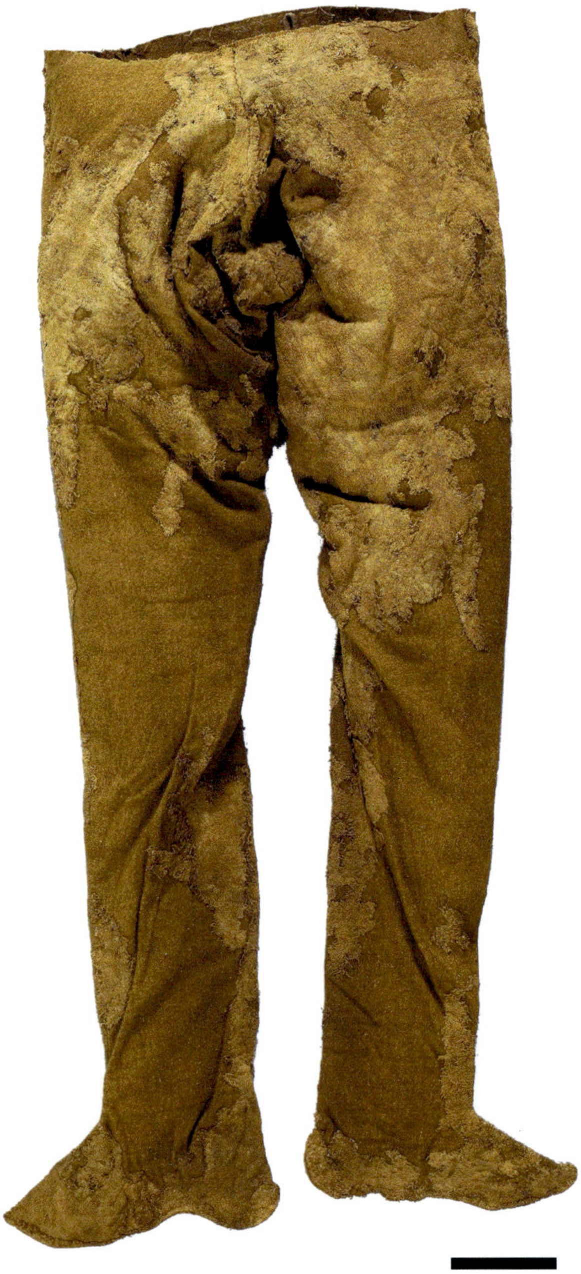

Fig. 7.32 The wool trousers from Thorsberg in Germany as they are preserved today. The darker brown part is a modern support fabric. The find is dated to the 3rd century AD (scale = 10 cm) (National Museum of Denmark).

2010). The Søgård leg wrappers were found together with a male bog body, while most other examples of trousers and leg wrappers do not have a clear gender identification due to the find context. The present evidence suggests that leg wrappers constituted a general male type, but it cannot be ruled out that females also used this clothing item, although, it does not seem obvious to combine them with a long dress or skirt (Croom 2000, 55–56).

Altogether, the archaeological costume finds attest to the presence of three different types of trousers appearing in the clothing finds in the Late Roman Iron Age: short trousers and long trousers with or without feet. From the onset of this period, but not before, trousers seem to be an essential part of Scandinavian male clothing, although they are extremely difficult to identify among the often fragmented archaeological textile finds. Nevertheless, the textile fragments from, *e.g.* Högom and Evebø/Eide attest that, long trousers were definitely worn in the Early Germanic Iron Age and reveal their utilisation at least in this period and to the end of the Viking Age (Nockert 1991, 34–35, 50–53, Magnus 1982, 301). A characteristic feature of these trousers is the use of tablet-woven cuffs and borders attached to the bottom of the trouser legs. A slit, often fastened with hooks and eyes or other kinds of fasteners further indicates that, the trouser legs were relatively tight-fitting (Nockert 1991, 41, 62, 127–128). The tunic-clad man from Bernuthsfeld wore leg wrappings on his lower legs, testifying that this item of clothing continued in use simultaneously with the trousers (Schlabow 1976, 16; Sanden 1996, 127).

Remains of several types of Viking Age leg wrappings or bindings are also known among the finds from Haithabu and in the male burials in Birka (Hägg 1986, 72; 1991a, 32–35). For instance, the textile fragments from Birka grave 905 suggest the presence of a pair of knee-length trousers with linen borders and a pair of long-legged trousers in wool twill (Geijer 1938, 145; Hägg 1986, 86). Usually however, fragments of this type of attire are rare. In Haithabu, remains of trousers resembling the Thorsbjerg type were recovered. The trouser legs are, however, not completely preserved, and thus, it is impossible to know whether or not they had feet attached.

Another type of trousers is the wide breeches, and also long stockings which make their appearance in this period (Hägg 1984a, 24–30). A striking detail of a voluminous pair of wide breeches found in Haithabu is that it is made of a thin and open tabby in two colours: red in front and yellow-brown at the back of the breeches (Hägg 1984a, 34, 28). In the male grave from Mammen, the reconstructed outfit from 1990 included a pair of Thorsberg type trousers (Østergård 1991), but most likely, the grave contained a pair of wide blue breeches made in a very openly woven tabby textile, that has until now not been linked with any other use.

It is not known if the long stocking from Haithabu had a foot attached, but it presumably covered the thigh and was held in place by a leather string (Hägg 1984a, 20–22). Close parallels to this long stocking are found in the Old Norse burials from South Greenland, where the introduction of the long tunics made trousers superfluous (Fig. 7.36). The stockings from this period reached just above the knees (Nørlund 1935, 104; Gjessing 1938, 51; Nockert 1985, 96–98; Hägg 1984a, 158–159; Østergård 2003).

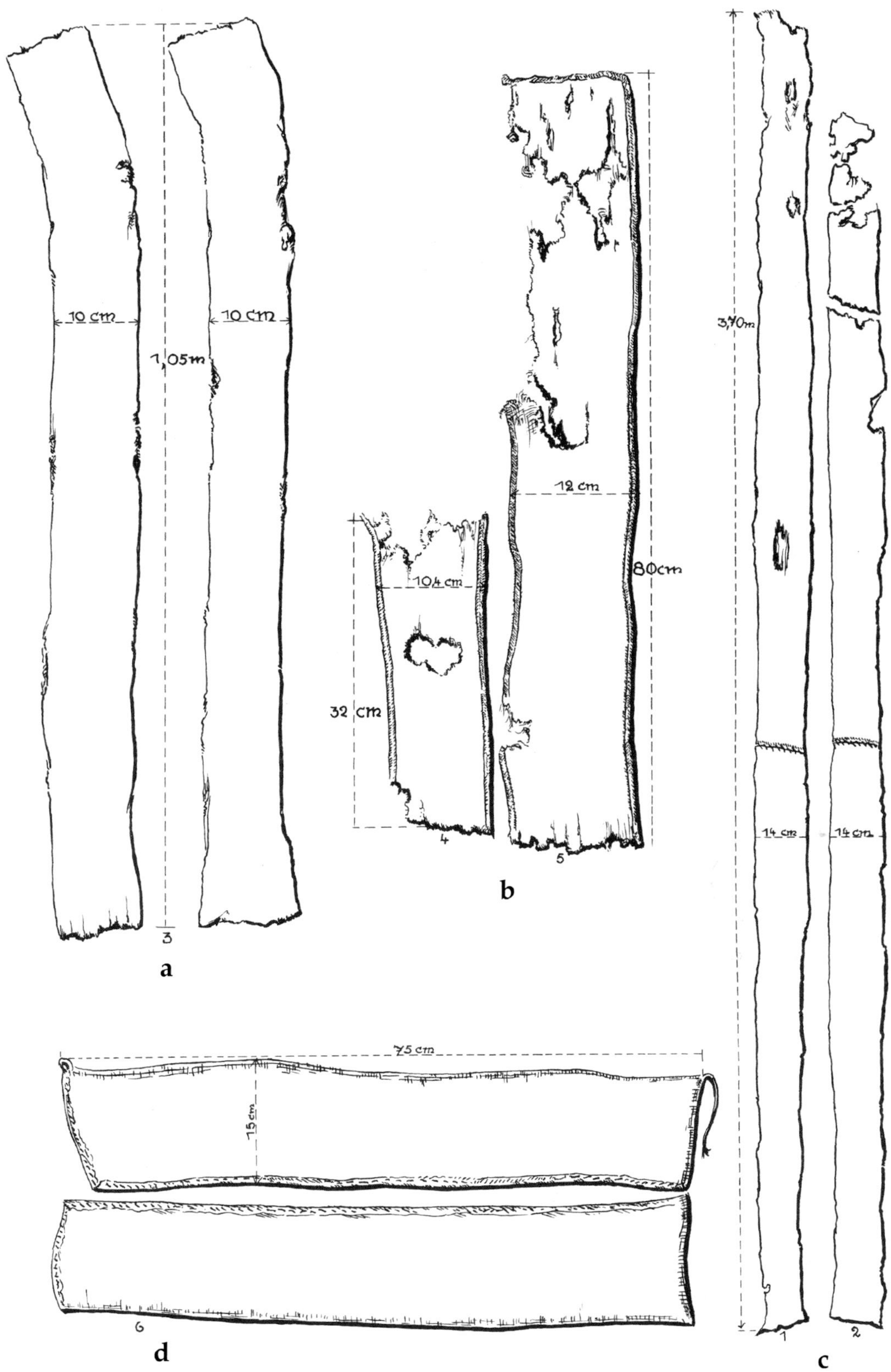

Fig. 7.33 Reconstruction drawings of various Iron Age leg wrappings found in Germany: a) Damendorf, b) Thorsberg, c) Bernuthsfeld and d) Obenaltendorf (after Schlabow 1976).

Fig. 7.34 The wool leg wrappers from Søgård in Denmark with the leg and foot of a man still preserved. The find is radiocarbon dated to AD 130–340 (National Museum of Denmark).

Fig. 7.35 The wool leg wrappers from Thorup in Denmark. The find is radiocarbon dated to 360–50 BC (scale = 10 cm) (National Museum of Denmark).

In conclusion, legwear in the archaeological textile record throughout the Iron Age shares many similarities with the depicted forms. Some seem to occur throughout the entire period in question, for example leg wrappings, while others seem to belong to a specific chronological period. It is likely, although it cannot be fully documented, that the long-legged trousers too, were used throughout the Late Iron Age, while it is very difficult to document when the breeches first appear. Especially the knee-length trousers seen on the bracteates and the gold-foil figures correspond well to finds from the Roman Iron Age, while wide breeches are encountered both among the depictions and in the archaeological textile finds dated to the Viking Age. The only type of leg wear encountered in the archaeological costume finds that are not

Fig. 7.36 The long wool stocking from Herjolfsnæs in South Greenland. The stocking measures 88 cm in length, and is dated to the 13th–15th century AD (National Museum of Denmark).

seen in the depictions, are leg wrappings. In iconography, they are known from contemporary depictions from the Continent (Müller 2003, 71–74), and why these were not included in the otherwise highly detailed Scandinavian depictions is difficult to explain. Presumably, this signifies that this detail was not so clearly visible or was considered unimportant to the appearance and understanding of the figures.

Textile textures and patterns

A final key feature on the depictions worth mentioning in comparison to the archaeological textile finds is the different decorations of the costume items. In general, male and female costumes are decorated with the same characteristic patterns constituting different stripes and checks that can be combined with different kinds of borders.

In Scandinavia, a wide range of textile types serve as markers for the characterisation of the different prehistoric chronological periods. The textile types are defined within a framework of fairly wide margins which also embraces the production of individually and locally produced goods. Each of the textile types has its own characteristics, and based on the surface structure, is easy to distinguish (Fig. 7.37, Tables 7.1 and 7.2) (Bender Jørgensen 1986).

In the Early Germanic Iron Age, the grave finds only contain a few characteristic textile types, such as the Haraldskær type. In general, the textiles are rather coarse, in comparison to both earlier and later periods. Diamond twill and ribbed twill do occur but only in a very limited number. However, this period is characterised by an evident preference for checked textiles in bright and contrasting colours, and the clothing is often composed of several different patterned and coloured textiles. It is obvious that, in the Early Germanic Iron Age, other aspects than the weaving technology was prioritised. For example the application of, and decoration with, tablet-woven bands is a characteristic element in this period, often together with the earlier mentioned bronze clasps. The tablet weaves are often unique and complex in their patterning, including the use of horsehair weft patterns. Especially the horsehair tablet-woven bands were immensely complex and time-consuming to produce, and represent a craft executed by specialists, perhaps children, as the technique demands good eyesight and small fingers.

In the Late Germanic Iron Age, new trends occur in the selection of textiles used in the burials, and the range of textile types expands markedly. Among the best known types are the Birka type, the Gerlev-Dråby type, the Gudmingegård type and the Nørre Sandegård type, which has been separated from the Gerlev-Dråby type (Mannering 1998b). The two types represent tabbies in wool and linen respectively (Tables 7.1 and 7.2). The Hessens/Elisenhof type is, like the ribbed twill, rather unusual. Especially the Birka type and other weaves with a clear texture are typical for the Late Germanic Iron Age. Likewise, new and improved raw materials are introduced, and textiles in plant fibres are much more common than in earlier periods. In the female clothing seen in the graves from Nørre Sandegård Vest on Bornholm it is possible to see a preference for dark blue wool textiles combined with undyed linen. At the same time, the textile finds demonstrate that colour patterns like checks and stripes are no longer used and neither are edgings with tablet-woven borders (Mannering 1997).

The Viking Age, seen from a textile perspective, is a continuation of the textile trends seen in the Late Germanic Iron Age, and there are many similarities between the textile types used in the two periods. A more detailed picture of the chronological differences emerges when the two periods are compared. New features for Viking Age Scandinavia are pleated fabrics, embroidery, applications and tablet weaves in silk, silver and gold threads. Another common feature for the Late Germanic Iron Age and the Viking Age is that, in these periods, it is no longer fashionable to produce and use textiles woven in colour patterns. Instead, single-coloured textiles with different surface structures and textures, providing contrasts between the different clothing items, dominate the visual appearance. This conclusion is further elaborated by dye analyses made on archaeological textiles from Dublin, York and London which were compared to a range of contemporaneous finds from Norway and Denmark. Here, the interpretation is that the Scandinavians preferred blue fabrics while the Dubliners had a passion for purples and the English a preference for red hues (Walton 1989).

The overall information that can be extracted from the textile technological analyses implies that, visually, there was a great difference between textiles produced in the Early Germanic Iron Age compared to the Late Germanic Iron and Viking Ages, while the latter two periods share great similarities. There can hardly be any doubt that each chronological period has its own visual preferences concerning choice of colours, texture and decorations, which makes it possible to differentiate between textiles and clothing produced in the three chronological periods respectively.

In conclusion, compared to the depictions, it is possible to identify some of the same characteristics in terms of the preference for stripes, checks and diamond patterns. Most likely, these exaggerated patterns were designed to represent some of the textiles types that were the most preferred during the Late Iron Age. Similarly, the presence of bands on much of the innerwear is striking, when the

Fig. 7.37 Textile reconstructions of three of the most characteristic textile types produced in the Scandinavian Late Iron Age, tabby (left), 2/2 twill (middle) and 2/2 diamond twill (right). The surface structures are so different that they are very easy to distinguish from one another (photo: Sidsel Frisch).

dating of the gold-foil figures and the helmet plates are taken into consideration. In the Early Germanic Iron Age, especially high-ranking male clothing was exquisitely embellished with tablet-woven bands constructed in intricate and time-consuming techniques (Nockert 1991; Sundström 1995; 1997), while in the Late Germanic Iron Age, there are no finds of tablet-woven bands from burials (Bender Jørgensen 1986). As far as the archaeological textile finds document, tablet-woven bands were not intrinsic to Late Germanic Iron Age clothing design, and this difference compared to the depictions, is indeed striking. Precisely on this point, there is an evident incongruity between the archaeological textile finds and the depictions. In Viking Age Scandinavia, the production of tablet-woven bands of silk, silver and gold thread is an entirely new development, which in the Danish area, is restricted to the upper social strata, while it is a more commonly used embellishment in the burials from Birka in Sweden (Geijer 1938; Hald 1980; Hägg 1986; Hansen 1991; Østergård 1991; Christensen *et al.*, 1992). Thus, here too, there are both differences and similarities between the iconographical costumes and archaeological clothing.

Conclusion

The vast majority of the iconographical costumes recorded are encountered in the archaeological textile and clothing material. At the same time, there is information on the depictions that is not seen in the archaeological textile finds and vice versa.

Clothing made of textile in the Late Iron Age was generally sewn together of several parts. The custom of cutting and sewing clothing items seems to come into the Scandinavian clothing tradition in the latter part of the Early Iron Age and continues to be used from the Late Iron Age onwards. In the Late Iron Age, clothing also becomes more gender specific compared to earlier periods, which can also be seen in the depictions.

Fig. 7.38 The piled wool hat found in the Early Bronze Age male oak coffin from Muldbjerg in Denmark. The find is dated to 1365 BC (National Museum of Denmark).

Fig. 7.39 Wool textile fragment 56A from Haithabu harbour in Germany with a brushed textile surface possibly imitating fur (Schloss Gottorf; photo: Charlotte Rimstad).

In the Late Iron Age, single coloured textiles with a clear texture were preferred. The textiles were adjusted to the function of the clothing and were suitable for cutting and sewing. Skin garments are no longer prominent in the archaeological record, although they must have been used as a resource. However, a new trend was the use of wild animal fur for trimmings, which was regarded as a prestigious material. Fur was occasionally imitated in piled textiles, as in the Early Bronze Age (Fig. 7.38), or by brushing the fabric to create a hairy surface (Fig. 7.39). It is possible that some of the patterns in the depictions represent furry surfaces (see Fig. 2.11 TØ1).

A typical male outfit in the Late Iron Age consisted of trousers, a tunic with sleeves, a rectangular cloak, belt, and shoes, while a typical female outfit included various dresses, skirts, blouses, cloaks, hairnets, and shoes, demonstrating both continuity from previous periods and new trends. It may be concluded that, the depictions selected for this study in most of the object categories are comparable to their original models, also when taking into consideration that depictions generally work with fewer, simplified or clearly exaggerated details. Precisely the deliberate and consequent utilisation of clear costume shapes enables us to understand and interpret these depictions today.

Generally, the investigation shows that the depictions represent clothing items that occur in the archaeological record. The only possible examples of archaic clothing forms could be the frequent occurrence of borders on the costumes portrayed on the gold-foil figures and the helmet plates. Thus, the depictions reveal costume items, *e.g.* kaftans, female jackets, and skirt and blouse ensembles that most likely were present in the archaeological clothing record in the Late Germanic Iron Age, but that have not been positively identified yet. They also demonstrate that the female dress most likely was long-sleeved, something which has not yet been securely documented via the archaeological textile record.

It is crucial to remember that, what is not known is difficult to recognise iconographically, and based on this paradox, the depictions are potentially of paramount interest for archaeological clothing research.

Notes

1 In Hägg's terminology this is termed a *livrock* in Swedish.

2 In Hägg's terminology this is termed a *tröje* in Swedish.

3 The example from Valsgärde grave 7 is a tabby weave, but several of the other examples refer to 2/2 twill textiles. See also Valsgärde graves 6 and 13 (Bender Jørgensen 1992a, 261–262).

4 A DNA analysis has rendered it probable that this is, indeed, a female rather than male burial, making the trousers worn by the deceased remarkable. Therefore, it has been suggested that this is a Sami costume, but as knowledge of the Sami clothing tradition is still in its infancy, it is not possible to verify this suggestion (Nockert and Possnert 2002, 60).

8

Conclusion

The analysis of the costumed figures on the object categories dated to the Late Iron Age and Scandinavian Middle Ages selected for this study is based on a material comprising 1221 gender determined figures. Of these, 998 were eventually ascertained as having a costume suitable for analysis. Although most object categories contain depictions of both men and women, there are twice as many male as female figures (Table 8.1). An exception is the helmet plates that only depict males. As demonstrated in the five individual costume analyses, this large amount of clothed figures can be grouped into a simple and homogenous system comprising 63 costume types.

The bracteates, as the earliest object category, are dated to the Early Germanic Iron Age. They primarily depict males in very simple costumes, and in several cases, it is difficult to see if the figures are clothed at all. It is evident that, on these depictions, the costume does not play a prominent role, but they are, nevertheless, germane to the costume analysis.

The costumes recorded on the full-figured bracteates (Group 1) primarily feature blouse and trouser ensembles, while in the group displaying only upper bodies (Group 2), an outer garment in the shape of a cloak or perhaps a jacket also occurs. In a few cases, bracteate figures are depicted in unusual costumes, *e.g.* men in kilts. This is probably due to the use of features belonging to the foreign sources of inspiration in an otherwise local costume context. The few females portrayed on the bracteates are all depicted homogenously in a skirt, while it is more difficult to ascertain if their upper body is covered in a blouse or is simply left bare. In this object category, the iconographic content is, thus, more important than the appearance of the costume and its recognisability, which is in accordance with the most obvious symbol-bearing elements of the bracteates being placed on the head and the neck of the figures. The simplest of the full figures (Group 2.4) can thus be compared with the relatively large group of singly represented male gold-foil figures primarily from Sorte Muld on Bornholm without specific costume details that are excluded from this analysis.

In the case of the gold-foil figures, the compilation of the costumes demonstrates that the Scandinavian gold-foil figures were produced within a shorter period than traditional archaeological dating suggests: presumably in the Late Germanic Iron Age with the emphasis on the 7th century AD. Thereby, the results of the costume analysis concur with well-accepted datings of the south Scandinavian gold-foil figures, and, on this basis, it is no longer valid to consider the gold-foil figures as an object category that was stagnant over a period of more than half a millennium. Moreover, based on the costumes, the Swedish and Norwegian gold-foil figures traditionally attributed to the 8th–10th centuries AD by conventional stylistic archaeological dating methods must have been produced contemporaneously with the southern Scandinavian gold-foil figures. The four costume groups that the gold-foil figures are divided into through a correspondence analysis thus show that this object category was sensitive to changes in the appearance, but do not represent a long chronological sequence. This result is further substantiated by the fibulae encountered on the gold-foil figures being either long-lived, enduring types or specific for the Late Germanic Iron Age, and gold-foil figures from different geographical areas in southern Scandinavia, Sweden and Norway being present in the same costume groups. Therefore, the costume groups are interpreted as representing various depiction strategies that depend on the composition of the depiction, *i.e.* whether they aim to represent single or double figures. It is even possible that these groupings can be seen as an expression of specific known individuals. Attributes that are often associated with Scandinavian gods, such as the one-eyed figure, the spear, the ring and the birds with Odin, and the disc-on-bow fibula and feathered pelt with Freyja, do appear on the gold-foil figures, but only to a very limited extent, although these features are the ones most often highlighted and used for interpretational purposes.

Of all the analysed object categories, gold-foil figures have the most varied selection of costumes and costume combinations in both genders. Among the males, it is the

Table 8.1 Male and female figures with a recorded costume, and their distribution among the object categories selected for this study.

Object category	*Females*	*Females %*	*Males*	*Males %*	*Total*	*Total %*
Bracteates	5	2%	146	20%	151	15%
Gold-foil figures	193	71%	317	43%	510	51%
Helmet plates	–	0%	48 (54)	7%	48	5%
Jewellery	23	9%	14	2%	37	4%
Tapestries	48 (265)	18%	204	28%	252	25%
Total	(486) 269	100%	(735) 729	100%	(1221) 998	100%

The figures in parentheses show the total number of gender determined figures. The numbers are based on information extracted from Tables 2.13, 2.15, 4.5, 5.3 and 6.8.

cloak or kaftan as outerwear, combined primarily with tunics and various trousers or leg types that are most often depicted. Blouse and trouser ensembles are less common. Among the females, the outerwear consists of various types of cloaks and jackets that are primarily combined with a dress, and, on rare occasions, with a skirt and blouse ensemble. A train or an apron may feature as an extra female adornment.

Helmet plates are dated to the Late Germanic Iron Age. This object category only consists of male figures. Outerwear of the kaftan type is the most frequent followed by trousers combined with a tunic, a blouse or a bare upper body. Compared to the much more variedly clad male gold-foil figures, the helmet plates contain only a limited selection of costume types and combinations, although there is a general correspondence between male costume types found in these two object categories. Gold-foil figures and helmet plates are the only object categories in which kaftans occur, and on the basis of the iconographical composition of the helmet plates, these figures most likely have a military function or connotation. That kaftan-clad males on the gold-foil figures are exclusively found as singly represented figures, and never together with female figures, while on the helmet plates, they do occur together with males in other costume types, makes their interpretation as warriors even more certain. The decisive iconographical factor for the occurrence of this costume is presumably that these depictions contain no women or any relation to typically female spheres.

Although the jewellery category is not as numerous as some of the other categories, it comprises a varied selection of both male and female figures. Male figures are dated to all three chronological periods (Early and Late Germanic Iron Age and the Viking Age), while the female figures, with a single exception, are dated to the Viking Age.

The earliest depictions are found on the relief brooches dated to the Early Germanic Iron Age. These are male figures wearing a costume consisting of a blouse and trouser or tunic and trouser ensemble. The range of costume types in this category corresponds closely to those of the bracteates. The two male depictions dated to the Late Germanic Iron Age portray a man in a tunic and another in a blouse and narrow breeches respectively. The former costume type is widespread among the gold-foil figures, while the latter, apart from the bracteates, is only found once among the gold-foil figures. The few male figures dated to the Viking Age are primarily clad in tunic and trouser combinations while a single figure wears an ensemble of a blouse and wide breeches. The breeches are a costume item that does not feature in any of the aforementioned object categories. However, breeches are also worn by the male figures on the tapestries, which further strengthens the connection of this costume item to the Viking Age.

Female figures in the jewellery category contain a varied selection of costume combinations, generally corresponding to those recorded on the gold-foil figures, although they also hold important differences. Those that occur most frequently are various combinations of the cloak and dress, either the closed cloak, the unusual female cloak with pointed corners, or the teardrop-shaped cloak, but never the open cloak. This is significant, as the open cloak is characteristic of Costume Group 2 among the gold-foil figures. Moreover, the open cloaks are especially frequently encountered among the Norwegian find material, which can be taken as a further argument for the gold-foil figures not reaching chronologically into the Viking Age. Furthermore, the female cloak with pointed corners and the teardrop-shaped cloak are not encountered in any female figures dated before the Viking Age, and as the teardrop-shaped cloak is also found on the tapestries, this type can be seen as indicative of a specific Viking Age clothing item. The jacket and dress combination appears only once on a single jewellery item, and except for the gold-foil figures, this is the only instance when a jacket is recorded among the object categories selected for this study.

The tapestries selected for this study are all dated to the Viking Age and the Scandinavian Middle Ages, with male figures being predominant. Further, the male and female costume ensembles depicted are relatively limited and have most similarities with those in the jewellery category.

Two typical male costume combinations are encountered: an ensemble of a tunic with narrow-legged trousers or bare legs/no specific leg wear, which also occurs in all the other object categories, and a more specific costume ensemble consisting of a teardrop-shaped cloak, a tunic worn with wide breeches or a blouse and wide breeches. Both in the early and later dated tapestries, wide breeches are attested, while the long loose tunic is only encountered once, on the Skog tapestry, dated to the 13th century AD (Nockert and Possnert 2002, 76). As in the jewellery category, male costumes show few similarities with the costume combinations attributed to the Germanic Iron Age, while there are many similarities with those dated to the Viking Age.

The female costumes in the tapestry category primarily consist of a dress, or ensembles of a closed cloak and dress, or a skirt and blouse combination, which are also common among the gold-foil figures and in the jewellery category. The majority of the female costumes consist of a combination of a teardrop-shaped cloak and a dress, which seems to represent the visual essence of Viking Age female appearance. As these female representations are only encountered on the Oseberg tapestry dated to the early part of the Viking Age (Christensen *et al.*, 1992, 9; Nockert and Possnert 2002, 69), it is not possible to make any further typological and chronological differentiation of the female costume recorded on the tapestries due to the more simplistic depictions used in the latest part of the examined period. Compared to the variations that the corresponding tapestry male costumes display, the simplified shape of the female costumes in this period is remarkable, and may possibly be rooted in a different, altered perception of the female role in contemporary society and narratives.

Overall costume combinations

The almost 1000 male and female figures and their costume items consisting of outerwear and/or innerwear can altogether be grouped into 253 female and 559 male costume combinations, excluding the male figures from the bracteate object category. The chronological sequence that can be extracted for these male (Table 8.2) and female (Table 8.3) costume ensembles can also be expressed graphically, as seen in Figure 8.1. This shows that male costumes, such as the kaftan (Y1), cloak with pointed ends combined with various items of innerwear (Y2–Y3), jackets (Y5) and narrow breeches (Y3b, Y6c and Y7b) appear in the Germanic Iron Age, while the teardrop-shaped cloak (Y4), the long wide tunic (Y6f) and wide breeches (Y7c) are chronologically later elements that belong to the Viking Age and Scandinavian Middle Ages.

In female costume, the open cloak (X3), the bell-shaped cloak (X6), the straight-edged jacket (X7b) and the closed cloak with a skirt and blouse ensemble are garments that appear in the Germanic Iron Age, while the teardrop-shaped cloak (Y4) and the pointed-ended female cloak (X5) are specific to the images that belong to the Viking Age and the Scandinavian Middle Ages. Basic costume combinations encountered throughout the entire period under review constitute elements, such as a tunic without an outer garment (Y6a–b and Y6d–e) as well as some blouse and trouser ensembles (Y7a and Y7d–f) for males, and the closed cloak (X2), the curved jacket (X7a), a dress without outerwear (X8), as well as a skirt and blouse ensemble for females.

In the costume analysis, outerwear can be described by a system comprising six different groups, OG1–OG6, half the groups consisting of cloaks, and the other half of jackets. The vast majority of the groups consist of female costumes, with 80% of all female figures wearing an outer garment. Of these, the cloak is the most frequent type. In contrast, only 24% of the males wear outer garments, most of which are of the kaftan type, with the remaining cloak types occurring more rarely. Thus, outerwear is utilised differently in male and female costume depictions. Only a single example each of a male in female outerwear and vice versa is encountered in the entire collection of material analysed.

Based on the costume analysis in each object category, it may be concluded that the male cloak with pointed corners is primarily encountered among the gold-foil figures dated to the Late Germanic Iron Age as well as a few figures from the Rolvsöy tapestry dated to the Viking Age. In the latter case, the cloak is, however, combined with narrow-legged trousers, which is not encountered among the gold-foil figures except in one case from Sorte Muld on Bornholm. The closed female cloak primarily found on the gold-foil figures and in the jewellery category is seen as a general type, while the open cloak is primarily connected to Costume Group 2 of the gold-foil figures (Fig. 8.2). The bell-shaped cloak is only found on the gold-foil figures and is also confined to the Late Germanic Iron Age. The teardrop-shaped cloak is limited to the Viking Age and is most widespread among the male costumes where it is found in a short version, while the long version belongs to the female costume. The kaftan is only encountered among the representations from the Late Germanic Iron Age, and only worn by males. The same chronological distribution applies to the jacket with a curved edge found among both male and female gold-foil figures, but also seen on two figures from the jewellery category dated to the Viking Age. The jacket with the straight lower edge is limited to the gold-foil figures and thereby to the Late Germanic Iron Age.

Innerwear is both simpler in shape and more uniform in appearance than outerwear. Female garments constitute dresses, and a skirt and blouse combination, while male garments constitute tunics, and blouse and trouser ensembles. The vast majority of females are represented in a dress, and only rarely in a skirt and blouse, while 68% of the males are represented in a tunic and less frequently in blouse and trouser ensembles. The greatest difference between the female dress and the male tunic lies in the length of the costume, while more specific details are rarely depicted.

Although clear costume type conventions are easily identifiable, the costumes also contain more individual characteristics. This is, for instance, seen in the various ways the costumes are embellished and patterned. The patterning comprises plain, vertically-striped, horizontally-striped and

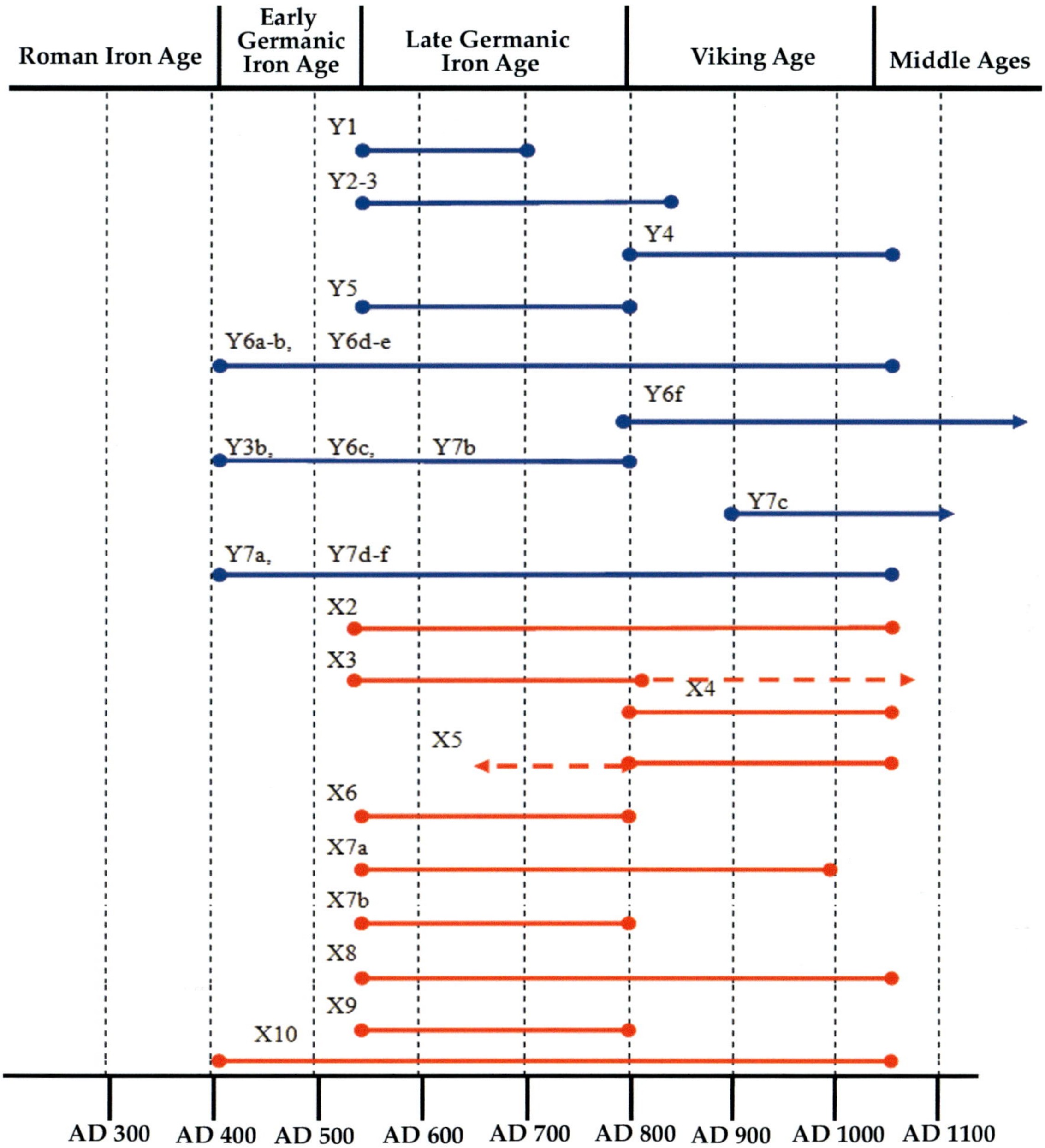

Fig. 8.1 The various costumes ensembles occurring in the selected object categories seen in relation to their individual dating, male costume ensembles (Y) and female ensembles (X). The numbering of the costume ensembles refers to the definitions given in Tables 8.2 and 8.3.

crisscross-patterned garments with or without borders, as well as checked, dotted or combinations of these elements. The most varied patterning is seen on dresses and tunics, although generally, the same types of patterning are encountered throughout the entire period under consideration.

It may be concluded that innerwear primarily consists of very simple and standardised types of garments that are found in all the object categories selected for this study. Among the male costumes, the narrow knee-length tunics are the most widespread, while the long, loose tunic seen on the Skog tapestry is a type exclusively found in the Viking Age. The same applies to the dress which is the most frequently occurring item of female innerwear in all object categories, except the bracteates, while the skirt and blouse combination is primarily encountered in the early part of the period and on figures from the later tapestry category, *e.g.* Oseberg and Rennebu.

A further significant result of the costume analysis is that some trouser types reflect chronological changes. Narrow breeches are relatively rare, but are exclusively connected to Germanic Iron Age depictions on the gold-foil figures and on a single male figure from the jewellery category. The same

Table 8.2 Distribution of male costume ensembles in the five object categories.

Costume ensemble	*Costume description*	*Males*					*Total*	*Total %*		
		BR	*GF*	*HP*	*JL*	*TP*				
XY	Closed cloak and bare legs (A5/L3,L5)		1				1	0%	0%	
Y1a	Kaftan and narrow trouser legs (D/L1)		3	8			11	2%		
Y1b	Kaftan and patterned trouser legs (D/L9)			1			1	0%	13%	
Y1c	Kaftan and bare legs (D/L3)		51	9			60	11%		
Y2a	Cloaks with pointed corners, tunic and wide trouser legs (A11-14/L2)		1				1	0%		
Y2b	Cloaks with pointed corners, tunic and bare legs (A11-14/L3,L5)	X	10				10	2%		
Y3a	Cloaks with pointed corners, and narrow trouser legs (A11-14/L1)					5	5	1%	6%	24%
Y3b	Cloaks with pointed corners and narrow breeches (A11-14/L4)		1				1	0%		
Y3c	Cloaks with pointed corners and bare legs (A11-14/L3)		14				14	3%		
Y4	Teardrop-shaped cloaks and wide breeches (A19/L7)					23	23	4%	4%	
Y5a	Slanted jackets, tunics and narrow trouser legs (B1/L1)		3				3	1%	1%	
Y5b	Slanted jackets, tunic and bare legs(B1/L3)	?	2				2	0%		
Y6a	Tunic and narrow trouser legs (E/L1)	X	21	8	1	9	39	7%		
Y6b	Tunic and wide trouser legs (E/L2)		6		1		7	1%		
Y6c	Tunic and narrow breeches (E/L4)		1				1	0%	65%	65%
Y6d	Tunic and patterned trouser leg (E/L9)			1			1	0%		
Y6e	Tunic with bare legs (E/L3,L5)		182	7	5	93	287	51%		
Y6f	Long, wide tunic with bare legs (E/L5)					30	30	5%		
Y7a	Blouse and wide trousers (BB/L2)		2				2	0%		
Y7b	Blouse and narrow breeches (BB/L4)	X	1		1		2	0%		
Y7c	Blouse and wide breeches (BB/L7)				1	36	37	7%	11%	11%
Y7d	Blouse and narrow trousers (BB/L8)	X		1	5	8	14	3%		
Y7e	Blouse and patterned trousers (BB/L9)		2	1			3	1%		
Y7f	Bare upper body and trousers (ND/L8,L9)	X		4			4	1%		
Total		-	301	40	14	204	559			
Total %			54	7	3	36		100%	100%	100%

Bracteates (BR), gold-foil figures (GF), helmet plates (HP), jewellery (JL), tapestries (TP).

applies to wide trouser legs. In contrast, wide breeches are a leg-covering only recorded among the depictions dated to the Viking Age, *i.e.* the tapestries where they predominate, and on a single male figure from the jewellery category.

Costume versus clothing

Based on the comparison of the iconographical costumes with archaeological textile and clothing finds, it may be concluded that, for the females figures, there are many similarities between the depictions and the archaeological textile and clothing finds, which are primarily manifested through the general outline of the costumes and the manner in which the outer garment was worn. The costume analysis further emphasises that outerwear constituted an essential component of female attire, with a varied selection of different modes of wear being attested to. The archaeological clothing finds from the Late Iron Age neither directly confirm nor reject the existence of the many different iconographical cloak types, but the overwhelming number of depictions in the costume analysis supports the assumption that the costume types represent actual female attire.

Female jackets constitute a special costume item primarily encountered on depictions dated to the Late Germanic Iron Age, although it cannot be documented in the archaeological textile and clothing finds until the Viking Age.

In the archaeological textile and clothing material, it is possible to differentiate among several types of innerwear constructed of different materials and worn in a specific order. This holds true for both male and female clothing. The same

Table 8.3 Distribution of female costume ensembles in the four object categories.

Costume ensemble	*Costume description*	*Females* *BR*	*GF*	*JL*	*TP*	*Total*	*Total %*		
XY	Closed cloak with visible feet (A5/L3,L5)		5			5	2%	2%	
X1	Open cloak with no feet visible (A3/L6)		1			1	0%	0%	
X2a	Closed cloak, dress with visible feet (A1-2,5-8,15-17/L5)		61	5		66	26%		
X2b	Closed cloak, dress and apron with visible feet (A1-2,5-8,15-17/L5)		3			3	2%	37%	
X2c	Closed cloak, dress with no feet visible (A1-2,5-8,15-17/L6)		15	2	1	18	7%		
X2d	Closed cloak, dress and apron with no feet visible (A1-2,5-8,15-17/L6)		3	1		4	2%		
X3a	Open cloak, dress with visible feet (A3-4/L5)		1			1	0%		
X3b	Open cloak, dress with no feet visible (A3-4/L6)		13		2?	15	6%	8%	80%
X3c	Open cloak, dress and apron with no feet visible (A3/L6)		5			5	2%		
X4a	Teardrop shaped cloak, dress with visible feet (A20-21/L5)			1		1	0%	13%	
X4b	Teardrop shaped cloak, dress with no feet visible (A20-21/L6)			1	32	33	13%		
X5a	Cloak with pointed ends, dress with visible feet (A18/L5)			2		2	1%	1%	
X5b	Cloak with pointed ends, dress with no feet visible (A11/L6)		1			1	0%	0%	
X6	Bell-shaped cloak, dress and apron with visible feet (A9-10/L5)		3			3	2%	2%	
X7a	Slanted jacket, dress with visible feet (B1-3/L5)		7	2		9	4%	6%	
X7b	Straight-edged jacket and dress (B4-5/L5)		5			5	2%		
X8a	Dress (C/L3,L5)		40	1	2	43	17%		
X8b	Dress (C/L6)		13	4	7	24	9%	16%	16%
X8c	Dress, apron and no feet visible (C1-12/L6)		1			1	0%		
X9	Closed cloak, blouse and skirt with visible feet (A1/SB/L5)		2			2	1%	1%	
X10a	Blouse and skirt with visible feet (SB/L5)	4	2			6	2%	4%	4%
X10b	Blouse and skirt and no visible feet (SB/L6)	1			4	5	2%		
Total		5	181	19	48	253			
Total %		2	72	7	19		100%	100%	100%

Bracteates (BR), gold-foil figures (GF), helmet plates (HP), jewellery (JL), tapestries (TP).

differentiation of innerwear cannot be seen in the depictions. At first glance, only single-layered costumes appear to be displayed. However, it cannot be fully excluded that dresses with borders positioned above the lower edge or hem indicate a double-layered costume. The same interpretation may also be valid for dress types recorded with an apron.

Most likely, the female dresses depicted have long sleeves. However, it cannot be excluded that some of the depicted dresses are sleeveless, as is known from the tubular dress type from the archaeological clothing finds of the Early Bronze and Early Iron Ages. Often, these kinds of details are not depicted systematically, and for the producer and the user, this was obviously not a feature that was considered necessary to depict in detail. Further, it has not been possible to identify the change in the positioning of the female dress fibulas from the shoulders to the chest, which is seen as the major proof for a change in the design of the Scandinavian female dress type during the 6th century AD. One explanation for this discrepancy could be that the design of the dress and the use of metal dress fibulae did not have a strong status-denoting function in the iconography, but equally important is the documented focus on female overwear, which often covers the entire upper body. If a fibula is depicted, it is always linked to the latter costume item. That archaeological textiles in burials are primarily preserved in connection with metal fibulae may also have given greater emphasis to the importance of this feature, and perhaps, on this basis, it may be possible to look for other alternative dress designs, just as is seen in the Viking Age textile finds from Haithabu.

Skirts and blouses are rarely depicted and are limited to a few instances on bracteates, gold-foil figures and tapestries, and it is evident that this costume combination is not as iconic as the dress. The skirt and blouse ensemble cannot be documented in archaeological textile material from the Late Iron Age, although it constitutes an important part of the preceding clothing tradition, as documented in the bog finds from Huldremose and Krogens Mølle in Denmark. Thus, the depictions reveal a more varied selection of clothing items than the archaeological textile and clothing finds do.

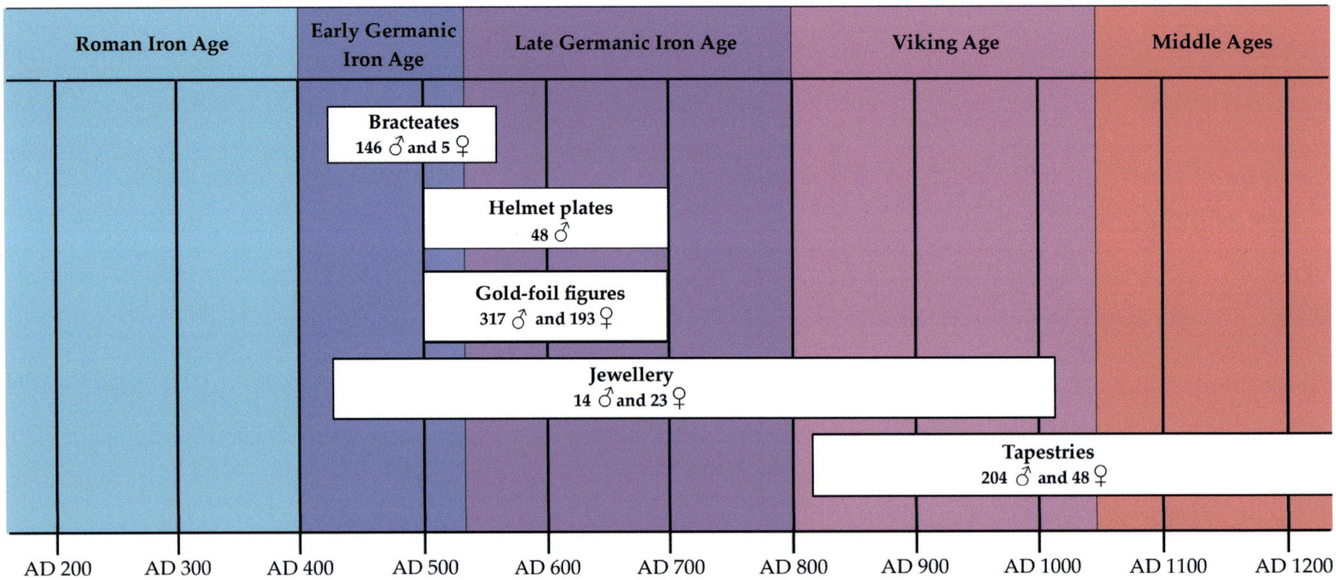

Fig. 8.2 Revised chronological overview of the five object categories, bracteates, gold-foil figures, helmet plates, jewellery and tapestries, selected for this study, with the gender distribution and total number of the analysed figures.

Similarly, the train-like ending of the dress and the skirt is a well-known and quite frequently occurring detail on the depictions, while due to the preservation conditions in the graves, it is difficult to document in the archaeological textile and clothing finds. Nevertheless, this element is so prominent in the depictions that it must have constituted a well-known feature of Late Iron Age female clothing. Further, the depictions support the assumption that female innerwear in the Late Iron Age in general was very long, almost floor-length, although it has not been possible to document any secure chronological differences in whether females are depicted with their feet visible or not.

A high level of correspondence is observed for males between the depictions and the archaeological textile and clothing finds, while outerwear is not as prominent as in the female costume. Whether this reflects a true relationship in the use of outerwear can only be surmised. The male cloaks encountered in the depictions are types also known from the archaeological textile and clothing finds, as indicated by the fibulae in the burials, but as these are most often highly fragmented, it is difficult to compare the two sources in detail.

The kaftan is an item of outerwear that has not yet been identified in the archaeological textile and clothing finds. It appears exclusively on the depictions from the Late Germanic Iron Age (singly represented male gold-foil figures and on helmet plates), and it is seen as a garment that plays a special role in the depictions. However, it is not unlikely that with a renewed focus on this clothing item, it will be easier to identify it among the archaeological textile and clothing finds, and the large number of kaftans depicted supports the assumption that it represents an actual clothing item.

Based on the iconographical context, it is further possible to see the use of the kaftan in depictions as a manifestation of a warrior elite in Late Germanic Iron Age society. However, when the sleeved jacket finally appears in the archaeological textile and clothing finds in the Viking Age, it is in different designs, with many common features between the male and female jacket. Thus, it seems that when the military significance of the kaftan waned during the latter part of the Late Germanic Iron Age, it eventually became a clothing item available to both genders. This could perhaps also explain why the warriors on the Oseberg tapestry are not attired in kaftans, but instead depicted in the short teardrop-shaped cloak, which iconographically is a new cloak type occurring late in the period under study. The depictions can thus provide evidence to support the interpretation that Viking Age jackets were not only inspired by Byzantine sources, but were indeed part of a well-known and much older Scandinavian clothing tradition.

In terms of innerwear, the vast majority of the males are clad in close-fitting tunics. In some cases, the tunic is thigh-length, but is most often knee-length. Similar to the females, it is single-layered. The model for this costume item is most likely tunics, such as those from Thorsbjerg, Lendbreen, Högum and Bernuthsfeld. It is also feasible to link the slightly longer tunic seen on the tapestry from Skog (dated to the 13th century AD) in comparison with the Old Norse clothing finds from Greenland. The costume analysis thus shows that, it is possible to follow a chronological development in the tunic types in the depictions, similar to that seen in the archaeological textile and clothing finds.

Another key characteristic of both male and female innerwear, especially encountered in the depictions from the Late Germanic Iron Age, are the patterned bands that, in most cases, are interpreted as attached or in-woven borders. Compared to the archaeological textile finds, this is possibly an archaic element, as decorations with tablet-woven bands are a feature which belongs to earlier periods and not later than the Early Germanic Iron Age.

The types of legwear encountered in the depictions, too, were shown to include both general types as well as those typical for a specific period, corresponding to the development seen in the archaeological textile and clothing finds. An example of this is the wide breeches found exclusively in the depictions dated to the Viking Age. Narrow breeches, however, only occur in the depictions dated prior to the Viking Age, and here, there are clear links to the Roman Iron Age trouser finds, such as those from Thorsberg and Damendorf.

Concluding remarks

Through the systematic analysis of the five object categories selected for this study, bracteates, gold-foil figures, helmet plates, jewellery and tapestries, it is now possible to determine where, and to what degree, the selected depictions of costumed figures complement the extant archaeological sources. This provides new insights into the chronological, regional and social variations in the clothing tradition in Late Iron Age Scandinavia.

The analyses of the five object categories demonstrate that a transformation of the costume types depicted occurred in the course of the Late Iron Age and Scandinavian Middle Ages, probably due to a combination of changes in clothing traditions, textile technology and shifting iconographical strategies.

It may further be concluded that the differences in the raw material used for the different iconographical sources did not create significant iconographical differences between those produced in an inorganic (metal) or an organic (textile) material. From a craft perspective, the depictions are created within very different traditions, yet, they nevertheless express a similar iconographical tradition. The costume combinations identified are not restricted to a specific craft or material, but rather controlled by chronological and narrative factors, and the analysis demonstrates that a distinct iconographical tradition for how to depict male and female costumes existed in Scandinavia in the Late Iron Age.

Further, the costume analysis demonstrates a greater variety of male rather than female costumes in the object categories selected for this study. This indicates a differentiated and more diverse attitude towards the male costume than to the female costume in the Late Iron Age. There can be no doubt that the producers of the depictions used contemporaneous archaeological textile and clothing as their models. Archaic clothing forms are difficult to pinpoint, while characteristic contemporary features, such as typical embellishments, fibulae and textile textures are clearly displayed in exaggerated forms.

Another key finding is that iconography can be used for investigating clothing traditions, and the fact that some costume items can only be documented through iconography makes this an important source in the interpretation of old and new archaeological textile and clothing finds. The analysis of the almost 1000 different male and female figures demonstrates that the iconographical costumes, both in the individual object categories and collectively, contribute with new and important knowledge to the clothing tradition in the period under consideration. It is evident that, the depictions are reliable sources for both iconographical and archaeological costume research.

The importance of clothing for the understanding of social systems in prehistory cannot be underestimated, and that the costumes assume such a prominent role in the iconography is a sign of their significance. Although the social significance of both iconographical costume and archaeological textile and clothing finds has long been known, this study provides a methodological tool that makes it possible to include this dimension in archaeological research in a new and more systematic way.

The costumes described in this study undoubtedly belong to the upper echelons of society, but this is also true for the main bulk of preserved archaeological textile and clothing finds, and all archaeological sources must be used with care in relation to content and context. The constraints of the preserved material must be kept in mind when reconstructing or disseminating knowledge of all kinds. The development of clothing through time is a phenomenon that cannot be discussed in isolation, but benefits from an interdisciplinary and cross-cultural perspective. It is crucial to base our discussions on systematic, comprehensive studies, thereby rendering it possible to describe the special, the innovative and the general aspects, and thereby enhance our knowledge.

Bibliography

This bibliography contains all the reference material from my original PhD thesis (2006) as well as pertinent new references published since then.

Ahrens, C., ed., (1979) *Sachsen und Angelsachsen*. Veröffentlichungen des Helms-Museum 32. Hamburg.

Almgren, B. ed., (1967) *The Viking*. Göteborg.

Almgren, B. (1983) Helmet, crowns and warriors' dress – from the Roman emperors to the chieftains of Uppland, in J.P. Lamm and H.-Å. Nordström (eds), *Vendel Period Studies. Transactions of the Boat grave symposium in Stockholm, February 2–3, 1981*, 11–16. Stockholm.

Almgren, O. (1901) En egendomlig båtgraf vid Ulltuna. *Kungliga Vitterhets-, Historie- och Antikvitetsakademien månadsblad*, vol. 30/31, 147–162. Stockholm.

Andersen, H. (1991) Dendrokronologisk datering af Mammengravens indhold, in M. Iversen (ed.), *Mammen. Grav, kunst og samfund i vikingetid*. Jysk Arkæologisk Selskabs Skrifter XXVIII, 43–44. Højbjerg.

Andersson, E. (1996a) Invisible Handicrafts. The General Picture of Textile and Skin Crafts in Scandinavian Surveys. *Lunds Archaeological Review* 1995, 7–20. Lund.

Andersson, E. (1996b) *Textilproduktion i arkeologisk kontext. En metodstudie av yngre järnåldersboplatser i Skåne.* University of Lund Institute of Archaeology Report series No. 58. Lund.

Andersson, E. (1999a) *The Common Thread. Textile Production during the Late Iron Age–Viking Age.* University of Lund Institute of Archaeology Report series No. 67. Lund.

Andersson, E. (1999b) *Textilproduktion i Lödde: köpinge endast för husbehov?* Lund.

Andersson, E. (1999c) *Textilproduktion i Birka med en jämförande analys av den vikingatida handelsplatsen Hedeby och samtida bosättningar i mälarområdet*. Lund.

Andersson, E. (2003a) *Tools for Textile Production from Birka and Hedeby*. Birka Studies 8. Stockholm.

Andersson, E. (2003b) Textile Production in Scandinavia during the Viking Age, in L. Bender Jørgensen, J. Banck-Burgess and A. Rast-Eicher (eds), *Textilien aus Archäologie und Geschichte. Festschrift für Klaus Tidow,* 46-62. Neumünster.

Andersson Strand, E. (2011). The organization of textile production in Birka and Hedeby, in S. Sigmundsson (ed.), *Viking Settlements and Viking Society, Papers from the Proceedings of the Sixteenth Viking congress Reykjavik 2009,* 1–17. Reykjavik.

Andersson Strand, E. (2015) Textile Exchange, Trade and Production in Scandinavian Viking Age and Early Medieval Societies, in C. Jahnke and A. Huang (eds), *Conference proceeding – Textiles and economy. Production and trade of textiles. The central and northern Europe as example, Copenhagen*, 8–22. Ancient Textiles Series, vol. 16. Oxford.

Andersson, E. and Batzer, A. (1999) *Sländspinning i Vikingatid och nutid*. Lund.

Andersson Strand, E., Frei, K., Gleba, M., Mannering, U., Nosch, M.-L. and Skals, I. (2010) Old textiles – new possibilities, *European Journal of Archaeology* 13:2, 149–173.

Andréasson, A. (1995) *Skandinaviens guldgubbar. Kända fynd och fyndplatser från 1740 till 1994.* Upubliceret C-uppsats i Arkeologi. Göteborgs Universitet.

Anon (1998) Amatørarkæolog gjorde årets fund. *Bornholms Tidende*, 21st January, p.1. Rønne.

Anon (1998) Sjældent fund af stempel til guldgubber: Han følte det som husets loft lettede. *Bornholms Tidende*, 21st January, p.5. Rønne.

Arbman, H. (1940) *Birka I. Die Gräber. Tafeln*. Stockholm.

Arbman, H. (1943) *Birka. Untersuchungen und Studien I. Die Gräber. Text*. Stockholm.

Arne, T. J. (1932) Ein Bemerkenswerter Fund in Östergötland. *Acta Archaeologica* 3, 67–112. København.

Arne, T. J. (1934) *Das Bootgräberfeld von Tuna in Alsike, Uppland.* Stockholm.

Arrhenius, B. (1962) Det flammande smycket. *Fornvännen* 57, 79–101. Stockholm.

Arrhenius, B. (1971) *Granatschmuck und Gemmen aus nordischen Funden des Frühen Mittelalters.* Acta Universitatis Stockholmiensis. Studies in North-European Archaeology. Series B. Stockholm.

Arrhenius, B. (1980) *The Chronology of the Vendel graves.* Rapport från Stockholms Universitets Arkeologiska forskningslaboratorium Nr. 2. Stockholm.

Arrhenius, B. (1983) The chronology of the Vendel graves, in J. P. Lamm and H.-Å. Nordström (eds), *Vendel Period Studies. Transactions of the Boat-grave symposium in Stockholm, February 2–3, 1981,* 39–70. Stockholm.

Arrhenius, B. (1986) Einige christliche Paraphrasen aus dem 6. Jahrhundert, in H. Roth (ed.), *Zum Problem der Deutung frühmittelalterlicher Bildinhalte. Akten des 1. Internationalen Kolloquiums in Marburg a. d. Lahn, 15. bis 19. Februar 1983,* 129–152. Sigmaringen.

Arrhenius, B. (1995) Regalia in Svealand in early medieval times. *Tor* 27. Uppsala.

Arwidsson, G. (1954) *Valsgärde 8*. Lund.

Arwidsson, G. (1977) *Valsgärde 7*. Lund.

Arwidsson, G. (1983) Valsgärde, in J. P. Lamm and H.-Å. Nordström (eds), *Vendel Period Studies. Transactions of the Boat. grave symposium in Stockholm, February 2–3, 1981*, 71–82. Stockholm.

Arwidsson, G. (1989a) Verschiedene Schmuckgegenstände in Form Menschenfiguren, Reitern, Pferden, Vierfüsslern, Vögeln und Schlangen, in G. Arwidsson (ed.), *Birka II.3. Systematische Analysen der Gräberfunde*, 55–59. Stockholm.

Arwidsson, G. ed., (1989b) *Birka II.3. Systematische Analysen der Gräberfunde*. Stockholm.

Arwill-Nordbladh, E. (1998) *Genuskonstruktioner i nordisk vikingatid. Förr och nu.* GOTARC. Serie B. Gothenburg Archaeological Theses no. 9. Göteborg.

Arwill-Nordbladh, E. (2013), Negotiating normativities – 'Odin from Lejre' as challenger of hegemonic orders. *Danish Journal of Archaeology* 2:1, 87–93.

Axboe, M. (1983) To tidlige fund af guldgubber. Kongsvik og "Norge". *Aarbøger for Nordisk Oldkyndighed og Historie 1981*, 97–108. København.

Axboe, M. (1986) Brísingamen – også et arkæologisk problem. *Danske Studier*, 116–121. København.

Axboe, M. (1999) The chronology of the Scandinavian gold bracteates, in J. Hines, K. Høilund Nielsen and F. Siegmund (eds), *The Pace of Change. Studies in Early-Medieval Chronology*, 126–147. Exeter.

Axboe, M. (2002) Sølvkræmmerhuset og Balders død – nye brakteatfund fra Bornholm, in J. Pind, A. Nørgård Jørgensen, L. Jørgensen, B. Storgård, P. O. Rindel and J. Ilkjær (eds), *Drik – og du vil leve skønt. Festskrift til Ulla Lund Hansen på 60-årsdagen 18. august 2002*. PNM. Studies in Archaeology & History vol. 7, 295–304. København.

Axboe, M. (2007) Brakteatstudier. *Nordiske Fortidsminder Serie B*. Bind 25. Copenhagen.

Axboe, M. (2011) Die Chronologie der Inschriften-Brakteaten, in W. Heizmann and M. Axboe (eds). *Die Goldbrakteaten der Völerwanderungszeit – Auswertung und Neufunde.* Ergänzungsbände zum Reallexikon der Germanischen Altertumskunde 40, 279–296. Berlin and New York.

Axboe, M. and Kromann, A. (1992) DN ODINN P F AUC? Germanic "Imperial Portraits" on Scandinavian Gold Bracteates. *Acta Hyperborea 4*, 271–305. København.

Axboe, M. with C. Behr and K. Düwel (2011) Katalog der Neufunde, in W. Heizmann and M. Axboe (eds). *Die Goldbrakteaten der Völerwanderungszeit – Auswertung und Neufunde.* Ergänzungsbände zum Reallexikon der Germanischen Altertumskunde 40, 893–999. Berlin and New York.

Bachrach, B. S. (1997) The imperial roots of Merovingian military organization, in A. Nørgård Jørgensen and B. Clausen (eds), *Military Aspects of the Scandinavian Society. In a European Perspective AD 1–1300.* PNM Studies in Archaeology & History, vol. 2, 25–31. København.

Bakka, E. (1968) Methodological Problems in the Study of Gold Bracteates. *Norwegian Archaeological Review* 1, 5–56. Oslo.

Banck-Burgess, J. (1999) *Die Textilefunde aus dem späthallstattzeitlichen Fürstengrab von Eberdingen-Hochdorf (Kreis Ludwigsberg) und weitere Grabtextilien aus hallstatt- und latènezeitlichen Kulturgruppen.* Forschungen und Berichte zur Vor- und Frühgeschichte in Baden-Würtemberg, Bd. 70. Stuttgart.

Banck-Burgess, J. (2003) Ein alamannischer Kleiderstoff, in L. Bender Jørgensen, J. Banck-Burgess and A. Rast-Eicher (eds), *Textilien aus Archäologie und Geschichte. Festschrift für Klaus Tidow*, 123–313. Neumünster.

Barfod Carlsen, E. (1999) Hængesmykker af guld. Den stilistiske udvikling af D-brakteaterne, in O. Høiris, H. J. Madsen, T. Madsen and J. Vellev (eds), *Menneskelivets mangfoldighed. Arkæologisk og antropologisk forskning på Moesgård*, 231–234. Moesgård.

Barfod Carlsen, E. (2002) Fabeldyr i udvikling. En analyse af D-brakteaterne. *Hikuin* 29, 119–142.

Barnes, R. and Eicher, J. B., eds, (1992) *Dress and Gender*. Oxford.

Bartel, A. and Knöchlein, R. (1993) Zu einem Frauengrab des sechsten Jahrhunderts aus Waging am See, Lkr. Traunstein, Oberbayern. Ein Beitrag zur Kenntnis der Frauentracht der älteren Merowingerzeit. *Germania* 71, 419–439.

Barthes, R. (1998) Billedets retorik, in B. Fausing and P. Larsen, (eds), *Visuel kommunikation* 1, 42–57. København.

Bau, F. (1981) Seler og slæb i vikingetid. Birkas kvindedragt i nyt lys. *Kuml*, 13–47.

Beck, H. (1968) Die Stanzen von Torslunda und die literarische Überlieferung. *Frühmittelalterliche Studien*, 2. Band, 237–250. Berlin.

Becker, C. J. (1990) *Nørre Sandegård. Arkæologiske undersøgelser på Bornholm 1948–52*. Historisk-filosofiske Skrifter 13. København.

Belting, H. (2002) *Bild-Anthropologie*. München.

Bender Jørgensen, L. (1986) *Forhistoriske tekstiler i Skandinavien. Prehistoric Scandinavian Textiles.* Nordiske Fortidsminder Serie B, Bind 9. København.

Bender Jørgensen, L. (1987) Metoder og kildekritik i textilforskningen Et svar på et debatoplæg. *Tor* 21, 263–281. Uppsala.

Bender Jørgensen, L. (1988) Textilfunde aus dem Mittelelbe-Saale-Gebiet (Eisenzeit bis frühes Mittelalter). *Jahresschrift für Mitteldeutsche Vorgeschichte*, Band 71, 91–123. Berlin.

Bender Jørgensen, L. (1991a) Textilteknologi i Oldtiden. *Arkeo. Nytt fra Historisk Museum i Bergen,* Nr. 2, 19–25. Bergen.

Bender Jørgensen, L. (1991b) The Textiles of the Saxons, Anglo-Saxons and Franks. *Studien zur Sachsenforschung* 7, 11–24. Hildesheim.

Bender Jørgensen, L. (1992a) *North European Textiles until AD 1000*. Århus.

Bender Jørgensen, L. (1992b) I Knud den Stores klæ'r. *Nationalmuseets Arbejdsmark*, 32–42. København.

Bender Jørgensen, L. (1993) Ancient costumes reconstructed, in G. Jaacks and K. Tidow (eds), *Archaeological Textiles. NESAT V. Textilsymposium Neumünster. Neumüster*, 109–113.

Bender Jørgensen, L. (1995) Karl den Store, Muhammed og Uldsækken. *Varia* 30, 73–83. Universitetes Oldsaksamling. Oslo.

Bender Jørgensen, L. (2000) The Mons Claudianus Textile Project, in D. Cardon and M. Feugère (eds), *Archéologie des textiles des origines au Ve siècle, Actes du colloque de Lattes, oct. 1999,* 253–264. Monographies Instrumentum 14. Montpellier.

Bender Jørgensen, L. (2003) Krigerdragten i folkevandringstiden, in P. Rolfsen and F.-A. Stylegar (eds), *Snartemofunnene i nytt lys*, 53–79. Oslo.

Bender Jørgensen, L. (2004a) A matter of material: Changes in textiles from Roman sites in Egypt's Eastern desert. *Antiquité Tardive* 11, 87–99.

Bender Jørgensen, L. (2004b) Team work on Roman textiles: the Mons Claudianus Textile Project, in C. Alfaro, J. P. Wild and B. Costa (eds), *Purpurae vestes: Actas del I Symposium Internacional sobre Textiles y Tintes del Mediterráneo en época romana (Ibiza, 8 al 10 de noviembre, 2002)*, 69–113. València.

Bender Jørgensen, L. and Walton, P. (1987) Dyes and fleece types in textiles from Scandinavia and Germany. *Journal of Danish Archaeology* 5, 1986, 177–188. Odense.

Bender Jørgensen, L. and Tidow, K., eds, (1988) *Archaeological Textiles in Northern Europe. Report from the 2nd NESAT Symposium.* Arkæologiske Skrifter 2. København.

Bender Jørgensen, L. and Munksgård, E., eds, (1992) *Archaeological Textiles in Northern Europe. Report from the 4th NESAT Symposium*. København.

Bender Jørgensen, L. and Rinaldo, C., eds, (1998) *Textiles in European Archaeology. Report from the 6th NESAT Symposium*. GOTARC serie A, bind 1. Göteborg.

Bender Jørgensen, L., Banck-Burgess, J. and Rast-Eicher, A., eds, (2003) *Textilien aus Archäologie und Geschichte. Festschrift für Klaus Tidow*. Neumünster.

Bennett, A. (1987) *Graven. Religiös och social symbol*. Theses and papers in North-European Archaeology 18. Tierp.

Bjowulf, in B. Wilmont (1983) *Bjowulf. Et heltedigt fra Danmarks sagntid*. Værløse.

Blindheim, C. (1947) Drakt og smykker. *Viking. Tidsskrift for Norrøn Arkeologi* 11, 1–139. Oslo.

Böhme, A. (1985) Tracht- und Bestattungssitten in den germanischen Provinzen und der Belgica. *Aufstieg und Niedergang der Römischen Welt*, Teilband II, 12,3, 423–455. Berlin.

Böhner, K. (1995) Die Frühmittelalterlichen Silberphaleren aus Eschwege (Hessen) und die Nordischen Pressblech-bilder. *Jahrbuch des Römisch-Germanischen Zentralmuseums Mainz*, 38. Jahrgang, 1991, Teil 2, 681–743.

Böhner, K. (1997) Eine Fränkische Simson-Fibel aus dem Xantener Dom. *Archäologisches Korrespondenzblatt* 27, Heft 2, 323–340.

Bonfante, L. (1989) Nudity as a Costume in Classical Art. *American Journal of Archaeology* 94:4, 543–570.

Bonfante, L. (2000) Classical nudity in Italy and Greece, in D. Ridgway (ed.), *Ancient Italy in its Mediterranean Setting. Studies in honour of Ellen Macnamara*, 271–293. Accordia Specialist Studies on the Mediterranean, vol. 4. London.

Brøgger, A.W. (1921) *Farmannshaugen og kong Bjørn, Harald Haarfangres søn. Aarbøger for nordisk oldkyngelighed og historie*. København.

Brøndsted, J. (1936) Danish Inhumation Graves of the Viking Age. A Survey. *Acta Archaeologica* 7, 81–229. København.

Broholm, H. C. and Hald, M. (1935) *Danske Bronzealders Dragter*. København.

Broholm, H. C. and Hald, M. (1939) *Skrydstrupfundet. En sønderjydsk Kvindegrav fra den ældre Bronzealder*. Nordiske Fortidsminder III. København.

Broholm, H. C. and Hald, M. (1940) *Costumes of the Bronze Age in Denmark*. København.

Broholm, H. C. (1943) *Danmarks Bronzealder*. København.

Callmer, J. (1997) Aristokratisk präglade residens från yngre järnåldern i forskningshistorien och deras problematik, in J. Callmer and E. Rosengren (eds), *"...gick Grendel at söka det höga huset...". Arkeologiska källor till aristokratiska miljöer i Skandinavien under yngre järnålder*, 11–18. GOTARC C. Arkeologiska Skrifter No. 17. Halmstad.

Callmer, J. (1998) Handelsplatser och kustplatser och deras förhållande till lokala politiska system. Ett bidrag till strukturen i den yngre järnålderns samhälle, in L. Larsson and B. Hårdh (eds), *Centrala Platser Centrala Frågor. Samhällsstrukturen under Järnåldern*, 27–37. Acta Archeologica Lundensia. Series 8°, No. 28, Lund.

Callmer, J. and Lundqvist, L. (1995) Slöingeprojektet 1993: fortsatt undersökning på en fyndplats med guldgubbar. *Fornvännen* 89, 257–263. Stockholm.

Callmer, J. and Rosengren, E., eds, (1997) *"...gick Grendel at söka det höga huset...". Arkeologiska källor till aristokratiska miljöer i Skandinavien under yngre järnålder*. GOTARC C. Arkeologiska Skrifter No. 17. Halmstad.

Capelle, T. (1968) *Der Metalschmuck von Haitabu. Studien zur wikingischen Metallkunst*. Neumünster.

Capelle, T. and Fischer, C., eds, (2005) *Ragnarok Odins Verden*. Silkeborg.

Capelle, T. and Vierck, H. (1971) Modeln der Merowinger- und Wikingerzeit. *Frühmittelalterliche Studien 5*, 42–100. Berlin.

Cardon, D. (2002) Chiffons dans le desert: textiles des praesidia de Maximianon, Krokodolô et Didymoi, in M. Durand and F. Saragosa (eds), Égypte, la trame de l'Histoire. *Textiles pharaoniques, coptes et islamiques*, 40–43. Paris.

Cardon, D. (2003) Chiffons dans le desert: textiles de Maximianon et Krokodolô, in H. Cuvigny (ed.), *La Route de Myos Hormos. L'armée dans le desert Oriental d'Égypte,* 619–659. Fouilles de l'IFAO, 48. Caire.

Cardon, D. and Feugère, M., eds, (2000) *Archéologie des textiles des origines au Ve siècle, Actes du colloque de Lattes, oct. 1999.* Monographies Instrumentum 14. Montpellier.

Cardon, D., Wouters, J., Vanden Berghe, I., Richard G. and Breniaux, R. (2004) Aperçus sur l'art de la teinture en Égypte Romaine: analyses de colorants des Textiles des *praesidia* du Désert Oriental. *Antiquité Tardive* 12, 101–111.

Carlsson, A. (1988) *Vikingatida ringspännen från Gotland*. Stockholm Studies in Archaeology 8. Stockholm.

Carus-Wilson, E. (1971) Haberget: a medieval textile conundrum. *Medieval Archaeology,* vol. XIII 1969, 148–166. London.

Christensen, A. E., Ingstad, A. S. and Myhre, B. (1992) *Osebergdronningens grav*. Oslo.

Christensen, A. E. and Nockert M., eds, (2006) *Osebergfunnet. Bind IV Tekstilene*. Oslo.

Christensen, T. (2013) A silver figurine from Lejre. *Danish Journal of Archaeology*, 2:1, 65–78.

Cleland, L. Davies, G. and Jones, L. L. (2007) *Greek and Roman Dress from A to Z*. London.

Cleland, L., Harlow, M. and Llewellyn-Jones, L., eds, (2005) *The Clothed Body in the Ancient World*. Oxford.

Comis, S. (2003) Garments from the Netherlands, in L. Bender Jørgensen, J. Banck-Burgess and A. Rast-Eicher (eds), *Textilien aus Archäologie und Geschichte. Festschrift für Klaus Tidow. Neumünster*, 193–204.

Crane, S. (2002) *The Performance of Self. Ritual, Clothing, and Identity During the Hundred Years War*. Philadelphia.

Croom, A. T. (2000) *Roman Clothing and Fashion*. Stroud and Charleston.

Crowfoot, E., Pritchard, F. and Staniland, K. (1992) *Textiles and Clothing c. 1150–c.1450*. Medieval finds from excavations in London 4. Museum of London. London.

Cumont, F. (1926) *L'Uniforme de la Cavalerie Orientale et le Costume byzantin. Byzantion. Revue Internationale des Études Byzantines*. Tome II, 1925, 181–191. Paris.

D'Amato, R. (2005) *Roman Military Clothing (3) AD 400–640*. Men-at-Arms. Oxford.

Danielsson, I.-M. B. (2007) *Masking Moments. The Transitions of Bodies and Beings in Late Iron Age Scandinavia*. Stockholm Studies in Archaeology 40. Stockholm.

Dedekam, H. (1924/25) To tekstilfund fra folkevandringstiden, Evebø og Snartemo. *Bergen Museums Aarbok 1924–25*, 3, 22–79. Bergen.

Deichmann, F. W. (1995) *Frühchristlichen Bauten und Mosaiken von Ravenna*. Wiesbaden.

dell'Orto, L. F. (1996) Problems of style: reflections on Scandinavian art (first–sixth centuries A.D.), in *Roman Reflections in Scandinavia*, 30–34. Rom.

Demant, I. (1996) Konstruktionen af jernalderdragterne til Hjemsted Oldtidspark. *Arkæologi i Slesvig* 5, 47–54.

Demant, I. (2000) Mikrostratigrafisk analyse af tekstilpræparaterne fra Lerdal grav 101. *Arkæologi i Slesvig* vol. 8, 75–87.

Demant, I. (2007) The poor people from Lønne Hede – Presentation of first century-graves with preserved textiles in A. Rast-Eicher and R. Windler (eds), *NESAT IX Archäologische Textilfunde – Archaeological Textiles*, 86–91. Ennenda.

Eicher, J. B., ed., (1995) *Dress and Ethnicity*. Oxford.

Eles, von P., ed., (2002) *Guerriero e sacerdote – Autorità e comunità nell`età del ferro a Verucchio. La Tomba del Trono*. Florence.

Elsner, H. (1989) *Wikinger Museum Haithabu: Schaufenster einer frühen Stadt*. Schleswig.

Engelhardt, C. (1863) *Thorsbjerg Mosefund*. København.

Engström, J. (1997) The Vendel Chieftains a study of military tactics, in A. Nørgård Jørgensen and B. Clausen (eds), *Military Aspects of the Scandinavian Society. In a European Perspective AD. 1–1300*, 248–255. PNM Studies in Archaeology & History, vol. 2. København.

Enright, M. (1988) Lady With a Mead-Cup. Ritual, Group Cohesion and Hierarchy in the Germanic Warband. *Frühmittelalterliche Studien* 22, 170–203. Berlin.

Enright, M. (1990) The Goddess Who Weaves. Some Iconographic Aspects of Bracteates of the Fürstenberg Types. *Frühmittelalterliche Studien* 24, 54–70.

Es, W.A. van and Ypey, J. (1977) Das Grab der "Prinzessin" von Zweeloo und seine Bedeutung im Rahmen des Gräberfeldes. *Studien zur Sachsenforschung* 1, 97–126. Hildesheim.

Ethelberg, P. (1986) *Hjemsted – en gravplads fra 4. og 5.årh. e.Kr*. Haderslev.

Ethelberg, P. (1990) *Hjemsted 2 – tre gravpladser fra 3. & 4.årh. e.Kr*. Haderslev.

Evans, A. C. (1986) *The Sutton Hoo Ship Burial*. London.

Evison, V. I. (1965) The Dover, Breach Downs and Birka Men. *Antiquity* 39, 214–217. Cambridge.

Ewing, T. (2006) *Viking Clothing*. Stroud.

Fabech, C. (1992) Edsten. Ett möjligt herresäte och kultcentrum i yngre järnålder. *Bohuslän Årsbok 1992*, 53–60. Årgang 14. Uddevala.

Fabech, C. and Ringtved, J., eds, (1991) *Samfundsorganisation og Regional Variation. Norden i romersk jernalder og folkevandringstid. Beretning fra 1. nordiske jernaldersymposium på Sandbjerg Slot 11–15 april 1989*. Jysk Arkæologisk Selskabs Skrifter XXVII. Århus.

Fabech, C. and Ringtved, J. (1995) Magtens geografi i Sydskandinavien – om kulturlandskap, produktion og bebyggelsesmønster, in H. G. Resi (ed.), *Produsjon og samfunn. Beretning 2. nordiske jernaldersymposium på Granvolden 7.–10–maj 1992*. Varia 30. Universitets Oldsakssamling. Oslo.

Falk, H. (1919) Altwestnordische Kleiderkunde. Mit besonderer Berücksichtigung der Terminologie. *Videnskapsselskapets Skrifter*, II. Hist.-Filos. Klasse, 1918. No. 3, 1–234. Kristiania.

Farke, H. (1991) Textile Reste an zwei Völkerwanderungszeitlichen Vogelfibeln. Zur Bergung und Präparation archäologischer Textilfunde. *Alt-Thüringen* 26, 197–206. Weimar.

Farke, H. (1998) Der Männerkittel aus Bernuthsfeld – Beobachtungen während einer Restaurierung, in L. Bender Jørgensen and C. Rinaldo (eds), *Textiles in European Archaeology. Report from the 6th NESAT Symposium*, 99–106. GOTARC serie A, bind 1. Göteborg.

Farke, H. (2001) A typical costume of the North German Iron Age? Some observations during conservation of the Bernuthsfeld 'plaid', in P. Walton Rogers, L. Bender Jørgensen and A. Rast-Eicher (eds), *The Roman Textile Industry and its influences. A Brithday Tribute to John Peter Wild*, 129-136. Oxford.

Fausing, B. and Larsen, P., eds, (1998) *Visuel kommunikation* 1. København.

Fentz, M. (1987) En hørskjorte fra 1000-årenes Viborg. *Kuml*, 23–46.

Fentz, M. (1992) An 11th Century Linen shirt from Viborg Søndersø, Denmark, in L. Bender Jørgensen and E. Munksgård (eds), *Archaeological Textiles in Northern Europe. Report from the 4th NESAT Symposium*, 83–92. København.

Fentz, M. (1994) Viking Age replicas in research and communication, in G. Jaacks and K. Tidow (eds), *Archaeological Textiles. NESAT V. Textilsymposium Neumünster*, 95–108. Neumüster.

Fischer, C. (1974) Guldgubbe. *Skalk* 1974/5, 29–30. Højbjerg.

Fischer, C. (1979) Moseligene fra Bjæverskovdal. *Kuml*, 7–44.

Frandsen, L. B. (1996) Nye grave med tekstiler fra Lønne Hede i Sydvestjylland. *Arkæologi i Slesvig 5*, 41–46.

Frandsen, L., Nørgård, A. and Østergård, E. (2010) *Nordbomønstre. Dragtsnit fra Middelalderen*. Århus.

Fredell, Å. (2003) *Bildbroar. Figurativ bildkommunikation av ideologi och kosmologi under sydskandinaviens bronsålder och förromersk järnålder*. GOTARC. Serie B. Gothenburg Archaeological Thesis No. 25. Göteborg.

Frei, K. M., Skals, I., Gleba, M. and Lyngstrøm, H. (2009) The Huldremose Iron Age textiles, Denmark: an attempt to define their provenance applying the strontium isotope system. *Journal of Archaeological Science*, V 36, 1965–1971.

Gärtner, B. and Ulriksen, J. (1997) Vester Egesborg-pladsen – en anløbsplads fra vikingetid ved Dybsø Fjord. *Liv og Levn* 11, 17–23. Næstved.

Gärtner, B. and Ulriksen, J. (2000) Mytens ornament. *Skalk* 2000/5, 5–9. Højbjerg.

Gaimster, M. (1998) *Vendel period bracteates on Gotland. On the significance of Germanic art*. Lund.

Garbsch, J. G. (1965) *Die Norische-pannonische Frauentracht im 1. und 2. Jahrhundert*. Münchner Beiträge zur Vor- und Frühgeschichte 11. München.

Garbsch, J. G. (1985) Die Norisch-pannonische Tracht. *Aufstieg und Niedergang der Römischen Welt, Teilband II* 12:3, 546–577. Berlin.

Geijer, A. (1938) *Birka III. Die Textilfunde aus den Gräbern*. Uppsala.

Geijer, A. (1994) *Ur textilkonstens historia*. Stockholm.

Gjessing, G. (1938) Skjoldehamndrakten. En senmiddelaldersk Nordnorsk mannsdrakt. *Viking. Tidsskrift for Norrøn Arkeologi* 2, 27–81.

Gjøl Hagen, K. (1992) *Solplissé – En reminisens av middelalderens draktutvikling? En kompartiv studie i plisserte stoffer fra Birka, Vangsnes, middelalderens Trondheim, Uvdal og Setesdal*. Varia 25. Oslo.

Glæsel, N. (2010) *Viking. Dress, Clothing, Garment*. CreateSpace Independent Publishing Platform.

Gleba, M. and Mannering, U. eds, (2012) *Textiles and Textile Production in Europe from Prehistory to AD 400*. Ancient Textiles Series, vol. 11. Oxford and Oakville.

Göransson, E.-M. (1999) *Bilder av kvinnor och kvinnlighet. Genus och kroppsspråk under övergången till kristendomen*. Stockholm Studies in Archaeology 18. Stockholm.

Goldhahn, J. (2004) Mångtydighetens tydlighet – till frågan om hällbilders mening och innerbörd, in G. Milstreu and H. Prøhl (eds), *Prehistoric Pictures. An Archaeological Source*, 121–135. Tanumshede.

Gotfredsen, L. (1998) *Billedets formsprog*. København.

Gotfredsen, L. and Frederiksen, H. J. (1988) *Troens Billeder. Romansk kunst i Danmark*. Herning.

Granger Taylor, H. (2000) The textiles from Kirbet Qazone (Jordan), in D. Cardon and M. Feugère (eds), *Archéologie des textiles des origines au Ve siècle, Actes du colloque de Lattes, oct. 1999*, 149–162. Monographies Instrumentum 14. Montpellier.

Gustafson, G. (1900) Et fund af figurerede guldplader. *Foreningen til norske Fortidsmindesmerkers bevaring. Aarsberetninger 1986–99*, 87–95. Bergen.

Hårdh, B. (1976) *Wikingerzeitliche Depotfunde aus Südschweden. Probleme und Analysen*. Acta Archaeologica Lundensia series in 8° minore No. 6. Bonn/Lund.

Hägg, I. (1968) Some Notes on the Origin of the Peplos-Type Dress in Scandinavia. *Tor. Meddelanden från Institutionen för nordisk fornkunskap vid Uppsala Universitet*. Vol. XII 1967–68, 81–127.

Hägg, I. (1971) Mantel och kjortel i vikingatidens dräkt. *Fornvännen* 3, 141–153. Stockholm.

Hägg, I. (1974) *Kvinnodräkten i Birka. Livplaggens rekonstruktion på grundlag av det arkeologiska materialet*. AUN 2. Archaeological Studies. Uppsala.

Hägg, I. (1983) Viking women's dress at Birka: A reconstruction by Archaeological methods, in N. B. Harte and K. G. Ponting (eds), *Cloth and Clothing in Medieval Europe: Essays in Memory of Prof. E. M. Carus-Wilson*, 316–350. Pasold Studies in Textile History 2. London.

Hägg, I. (1984a) *Die Textilreste aus dem Hafen von Haithabu*. Berichte Haithabu 20. Neumünster.

Hägg, I. (1984b) Birkas orientaliska praktplagg. *Fornvännen* 78, 204–220. Stockholm.

Hägg, I. (1985) Textilhistoria, statistik och källkritik. *Tor* XX, 259–278. Uppsala.

Hägg, I. (1986) Die Tracht, in G. Arwidsson (ed.), *Birka II:2. Systematische Analysen der Gräberfunde*, pp. 51–72. Stockholm.

Hägg, I. (1988) Textilfunds als Spiegel der Gesellschaft. Erwägungen über das Beispiel Haithabu, in L. Bender Jørgensen and K. Tidow (eds), *Archaeological Textiles in Northern Europe. Report from the 2nd NESAT Symposium*, 187–196. Arkæologiske Skrifter 2. København.

Hägg, I. (1991a) *Textilfunde aus der Siedlung und aus den Gräbern von Haithabu*. Berichte Haithabu 29. Neumünster.

Hägg, I. (1991b) Rangsymboliska element i vikingatida gravar. Hedeby – Birka – Mammen, in M. Iversen (ed.), *Mammen. Grav, kunst og samfund i vikingetid*, 155–162. Jysk Arkæologisk Selskabs Skrifter XXVIII. Højbjerg.

Hägg, I. (1997) Grabtextilien und christliche Symbolik am Beispiel der Funde unter dem Schleswiger Rathausmarkt, in V. Vogel (ed.), *Ausgrabungen in Schleswig. Berichte und Studien 12. Kirche und Gräberfeld des 11.–13. Jahrhunderts unter dem Rathausmarkt von Schleswig*, 88–146. Neumünster.

Hagberg, U. E. (1976) Fundort und Fundgebiet der Modeln aus Torslunda. *Frühmittelalterliche Studien*, 10, 323–349. Berlin.

Hald, M. (1950) *Olddanske Tekstiler*. Nordiske Fortidsminder Serie B, Bind 5. København.

Hald, M. (1972) *Primitive Shoes. An Archaeological-Ethnological Study Based upon Shoe Finds from the Jutland Peninsula*. Publications of the National Museum, Archaeological-Historical Series vol. XIII. København.

Hald, M. (1980) *Ancient Danish Textiles from Bogs and Burials*. Publications of the National Museum, Archaeological-Historical Series vol. XXI. København.

Hannestad, N. (1979) Rome – Ideology and Art, in M. T. Larsen (ed.), *Power and Propaganda. A Symposium on Ancient Empires*. Mesopotamia 7, Copenhagen Studies in Assyriology, 361–389. Copenhagen.

Hansen, E. H. (1991) Nålebinding og brikvævning fra Mammengraven, in M. Iversen (ed.), *Mammen. Grav, kunst og samfund i vikingetid*. Jysk Arkæologisk Selskabs Skrifter XXVIII, 145–148. Højbjerg.

Haraldsen, T. H. (1995) Gullgubben fra Vingrom – på Frigg og Odins Hov? Fåberg og Lillehammer. *Et lokalskrift utgitt av Fåberg Historielag*, 136–146.

Harboe, J. (1992) The Real Image: The Imperial Portrait and the Icon in Early Byzantinum. *Acta Hyperborea 4*, 249–270. København.

Harlow, M. (2004) Female Dress, Third–Sixth Century: The Messages in the Media? *Antiquité Tardive* 12, 203–215.

Haseloff, G. (1981) *Die germanischen Tierornamentik der Völkerwanderungszeit. Studien zur Salin's Stil I*. Berlin.

Haseloff, G. (1986) Bild und Motiv im Nydam-Stil und Stil I, in H. Roth (ed.), *Zum Problem der Deutung frühmittelalterlicher Bildinhalte. Akten des 1. Internationalen Kolloquiums in Marburg a.d. Lahn, 15. bis 19. Februar 1983*, 67–110. Sigmaringen.

Haseloff, G. (1991) *Die germanische Tierornamentik der Völkerwanderungszeit. Studien zu Salin's Stil I*. Band I–III. Berlin.

Hauck, K. (1978) Bildforschung als Historische Sachforschung. Zur vorchristlichen Ikonographie der figuralen Helmprogramme aus der Vendelzeit, in K. Hauck and H. Mordek (eds), *Geschichtsschreibung und Geistiges Leben im Mittelalter*, 27–70. Köln.

Hauck, K. (1986) Methodenfragen der Brakteatendeutung. Erprobung eines Interpretationsmusters für die Bildzeugnisse aus einer oralen Kultur, in H. Roth (ed.), *Zum Problem der Deutung frühmittelalterlicher Bildinhalte. Akten des 1. Internationalen Kolloquiums in Marburg a.d. Lahn, 15. bis 19. Februar 1983*, 273–298. Sigmaringen.

Hauck, K., ed., (1992) *Der historische Horizont der Götterbild-Amulette aus der Übergangsepoche von der Spätantike zum Frühmittelalter*. Berichte über das Colloquium vom 28.11.–1.12. 1988 in der Werner-Reimers-Stiftung. Bad Homburg.

Hauck, K. (1993) Die bremische Überlieferung zur Götter-Dreiheit Altuppsalas und die bornholmischen Goldfolien aus Sorte Muld (Zur Ikonologi der Goldbrakteaten, LII). *Frühmittelalterlichen Studien* 27, 409–479. Berlin.

Hauck, K. (1994) Altuppsalas Polytheismus exemplarisch erhellt mit Bildzeugnissen des 5. – 7. Jahrhunderts (Zur Ikonologie der Goldbrakteaten, LIII), in H. Uecker (ed.), *Studien zum Altgermanischen*. Festschrift für Heinrich Bech, 197–302. Berlin & NewYork.

Hauck, K. (2011) Die Bildformeln der Goldbrakteaten in ihren Leitvarianten (Zur Ikonologie der Goldbrakteaten, LV), in W. Heizmann and M. Axboe (eds), *Die Goldbrakteaten der Völkerwanderungszeit – Auswertung und Neufunde*. Ergänzungsbände zum Reallexikon der Germanischen Altertumskunde 40, 61–152. Berlin and New York.

Hauck, K. Axboe, M., Düwel, K., Padberd, L. von, Smyra, U. and Wypior, C. (1985a) *Die Goldbrakteaten der Völkerwanderungszeit. 1,1 Ikonographischer Katalog. Einleitung*. München.

Hauck, K., Axboe, M., Clavadetscher, U., Düwel, K. and Padberd, L. von (1985b) *Die Goldbrakteaten der Völkerwanderungszeit. 1,2 Ikonographischer Katalog (IK 1, Text)*. München.

Hauck, K., Lange, H. and Padberg, L. von (1985c) *Die Goldbrakteaten der Völkerwanderungszeit. 1,3 Ikonographischer Katalog. (IK 1, Tafeln)*. München.

Hauck, K., Axboe, M., Düvel, K., Padberg, L. von and Wypior, C. (1986a) *Die Goldbrakteaten der Völkerwanderungszeit. 2,1 Ikonographischer Katalog. (IK 2, Text)*. München.

Hauck, K., Lange, H. and Padberg, L. von (1986b) *Die Goldbrakteaten der Völkerwanderungszeit. 2,2 Ikonographischer Katalog. (IK 2, Tafeln)*. München.

Hauck, K., Axboe, M., Düvel, K., Padberg, L. von and Rulffs, H. (1989a) *Die Goldbrakteaten der Völkerwanderungszeit. 3,1 Ikonographischer Katalog. Text*. München.

Hauck, K., Axboe, M., Lange, H. and Padberg, L. von (1989b) *Die Goldbrakteaten der Völkerwanderungszeit. 3,1 Ikonographischer Katalog. (IK 3, Tafeln)*. München.

Hauck, K. and Mordek, H. (1978) *Geschichtsschreibung und Geistiges Leben im Mittelalter*. Köln.

Hawkes, S. C., Davidson, H. R. E. and Hawkes, C. (1965) The Finglesham Man. *Antiquity 39*, 17–32. Cambridge.

Hedeager, L. (1988) *Danernes Land. Fra ca. år 200 f.Kr. – ca. 700 e.Kr*. Gyldendal og Politikens Danmarkshistorie, Bind 2. København.

Hedeager, L. (1997) *Skygger af en anden virkelighed. Oldnordiske myter*. Haslev.

Hedeager, L. (1999) Myth and art: a passport to political authority in Scandinavia during the Migration Period. *Anglo-Saxon Studies in Archaeology and History* 10, 151–156.

Hedeager, L. (2000) Migration Period Europe: the formation of a political mentality, in F. Theuws and J. L. Nelson (eds), *Rituals of Power. From Late Antiquity to the Early Middle Ages*, 15–57. Brill, Leiden.

Hedeager, L. (2002) Scandinavian 'Central places' in a Cosmological Setting, in B. Hårdh and L. Larsson (eds), *Central Places in the Migration and the Merovingian Periods*. Uppåkrastudier 6. Acta Archaeologica Lundensia, Series 8^0, No. 39, 3–18. Lund.

Hedeager Krag, A. (1994) Reconstruction of a Viking Magnate Dress, in G. Jaacks and K. Tidow (eds), *Archaeological Textiles. NESAT V. Textilsymposium Neumünster*, 114–119. Neumüster.

Hedeager Krag, A. (1998) Dress and Power in Prehistoric Scandinavia, c. 550–1050 AD, in L. Bender Jørgensen and C. Rinaldo (eds), *Textiles in European Archaeology. Report from the 6th NESAT Symposium*, 125–130. GOTARC serie A, bind 1. Göteborg.

Hedeager Krag, A. (1999) Fränkisch-byzantinische Trachteinflüsse in drei Dänischen Grabfunden des 10. Jahrhunderts. *Archäologisches Korrespondenzblatt* 29, 425–444.

Hedeager Krag, A. (2005) Dress and Fashion in Denmark's Viking Age, in F. Pritchard and J. P. Wild (eds), *North European Symposium for Archaeological Textiles VII*, 29–35. Edinburgh.

Hedeager Krag, A., ed., (2003) *Dragt og magt*. København.

Hedeager Krag, A. and Ræder Knudsen, L. (1999) Vikingetidstekstiler. Nye opdagelser fra gravfundene i Hvilehøj og Hørning. *Nationalmuseets arbejdsmark 1999*, 159–168. København.

Hedeager Madsen, A. (1990) Women's dress in the Viking period in Denmark, based on the tortoise brooches and textile remains, in P. Walton and J.P. Wild (eds), *Textiles in Northern Archaeology. NESAT III: Textile Symposium in York*, 101–106. London.

Heizmann, W. and Axboe, M., eds, (2011) *Die Goldbrakteaten der Völerwanderungszeit – Auswertung und Neufunde*. Ergänzungsbände zum Reallexikon der Germanischen Altertumskunde 40. Berlin and New York.

Helgesson, B. (1998) Vad är centralt? – fenomen och funktion; lokalisering och person, in L. Larsson and B. Hårdh (eds), *Centrala Pladser Centrala Frågor. Samhällsstrukturen under Järnåldern*, 39–46. Lund.

Hellerström, S., ed., (2000) *Väg 11. Sträckan Östra Tommarp-Simrishamn*. Riksantikvarieämbetet. UV Syd Rapport 2000:25. Lund.

Helmbrecht, M. (2011) *Wirkmächtige Kommunikationsmedien. Menschenbilder der Vendel- und Wikingerzeit und ihre Kontexte.* Acta Archaeologica Lundensia Series Prima 4°, No. 40. Lund.

Henriksen, P. S. and Harild, J. A. (2005) *Hør-industrien på Fyn i yngre Jernalder*. NNU Rapport no. 13. Nationalmuseet, København.

Herschend, F. (1995) Hus på Helgö. *Fornvännen 90*, 221–228. Stockholm.

Hildebrand, H. (1880) Från jernåldern. *Kongl. Vitterhets Historie och Antiqvitets Akademiens Månedsblad,* Årg. 98, 17–23. Stockholm.

Hines, J. (1993) *Clasps, Hegtespenner, Agraffen. Scandinavian clasps of classes A–C of the 3rd to 6th centuries A.D.* Kungl. Vitterhets Historie och Antikvitete Akademien. Stockholm.

Hodder, I. (1992) *Theory and Practice in Archaeology*. London & New York.

Holck, P. (1988) Myrfunnet fra Skjoldehamn – Mannlig Same Eller Norrøn Kvinne? *Viking, Tidsskrift for Norrøn Arkeologi,* vol. 51, 109–115.

Høigaard Hofseth, E. (1998) Valgets kval. To rekonstruksjoner fra folkevandringstid. *Universitets Oldsaksamling Årbok* 1997/1998, 93–112. Oslo.

Høilund Nielsen, K. (1987) Zur Chronologie der jüngeren germanischen Eisenzeit auf Bornholm. Untersuchungen zu Schmuckgarnituren. *Acta Archaeologica* 57, 47–86. København.

Høilund Nielsen, K. (1988) Dyreornamentik. Social indikation eller modefænomen? in T. Madsen (ed.), *Bag Moesgårds maske. Kultur og samfund i fortid og nutid*, 137–144. Aarhus.

Høilund Nielsen, K. (1998) Animal Style – A Symbol of Might an Myth. Salin's Style II in a European Context. *Acta Archaeologica* 69, 1–52. København.

Høilund Nielsen, K. (1999) Ulvekrigeren. Dyresymbolik på våbenudstyr fra 6.–7. århundrede, in O. Høiris, H. J. Madsen, T. Madsen and J. Vellev (eds), *Menneskelivets mangfoldighed. Arkæologisk og antropologisk forskning på Moesgård*, 327–334. Moesgård.

Høilund Nielsen, K. (2002a) Fra antikristne symboler til "ophitisk kunstsmag". Dyrestil i oldtid og nutid. *Hikuin* 29, 7–14.

Høilund Nielsen, K. (2002b) Ulv, hest og drage. Ikonografisk analyse af dyrene i stil II–III. *Hikuin* 29, 187–218.

Høilund Nielsen, K. (2002c) Den anden germanske dyrestil – og dens vej til Norge. *Frå haug ok heidni* 1, 18–27.

Hölscher, T. (2000) Bildwerke: Darstellungen, Funktionen, Botschaften, in A. H. Borbein and T. Hölscher (eds), *Klassische Archäologie. Eine Einführung*, 146–165. Darmstadt.

Holmqvist, W. (1952) Ryttaren från Möjbro. *Fornvännen* 47, Häfte 1–2, 241–261. Stockholm.

Holmqvist, W. (1955) *Germanic Art During the First Millennium A.D.* Stockholm.

Holmqvist, W. (1960) The Dancing Gods. *Acta Archaeologica* 31, 101–127. København 1961.

Holmqvist, W., ed., (1961) *Excavations at Helgö I. Report for 1954–1956*. Stockholm.

Holmqvist, W., ed., (1970) *Excavations at Helgö III. Report for 1960–1964*. Stockholm.

Holmqvist, W. (1977) *Vår tidiga konst*. Stockholm.

Holmqvist, W. and Arrhenius, B., eds, (1964) *Excavations at Helgö II. Report for 1957–1959*. Stockholm.

Horneij, R. (1991) *Bonaderna från Överhogdal*. Östersund.

Hougen, B. (1933) Helgelandsfundet. Et myrfund av tekstiler fra eldre jernalder. *Stavanger Museums Årshefte 1930–32*, årg. 41–42, 55–75.

Hougen, B. (1935) *Snartemofunnene. Studier i folkevandringstidens ornamentikk og tekstil-historie*. Oslo.

Hougen, B. (1940) Osebergfunnets billedvev. *Viking IV*, 85–124. Oslo.

Hougen, B. (1967) *The Migration Style of Ornament in Norway*. Oslo.

Hundt, H.-J. (1987) Nachtrag zu den Textilresten aus dem fränkischen Knabengrab unter dem Kölner Dom. *Kölner Domblatt* 52, 237–243. Jahrbuch des Zentral-Dombau-Vereins. Köln.

Ingstad, A. S. (1988) Textiles from Oseberg, Gokstad and Kaupang, in L. Bender Jørgensen and K. Tidow (eds), *Archaeological Textiles in Northern Europe. Report from the 2nd NESAT Symposium*, 133–148. Arkæologiske Skrifter 2. København.

Ingstad, A. S. (1992) Oseberg dronningen-hvem var hun? in A. E. Christensen, A. S. Ingstad and B. Myhre (eds), *Oseberg dronningens grave: vår arkeologiske nasjonalskatt i nytt lys*, 224–257. Oslo.

Iversen, M. ed., (1991) *Mammen. Grav, kunst og samfund i vikingetid.* Jysk Arkæologisk Selskabs Skrifter XXVIII. Højbjerg.

Jaacks, G. and Tidow, K., eds, (1993) *Archaeological Textiles. NESAT V. Textilsymposium Neumünster*. Neumüster.

James, E. (1988) *The Franks*. Oxford.

James, E. (1997) The militarization of Roman society, 400–700, in A. Nørgård Jørgensen and B. Clausen (eds), *Military Aspects of the Scandinavian Society. In a European Perspective AD1–1300*, 19–24. PNM Studies in Archaeology & History, vol. 2. København.

Jansson, K. (2003) Tolv graver och en guldgubbe från Visingsö. *Fornvännen* 98, 127–130.

Jensen, C. K. and Høilund Nielsen, K. (1997) Burial data and Correspondence Aalysis, in C. K. Jensen and K. Høilund Nielsen (eds), *Burial and Society. The Chronological and Social Analysis of Archaeological Burial Data,* 29–61. Århus.

Jørgensen, A. Nørgård (1999) *Waffen und Gräber. Typologische und chronologische Studien zu skandinavischen Waffengräber 520/30 bis 900 n. Chr.* Nordiske Fortidsminder Serie B, vol. 17. København.

Jørgensen, L. (1990) *Bækkegård and Glasergård. Two cemeteries from the Late Iron Age on Bornholm*. Arkæologiske Studier vol. VIII. København.

Jørgensen, L. (1997) En stormandsgård fra vikingetiden. *Nyt fra Nationalmuseet* No. 77, 10. København.

Jørgensen, L. (1999) Fra nutidens pløjelag til jernalderens samfund – stormænd og hånedværkere i Gudme. Årbog for Svendborg og Omegns Museum 1998, 8–21. Svendborg.

Jørgensen, L. (2005) Hov og hørg ved Tissø, in T. Capelle and C. Fischer (eds), *Ragnarok Odins Verden*, 131–142. Silkeborg.

Jørgensen, L. and Jørgensen, A. N. (1997) *Nørre Sandegård Vest. A Cemetery from the 6th–8th Centuries on Bornholm*. Nordiske Fortidsminder Serie B, Bind 14. København.

Jørgensen, L., Bican, J. F., Thomsen, L. G. and Jensen, X. P. (2004) Stormænd, købmænd og håndværkere ved Tissø i det 6.–11. årh. *Fra Holbæk Amt 2003/2004*, 55–65.

Jørgensen, L. and Petersen, P. V. (1998) *Guld, Magt og Tro. Danske guldskatte fra oldtid og middelalder*. København.

Kaul, F. (1998) *Ships on Bronzes. A Study in Bronze Age Religion and Iconography*. Publications from the National Museum. Studies in Archaeology and History 3. Copenhagen.

Kaul, F. (2004) Bronzealderens ikonografiske motiver og deres fremkomst i en formativ fase. Skibet, det vifteformede symbol, hesten, hjulkorset, fisken og slangen, in G. Milstreu and H. Prøhl (eds), *Prehistoric Pictures. An Archaeological Source*, 85–119. Tanumshede.

Kippenberg, H. G. ed., (1990) *Genres in visual Representations. Visible Religion* vol. VII. Leiden.

Kjørup, S. (1998) Billedkommunikation, in B. Fausing and P. Larsen (eds), *Visuel kommunikation* 1, 58–82. København.

Kjærum, P. and Olsen, R.A., eds, (1990) *Oldtidens ansigt*. København.

Klindt-Jensen, O. and Wilson, D. M. (1965) *Vikingetidens kunst*. København.

Knauer, E. R. (1978) Towards the History of the Sleeved Coat. *Expedition* 21, no. 1, 18–36. Pennsylvania.

Knauer, E. R. (1985) *Ex oriente vestimenta*. Trachtgeschichtliche Beobachtungen zu Ärmelmantel und Ärmeljacke. Aufstieg und Niedergang der Römischen Welt, Teilband II 12,3, 578–741. Berlin.

Knauer, E. R. (2004) A Quest for the Origin of the Persian Riding-Coats: Sleeved Garments with Underarm Openings, in C. Fluck and G. Vogelsang-Eastwood (eds), *Riding Costume in Egypt. Origin and Appearance*, 7–28. Leiden & Boston.

Koch, R. (1985) Die Tracht der Alamannen in der Spätantike. *Aufstieg und Niedergang der Römischen Welt*, Teilband II 12,3, 456–577. Berlin.

Koch, U. (2001) Das Gold der Barbarenfürsten. *AiD* 1/2001, 62–65.

Kolstrup, I.-L. (1993) Det nøgne look. *Passepartout: skrifter for kunsthistorie*, Nr. 2, 1. årgang vol. 1, 1–2, 121–136.

Kolstrup, I.-L. (1995) *Magi og Kunst*. Skippershoved.

Kondakov, N. P. (1924) Les Costumes orientaux à la Cour Byzantine. *Byzantion. Revue Internationale des Études Byzantines*. vol. I, 7–49. Paris.

Krafft, S. (1955) *Osebergfunnets tekstiler. Fragmenter og rekonstruerte mønstre*. Oslo.

Kristoffersen, S. (2002) Den første germaniske dyrestilen i Rogaland. *Frå haug ok heidni* 1, 12–17.

Kromann, A, and Watt, M. (1984) Skattefundet fra Smørenge. En nedgravet skat fra folkevandringstid på Bornholm. *Nationalmuseets Arbejdsmark*, 29–41. København.

Lamm, J. P. (1997) Jackpot för Efterundersökningsprojektet. *En årskrönika från Statens historiska museer 1996*, 18–19. Borås.

Lamm, J. P. (2004) Figural Gold Foils Found in Sweden: a study based on the discoveries from Helgö, in B. Gyllensvärd, P. Harbison, M. Axboe, J. P. Lamm, T. Zachrisson and S. Reisborg (eds), *Excavations at Helgö XVI. Exotic and Sacral Finds from Helgö*, 41–142. Stockholm.

Lamm, J. P., Hydman, H., Axboe, M., Hauck, K., Beck, H., Behr, C. and Pesch, A. (2000) Der Brakteat des Jahrhunderts. *Frühmittelalterliche Studien* 34, 1–93.

Langner, L. (1991) *The Importance of Wearing Clothes*. Los Angeles.

Laporte, J. P. and Boyer, R. (1991) *Tresors de Celles: Sepultures et reliques de la Reine Bathilde et de l'abesse Bertille*. Chelles.

Larsson, L. ed., (2004) *Continuity for Centuries. A ceremonial building and its context at Uppåkra, southern Sweden*. Uppåkrastudier 10. Acta Archaeologica Lundensia, Series 8^0, No. 30. Lund.

Larsson, L. (2005) Kulthuset i Uppåkra, in T. Capelle and C. Fischer (eds), *Ragnarok Odins Verden*, 107–118. Silkeborg.

Larsson, A. (2007) Klädd krigare: skifte i skandinaviskt dräktskick kring år 1000. Occasional Papers in Archeology 39. Uppsala.

Larsson, L. and Hårdh, B. (1998a) Uppåkra – et hövdinga- eller kungasäte. *Fornvännen* 92, 139–154. Stockholm.

Larsson, L., and Hårdh, B. eds, (1998b) *Centrala Pladser Centrala Frågor. Samhällsstrukturen under Järnåldern*. Lund.

Larsson, L. and Lenntorp, K.-M. (2004) The Enimagic House, in L. Larsson (ed.), *Continuity for Centuries. A Ceremonial Building and its Context at Uppåkra, Southern Sweden*, 3–48. Uppåkrastudier 10. Acta Archaeologica Lundensia, Series 8^0, No. 30. Lund.

Lidén, H.-E. (1969) From Pagan Sanctuary to Christian Church. The Excavation of Mære Church in Trøndelag. *Norwegian Archaeological Review*, vol. 2, 3–32. Oslo.

Lindqvist, S. (1936) *Uppsala Högar och Ottarshögen*. Stockholm.

Lindqvist, S., von Post, L., von Walterstoff, E. and Meyer, E. (1925) *Bronsåldersmanteln från Gerumsberget*. Stockholm.

Lund Hansen, U. (1976) En af de ypperste – en rig romertidsgrav fra Bennebo Mark, in *Festskrift til K. G. Kaaber, Fra Holbæk Amt 69, Årbog for Historisk Samfund for Holbæk Amt 1976*, 63–76. Holbæk.

Lund Hansen, U. (2001) Jüngere Kaiserzeit und Völkerwanderungszeit. Heimisches Material – fremder Rahmen. Spätkaiserzeitliche Tradition und neue Tendenzen innerhalb der völkerwanderungszeitlichen Chronologie – oder wie sich das kontinentale Material auf die Dänische Chronologie auswirkt, in M. Meyer (ed.), *"...Trans Albium Fluvium" Forschungen zur vorrömischen, kaiserzeitlichen und mittelalterlichen Archäologie. Festschrift für Achim Leube zur 65. Geburtstag*, 349–357. Leidorf.

Lund Hansen, U. (1995) *Himlingøje – Seeland – Europa. Ein Gräberfeld der jüngeren römischen Kaiserzeit auf Seeland, seiner Bedeutung und internationalen Beziehungen*. Nordiske Fortidsminder Serie B, vol. 13. København.

Lundqvist, L. (1995) Gubber og granater. Skalk 1995/1, 10–13. Højbjerg.

Lundqvist, L. (1996a) Slöinge- en stormansgård från järnåldern, in L. Lundqvist, K. Lindeblad, A. Nielsen and L. Ersgård, *Slöinge och Borg. Stormansgårdar i öst och väst*. Arkeologiska undersökningar. Skrifter No. 18, 9–52. Linköping.

Lundqvist, L. (1996b) Slöingeprojekt 1994 och 1995 – stolpar, guldgubbar och bebyggelse. *Fornvännen* 91. Stockholm.

Lundqvist, L. (1996c) Om fyra års fältarbete utanför Slöinge. *Hallandsbygd 1995–1996*, 5–10. Varberg.

Lundqvist, L. (1996d) Ett 700-talshus med offrade guldgubbar. Om Slöinge-grävningarna 1994–95. *Fynd 1/95. Tidskrift för Göteborgs Stadsmuseum och Fornminnesforeningen i Göteborg*, 59–63. Göteborg.

Lundqvist, L. (1997) Slöinge: om et pågående projekt, in J. Callmer and E. Rosengren (eds), *"...gick Grendel at söka det höga huset...". Arkeologiska källor till aristokratiska miljöer i Skandinavien under yngre järnålder*. GOTARC C. Arkeologiska Skrifter No. 17, 89–98.

Lundqvist, L., Callmer, J. and Rosengren, E. (1993) En fynfplats med guldgubbar vid Slöinge, Halland. *Fornvännen* 88, 65–70. Stockholm.

Lundqvist, L., Lindeblad, K., Nielsen, A. and Ersgård, L. (1996) *Slöinge och Borg. Stormansgårdar i öst och väst.* Riksantikvarieämbetet. Arkeologiska undersökningar. Skrifter Nr. 18. Linköping.

Lyngstrøm, H. (1989) Lousgaard – éthundrede år efter J.A. Jørgensen og E. Vedel. *Aarbøger for Nordisk Oldkyndighed og Historie 1989*, 115–170. København.

MacGregor, A. (1982) *Anglo-Scandinavian Finds from Lloyds Bank, Pavement, and Other sites.* The Archaeology of York, vol. 17: The small finds. London.

Mackeprang, M. B. (1935) Menschendarstellungen aus der Eisenzeit Dänemarks. *Acta Archaeologica* 6, 228–249. København.

Mackeprang, M. B. (1943) Om de saakaldte "guldgubber". *Nationalmuseets Arbejdsmark*, 69–76. København.

Mackeprang, M. B. (1952) *De nordiske guldbrakteater.* Jysk Arkæologisk Selskabs Skrifter II. Århus.

Madsen, T. ed., (1988) *Bag Moesgårds maske. Kultur og samfund i fortid og nutid.* Aarhus.

Magnus, B. (1982) How was he dressed? New light on the garments from the grave at Evebø/Eide in Gloppen, Norway. *Studien zur Sachsenforschung* 4, 293–313. Hildesheim.

Maik, J. ed., (2004) *Priceless Invention of Humanity – Textiles, NESAT VIII.* Acta Archaeologica Lodziensia Nr. 50/1. Łódź.

Malmer, M. P. (1963) *Metodproblem inom Järnålderns konsthistoria.* Acta Archaeologica Lundensia Series 8°, No. 3. Lund.

Malmius, A. (1997) Textilfragmenten i båtgrav 5, Valsgärde, en första analys, in B. Arrhenius and G. Eriksson (eds), *SIV Svealand i vendel och vikingatid*: Studier från delprojekten vid Stockholms Universitet, 42–57. Stockholm.

Malmius, A. (1998) The textile fragments in boat-grave 5, Valsgärde, Old Uppsala parish, Uppland. *Laborativ Arkeologi* 10–11, 65–80.

Malmius, A. (2003) Textilfragmenten i båtgrav 5, Valsgärde, en tolkning av en mandsdräkt, in A. Hedeager Krag (ed.), *Dragt og magt*, 46–61. København.

Mannering, U. (1997) The Textiles from Nørre Sandegård Vest, in L. Jørgensen and A. N. Jørgensen., *Nørre Sandegård Vest. A Cemetery from the 6th– 8th Centuries on Bornholm*, 118–140. Nordiske Fortidsminder Serie B, Bind 14. København.

Mannering, U. (1998a) *Guldgubber. Et billede af yngre jernalders dragt.* Upubliceret cand.phil.-speciale fra Københavns Universitet (Unpublished Master's thesis, University of Copenhagen).

Mannering, U. (1998b) The Textiles from Nørre Sandegaard Vest, in L. Bender Jørgensen and C. Rinaldo (eds), *Textiles in European Archaeology. Report from the 6th NESAT Symposium*, 107–115. GOTARC serie A, bind 1. Göteborg.

Mannering, U. (1999) Sidste skrig. *Skalk* 1999/4, 20–27. Højbjerg.

Mannering, U. (2000) Roman Garments from Mons Claudianus, in D. Cardon and M. Feugère (eds), *Archéologie des textiles des origines au Ve siècle, Actes du colloque de Lattes, oct. 1999*, 283–290. Monographies Instrumentum 14. Montpellier.

Mannering, U. (2004) Dress in Scandinavian Iconography of the 5th–10th centuries AD, in J. Maik (ed.), *Priceless Invention of Humanity – Textiles, NESAT VIII*, 67–74. Acta Archaeologica Lodziensia Nr. 50/1. Łódź.

Mannering, U. (2008) Iconography and Costume from the Late Iron Age in Scandinavia, in M. Gleba, C. Munkholt and M.-L. Nosch (eds), *Dressing the Past,* 59–67. Ancient Textiles Series, vol. 3. Oxford.

Mannering, U. (2010) The Huldremose find. An Early Iron Age woman with an exceptional costume *Fasciculi Archaeologiae Historicae* XXIII, 15–24. Łódź.

Mannering, U. (2011) Early Iron Age Craftsmanship from a Costume Perspective, in L. Boye (ed.), *Arkæologi i Slesvig/ Archäologie in Schleswig,* Sonderband "Det 61. Internationale Sachsensymposion 2010", 85–94. Haderslev.

Mannering, U. (2013) Man or woman? – perception of gender through costume. *Danish Journal of Archaeology* 2:1, 79–86.

Mannering, U. and Andersson Strand, E. (2009) Dress Images on Gold-foil Figures, in C. Adamsen, U. Lund Hansen, F. O. Nielsen and M. Watt (eds), *Sorte Muld. Wealth, Power and Religion at an Iron Age Central Settlement on Bornholm*, 54–61. Bornholms Museum, Rønne.

Mannering, U., Possnert, G., Heinemeier, J. and Gleba M. (2010) Dating Danish textiles and skins from bog finds by means of ^{14}C AMS, *Journal of Archaeological Science* 37, 261–268.

Mannering, U., Gleba, M. and Bloch Hansen, M. (2012) Denmark, in M. Gleba and U. Mannering (eds), *Textiles and Textile Production in Europe from Prehistory to AD 400*, Ancient Textiles Series, vol. 11, 91–118. Oxford and Oakville.

Mannering, U. and Ræder Knudsen, L. (2013) Hammerum, The Find of the Century, in J. Banck-Burgess and C. Nübold (eds), *NESAT XI, The North European Symposium for Archaeological Textiles XI*, 157–160. Leidorf.

Marxen, I. and Moltke, E. (1978) Jellingmanden. Danmarks ældste figurmaleri. *Nationalmuseets Arbeijdsmark*, 111–118. København.

Melle, J. von (1725/1997) Commentatiuncula de simulachris aureis, quæ in Bornholmia, maris balthici insula, agris eruuntur, oversættelse, in J. Sørensen, E. Thykier and M. Watt, *Jacob von Melle og de Bornholmske guldgubber*. Rønne.

Milstreu, G. and Prøhl, H., eds, (2004) *Prehistoric Pictures. An Archaeological Source.* Tanumshede.

Mirzoeff, N. (2001) *An Introduction to Visual Culture*. London.

Mitchell, W. J. T. (1986) *Iconology: Image, Text, Ideology*. Chicago & London.

Møllenhus, K. R. (1967) Gullgubbene fra Mære Kirke. *Viking. Tidsskrift for Norrøn Arkeologi* 31, 163–165. Oslo.

Møller Hansen, K., Veien Christiansen, D. and Høier, H. (1999) Næs – Et nyt aspekt af vikingetidens erhverv. *Kulturhistoriske studier*, 7–35. Sydsjællands Museum.

Möller-Wiering, S. (2006) The Bog Body from Obenaltendorf, *Archaeological Textiles Newsletter* 43, 9–12.

Möller-Wiering, S. (2011) *War and Worship. Textile from 3rd to 4th-century AD weapon Deposits in Denmark and Northern Germany.* Ancient Textiles Series vol. 9. Oxford and Oakville.

Montelius, O. (1869) *Från Jernåldern*. Stockholm.

Mortensen, P. and Rasmussen, B. M., eds, (1991) *Fra Stamme til Stat i Danmark 2. Høvdingesamfund og Kongemagt*. Jysk Arkæologisk Selskabs Skrifter XXII:2. Højbjerg.

Müller, M. (2003) *Die Kleidung nach Quellen des frühen Mittelalters*. Berlin.

Munksgård, E. (1974) *Oldtidsdragter*. København.

Munksgård, E. (1976) Frisurer fra ældre jernalder. Mosefund og gravfund. Årbøger for nordisk oldkyndighed og historie, 5–21. København.

Munksgård, E. (1978) Brakteaten – Archäologisches. *Reallexikon der Germanischen Altertumskunde* 3, 338–341.

Munksgård, E. (1989) Vort ældste kongeportræt? Knud den Store i Liber vitae. Årbøger for nordisk oldkyndighed og historie, 171–177. København.

Munksgård, E. (1990a) The costumes depicted on gold-sheet figures (guldgubbar), in P. Walton and J. P. Wild (eds), *Textiles in Northern Archaeology. NESAT III: Textile Symposium in York* 97–100. London.

Munksgård, E. (1990b) A Tandem accelerator dating, in P. Walton and J. P. Wild (eds), *Textiles in Northern Archaeology. NESAT III: Textile Symposium in York*, 211–214. London.

Munksgård, E. (1991) Kopien af dragten fra Mammengraven, in M. Iversen (ed.), *Mammen. Grav, kunst og samfund i vikingetid*, 151–154. Jysk Arkæologisk Selskabs Skrifter XXVIII. Højbjerg.

Munksgård, E. and Østergård, E. (1988) Textiles and costume from Lønne Hede. An early Roman Iron Age Burial, in L. Bender Jørgensen and K. Tidow (eds), *Archaeological Textiles in Northern Europe. Report from the 2nd NESAT Symposium*, 53–64. Arkæologiske Skrifter 2. København.

Muthesius, A. (1982) The silk fragment from 5 Coppergate, in A. MacGregor, *Anglo-Scandinavian Finds from Lloyds Bank, Pavement, and Other sites*, 132–136. The Archaeology of York, vol. 17: The small finds. London.

Nielsen, H. (1985) Guldsmedekunsten. *Skalk* 1985/6, 14–15. Højbjerg.

Näsman, U. (1998) Sydskandinavisk samhällsstruktur i ljuset av merovingisk och anglosaxisk analogi eller i vad är det som scentralplatserna är centrala?, in L. Larsson and B. Hårdh (eds), *Centrala Pladser Centrala Frågor. Samhällsstrukturen under Järnåldern*, 1–26. Lund.

Nielsen, F. O. S. (1999) *Bornholms Museums antikvariske arbejde 1998–1999*, 57–92. Bornholms Museum.

Nielsen, H. (1997) Ett regionalt rigedomscenter i Sydvestsjælland, in J. Callmer and E. Rosengren (eds), *"...gick Grendel at söka det höga huset...". Arkeologiska källor till aristokratiska miljöer i Skandinavien under yngre järnålder*, 55–70. GOTARC C. Arkeologiska Skrifter No. 17. Halmstad.

Nilsen, M. V. (1997) Middelalderdrakt i lys av kunstog arkeologisk materiale. *Spor – fortidsnyt fra Midt-Norge* 1997/2, 20–21.

Nilsson, T. (1992) Stentinget. *Skalk* 1992/4, 9. Højbjerg.

Nilsson, T. (1994) Stentinget – en boplads med handel og håndværk fra germansk jernalder og vikingetid. *Vendsyssel Nu & Da 1991–1993*, 64–77. Vendsyssel.

Nockert, M. (1985/1997) *Bockstensmannen och hans dräkt*. Falkenberg.

Nockert, M. (1991) *The Högom Find and other Migration Period Textiles and Costumes in Scandinavia*. Archaeology and environment 9. Umeå.

Nockert, M. (1994) Textilerna från Tuna i Badelunda, in E. Nylén and B. Schönbäck, *Tuna i badelunda. Guld Kvinnor Båtar*, Part 1, 112–117.Västerås kulturnämnds skriftserie. Västerås.

Nockert, M. and Franzén, A. M. (1992) *Bonanderna från Skog och Överhogdal och andra medeltida väggbeklädnader*. Stockholm.

Nockert, M. and Possnert, G. (2002) *Att datera textilier*. Oslo.

Nørgård Jørgensen, A. (1999a) *Waffen und Gräber. Typologische und chronologische Studien zu skandinavischen Waffengräbern 520/30 bis 900 n.Chr.* Nordiske Fortidsminder Serie B, Bind 17. København.

Nørgård Jørgensen, A. (1999b) A peaceful discussion of a martial topic: the chronology of Scandinavian weapon graves, in J. Hines, K. Høilund Nielsen and F. Siegmund (eds), *The Pace of Change. Studies in Early-Medieval Chronology*, 148–159. Exeter.

Nørlund, P. (1935) *De gamle nordbobygder ved verdens ende. Skildringer fra Grønlands Middelalder*. Folkelæsning Nr. 361 årg. 1935. København.

Nørlund, P. (1941) Klædedragt i oldtid og middelalder. *Nordisk Kultur XV:B. Dragt*, 1–88. Stockholm.

Nordahl, E. (2001) Båtgravar i Gamla Uppsala. Spår av en vikingatida högreståndsmiljö. *AUN* 29. Uppsala.

Nordbladh, J. (2004) Hällristningar som allmän bild – och som speciell, in G. Milstreu and H. Prøhl (eds), *Prehistoric Pictures. An Archaeological Source*, 63–67. Tanumshede.

Nordén, A. (1938) Le problème des "Bonshommes en or". *Acta Archaeologica* 9, 151–163. København.

Nylén, E. and Schönbäck, B. (1994) *Tuna i badelunda. Guld Kvinnor Båtar*. Part 1. Västerås kulturnämnds skriftserie. Västerås.

Ørsnes, M. (1966) *Form og Stil i Sydskandinaviens yngre germanske jernalder*. København.

Østergård, E. (1987) Nordboernes tøj – dagligdragten i middelalderen, in B. Wittgren (ed.), *Textila tekniker i nordisk tradition. Rapport från nordiskt symposium om textila tekniker 1986*. Etnolore 6, 95–104.

Østergård, E. (1991) Tekstilfragmenterne fra Mammengraven, in M. Iversen (ed.), *Mammen. Grav, kunst og samfund i vikingetid*, 123–138. Jysk Arkæologisk Selskabs Skrifter XXVIII. Højbjerg.

Østergård, E. (2003) *Som syet til jorden. Tekstilfund fra det norrøne Grønland*. Århus.

Østergård, E. (2004) *Woven into the Earth: textile finds in Norse Greenland*. Aarhus.

Olsen, B. (1997) *Fra ting til tekst. Teoretiske perspektiv i arkeologisk forskning*. Oslo.

Olsen, M. (1909) Fra gammelnorsk myte og kultus. *Maal og minne*, 17–36.

Olsen, M. and Schetelig, H. (1909) De to runestener ra Tu og Klepp paa Jæderen. *Bergens Museums Aarbog* No. 11, 2–29. Bergen.

Opedal, A. (1998) *De glemte skipsgravene. Makt og myter på Avaldsnes*. Stavanger.

Oscarsson, U. (1994) *The Överhogdal Weaves*. Östersund.

Owen-Crocker, G. R. (1986/2004) *Dress in Anglo-Saxon England*. Manchester.

Panofsky, E. (1998) Ikonografi og ikonologi, in B. Fausing and P. Larsen (eds), *Visuel kommunikation* 1, 9–21. København.

Pentz, P., Baastrup, M. P., Karg, S. and Mannering, U. (2009) Kong Haralds vølve, *Nationalmuseets arbejdsmark*, 215–232.

Pesch, A. (2002a) Uppåkra im Licht der Formular-Familien der Völkerwanderungszeitlichen Goldbrakteaten, in B. Hårdh and L. Larsson (eds), *Central Places in the Migration and the Merovingian Periods*, 55–78. Uppåkrastudier 6. Acta Archaeologica Lundensia, Series 8⁰, No. 39. Lund.

Pesch, A. (2002b) Frauen und Brakteaten – eine Skizze, in R. Simek and W. Heizmann (eds), *Mythological Woman. Studies in Memory of Lotte Motz*, 33–80. Wien.

Pesch, A. (2005) På glatis med ikonografi!, in T. Capelle and C. Fischer (eds), *Ragnarok Odins Verden*, 25–32. Silkeborg.

Petersen, P. V. (1992) Danefæ 1991. Oldtid. *Arkæologiske Udgravninger i Danmark 1991*, 208–219. København.

Petersen, P. V. (1993a) Danefæ 1992. Oldtid. *Arkæologiske Udgravninger i Danmark 1992*, 226–237. København.

Petersen, P. V. (1993b) Valkyrier i Ribe. *By, marsk og geest 5. Årsberetning 1992*, 41–46. Den antikvariske Samling i Ribe.

Petersen, P. V. (2005) Odins fugle, valkyrier og bersærker, in T. Capelle and C. Fischer (eds), *Ragnarok Odins Verden*, 57–86. Silkeborg.

Plicht, J. v. d., Sanden, W. A. B. v. d., Aerts, A.T., and Streurman, H. J. (2004) Dating bog bodies by means of ^{14}C-AMS, *Journal of Archaeological Science* 31, 471–491.

Pochat, G. (1977) *Symbolbegreppet i konstvetenskapen*. Antikvariska Serien 30. Stockholm.

Ræder Knudsen, L. (1998) An Iron Age Cloak with Tablet-woven Borders: a New Interpretation of the Method of Production, in L. Bender Jørgensen and C. Rinaldo (eds), *Textiles in European Archaeology. Report from the 6th NESAT Symposium*, 79–84. GOTARC serie A, bind 1. Göteborg.

Ræder Knudsen, L. (2005) Brocaded Tablet Woven Bands: Same Appearance, Different Weaving technique, Hørning, Hvilehøj and Mammen, in F. Pritchard and J. P. Wild (eds), *North European Symposium for Archaeological Textiles VII*, 36–43. Edinburgh.

Ræder Knudsen, L. (2012) Case-Study: The Tablet-Woven Borders of Verucchio, in M. Gleba and U. Mannering (eds), *Textiles and Textile Production in Europe from Prehistory to AD 400*, 254–263. Ancient Textiles Series vol. 11, Oxford.

Ræder Knudsen, L. and Mannering, U. (2007) A Danish Early Germanic Iron Age Grave with Tablet Woven Cuffs, in A. Rast-Eicher and R. Windler (eds), *Archäeologische Textilfunde – Archaeological Textiles: North European Symposium for Archaeological Textiles IX, Braunwald 18.–20. Mai 2005*, 119–123. Ennenda.

Ringquist, P. O. (1969) Två vikingatida uppländska människofigurer i brons. *Fornvännen* 64, 287–296. Stockholm.

Robinson, D. E. (1993) Dyrkede planter fra Danmarks forhistorie. *Arkæologiske Udgravninger i Danmark 1993*, 20–39.

Robinson, D. E. (1994) *Et katalog over rester af dyrkede planter fra Danmarks forhistorie*. NNU Rapport Nr. 13. Nationalmuseet, København.

Robinson, D. E., Moltsen, A. and Harild, J. (1995) *Arkæobotaniske analyse af bronzealder gårdsanlæg og marksystemer ved Bjerre Enge, Hantsholm, Thy*. NNU Rapport Nr. 15. Nationalmuseet, København.

Roesdahl, E. (1987) *Vikingernes verden*. København.

Rolfsen, P. and Stylegar, F.-A., eds, (2003) *Snartemofunnene i nytt lys*. Oslo.

Roth, H. (1986a) Einführung in die Problematik, Rückblick und Ausblick, in H. Roth (ed.), *Zum Problem der Deutung frühmittelalterlicher Bildinhalte. Akten des 1. Internationalen Kolloquiums in Marburg a.d. Lahn, 15. bis 19. Februar 1983*, 9–24. Sigmaringen.

Roth, H. (1986b) Stil II – Deutungsprobleme. Skizzen zu Pferdemotiven und zur Motivkopplung, in H. Roth (ed.), *Zum Problem der Deutung frühmittelalterlicher Bildinhalte. Akten des 1. Internationalen Kolloquiums in Marburg a.d. Lahn, 15. bis 19. Februar 1983*, 111–128. Sigmaringen.

Roth, H., ed., (1986c) *Zum Problem der Deutung frühmittelalterlicher Bildinhalte. Akten des 1. Internationalen Kolloquiums in Marburg a.d. Lahn, 15. bis 19. Februar 1983*. Sigmaringen.

Rubinstein, R. P. (1995) *Dress Codes*. Oxford.

Rud, M. (1988) *The Bayeux Tapestry and the Battle of Hastings 1066*. København.

Salin, B. (1935) *Die Altgermanische Thierornamentik*. Stockholm.

Sanden, W. van der (1995) C14-Datierungen von Moorleichen aus Niedersachsen und Schleswig-Holstein. *Die Kunde*, vol. N. F. 46, 137–155.

Sanden, W. van der (1996) *Through nature to eternity. The bog bodies of northwest Europe*. Amsterdam.

Sanden, W. van der, and Capelle, T. (2001) *Mosens Guder. Antropomorfe træfigurer fra Nord- og Nordvesteuropas fortid*. Silkeborg.

Saxtorph, N. (1992) *Nordiske Gude- og Heltesagn*. Viborg.

Schlabow, K. (1976) *Textilfunde der Eisenzeit in Norddeutschland*. Neumünster.

Schmidt-Lornsen, J. (1986) Bilddarstellungen auf wikingerzeitlichen Mähnenstuhlpaaren, in H. Roth (ed.), *Zum Problem der Deutung frühmittelalterlicher Bildinhalte. Akten des 1. Internationalen Kolloquiums in Marburg a.d. Lahn, 15. bis 19. Februar 1983*, 297–302. Sigmaringen.

Sebesta, J. L. and Bonfante, L. eds, (1994) *The World of Roman Costume*. Wisconsin.

Shamir, O. (2005) Tunics from Kasr al-Yahud, in L. Cleland, M. Harlow and L. Llewellyn-Jones (eds), *The Clothed Body in the Ancient World*, 162–168. Oxford.

Simek, R. (2002) Goddesses, Mothers, Disir: Iconography and interpretation of the female deity in Scandinavia in the first Millenium, in R. Simek and W. Heizmann (eds), *Mythological Woman. Studies in Memory of Lotte Motz*, 93–123.Wien.

Sindbæk, S. (1998) Mammendyret, græsli-fuglen og den madjariske hest, Øst-Europa i vikingtidens kunst og symbolen. *Spor* 7, 8–17.

Sjöberg, N. (1905) Några ord om bilderna på Torslundaplåtarna. *Svenska Fornminnesföreningens Tidskrift* 12, 323–325. Stockholm.

Sjöborg, N. H. (1791) *Topographia parochiae Raflunda et monumentorum quae circa sunt*. Lund.

Sonesson, G. (1992) *Bildbetydelser. Inledning till bildsemiotiken som vetenskap*. Lund.

Stamsø Munch, G. (1990) Arkeologiske utgravninger på Borg 1989. Årboka Lofotar, 62–66. Værøy.

Stamsø Munch, G. (1991) Høvdinggården "Borg i Lofoten". *Ottar* Nr. 187, 4, 43–48. Tromsø Museum.

Stamsø Munch, G. and Johansen, O. S. (1988) Borg in Lofoten – An Inter-Scandinavian Research Project. *Norwegian Archaeological Review*, vol. 21, 2, 119–126. Oslo.

Stamsø Munch, G., Johansen, O. S. and Roessdahl, E., eds, (2003) *Borg in Lofoten : a chieftain's farm in North Norway*. Bøstod.

Stauffer, A. (2002) Tessuti, in P. Von Eles (ed.), *Guerriero e sacerdote – Autorità e comunità nell'età del ferro a Verucchio. La Tomba del Trono*, 192–219. Florence.

Stauffer, A. (2004) Villanovan garments from Verucchio (700 BC). *CIETA Bulletin 81,* 14–20.

Stauffer, A. (2012) Case Study: The Textiles from Verucchio, Italy, in M. Gleba and U. Mannering (eds), *Textiles and Textile Production in Europe from Prehistory to AD 400*, 242–253. Ancient Textiles Series vol. 11, Oxford.

Steinsland, G. (1989) *Det hellige bryllup og Norrøn kongeideologi. En undersøkelse av hierogami-myten i Skirnismál, Ynglingatal, Háleygjatal og Hyndluljód.* Oslo.

Steinsland, G. (1990) De nordiske gullblekk med parmotiv og norrøn fyrsteideologi. Et tolkningsforslag. *Collegium Medivale* 1, 73–94.

Steinsland, G. (1997) *Eros og død i norrøne myter*. Oslo.

Steinsland, G., Drobin, U., Pentikäinen, J. and Meulengracht, P., eds, (1991) *Nordisk Hedendom. Et symposium*. Odense.

Stenberger, M. (1964) *Det forntida Sverige*. Stockholm.

Stenberger, M. (1973) Öländska figurbleck av guld. Finska fornminnesföreningens tidsskrift 75, 218–220. Helsingfors.

Steuer, H. (1987) Helm und Ringschwert. Prunkbewaffung und Rangabzeichen germanischer Krieger. Eine Übersicht. Studien zur Sachsenforschung 6, 189–236. Hildesheim.

Stjernquist, B. (1951) *Vä under järnåldern*. Skrifter utgivna av Kungl. Humanistiska Vetenskapssamfundet i Lund, vol. 47. Lund.

Stjernquist, B. (1993a) *Gårdlösa. An Iron Age Community in its natural and Social Setting II. The archaeological Fieldwork, the Features and the Finds*. Skrifter utgivna av Kungl. Humanistiska Vetenskapssamfundet i Lund, vol. 80. Lund.

Stjernquist, B. (1993b) *Gårdlösa. An Iron Age Community in its natural and Social Setting III. Chronological, Economic and Social Analyses*. Skrifter utgivna av Kungl. Humanistiska Vetenskapssamfundet i Lund, vol. 81. Lund.

Stolpe, H. J. and Arne, T. J. (1912) *Gravfältet vid Vendel*. Stockholm.

Stone, S. (1994) The Toga: From National to Ceremonial Costume, in J. L. Sebesta and L. Bonfante (eds), *The World of Roman Costume*, 13–45. Wisconsin.

Sumner, G. (2002) *Roman Military Clothing (1) 100 BC–AD 400*. Men-at-Arms. Oxford.

Sumner, G. (2003) *Roman Military Clothing (2) AD 200–400*. Men-at-Arms. Oxford.

Sundkvist, A. (2001) *Hästarnas land. Aristokratisk hästhållning och ridkonst i Svealands yngre järnålder*. OPIA (Occasional Papers in Archaeology) 28. Uppsala.

Sundström, A. (1995) Reproduction of Horsehair Tablet Braids from Scandinavia's Migration Period. *Archaeological Textiles Newsletter*, No. 21, Autumn 1995, 24–26. Leiden.

Sundström, A. (1997) Brikkevevde bånd av hestehår – en reproduksjon av folkevandringstidens kunsthåndverk. *Spor – fortidsnyt fra Midt-Norge 1997/2*, 24–25.

Tauber, H. (1979) Kulstof-14 datering af moselig. *Kuml 1979,* 73–78.

Thomsen, P. O. (1990) Guldets billedverden. *Skalk* 1990/1, 3–7. Højbjerg.

Thomsen, P. O., Blæsild, B., Hardt, N. and Michaelsen, K. K. (1993) *Lundeborg – en handelsplads fra jernalderen*. Skrifter fra Svendborg & Omegns Museum. Bind 32. Ringe.

Thordeman, B. (1941) Stridsdräkten under forntid och medeltid. *Nordisk Kultur XV: B. Dragt*, 89–123. Stockholm.

Thrane, H. (1993) *Guld, guder og godtfolk*. Odense.

Tillväxten under år (1934) *Kungl. Vitterhets Historie och Antikvitetsakademiens Årsbok 1936–38*. Stockholm 1938.

Tornbjerg, S. Å. (1997) Fra gubbernes verden. *Skalk* 1997/3, 6–10. Højbjerg.

Ulriksen, J. (1998) *Anløbspladser. Besejling og bebyggelse i Danmark mellem 200 og 1100 e.Kr*. Odense.

Ulriksen, J. (2000) Anløbspladsen ved vester Egesborg. *Nyt fra Næstved Museum Nr. 51*, p. 6. Næstved.

Vedel, E. (1886) *Bornholms Oldtidsminder og Oldsager*. København.

Vedeler, M. (2007a) *Klær og formspråk i norsk middelalder*. Universitetet i Oslo.

Vedeler, M. (2007b) Medieval clothing in Uvdal, Norway, in A. Rast-Eicher and R. Windler (eds), *NESAT IX. Archäologische Textilfunde – Archaeological Textiles*, 141–147. Ennenda.

Vedeler, M. (2013) *Silk for the Vikings*. Ancient Textiles Series vol. 15. Oxford and Philadelphia.

Vedeler, M. and Bender Jørgensen, L. (2013) Out of the Norwegian glaciers: Lendbreen—a tunic from the early first millennium AD. *Antiquity* 87, 788–801.

Vensild, H. (1993) Bornholms Museumsforening 1893–1993. Træk af Foreningens og Bornholms Museums historie, in *Bornholms Museumsforening 1893–1993*, 7–88. Rønne.

Vierck, H. (1979a) Trachtenkunde und Trachtgeschichte in der Sachsenforschung, ihre Quellen, Zeile und Methoden, in C. Ahrens (ed.), *Sachsen und Angelsachsen. Veröffentlichungen des Helms-Museum* 32, 231–243. Hamburg.

Vierck, H. (1979b) Die anglische Frauentracht, in C. Ahrens (ed.), *Sachsen und Angelsachsen. Veröffentlichungen des Helms-Museum 32*, 245–252. Hamburg.

Vierck, H. (1979c) Zur angelsächsischen Frauentracht, in C. Ahrens (ed.), *Sachsen und Angelsachsen. Veröffentlichungen des Helms-Museum 32*, 255–262. Hamburg.

Vierck, H. (1979d) Zur Seegermanischen Männertracht, in C. Ahrens (ed.), *Sachsen und Angelsachsen. Veröffentlichungen des Helms-Museum* 32, 263–283. Hamburg.

Vierck, H. (1981) Die wikingische Frauentracht von Birka. *Offa*, 119–133. Neumünster.

Vogel, V. ed., (1997) *Ausgrabungen in Schleswig. Berichte und Studien 12. Kirche und Gräberfeld des 11.–13. Jahrhunderts unter dem Rathausmarkt von Schleswig*. Neumünster.

Vogelsang-Eastwood, G. (1992) *Patterns for Ancient Egyptian Clothing*. Leiden.

Vogelsang-Eastwood, G. (1993) *Pharaonic Egyptian Clothing*. Studies in Textile and Costume History vol. 2. Leiden.

Vogelsang-Eastwood, G. (2000) The Clothing of the "Shami prince". *PERSICA XVI*, 31–47. Leuven.

Voss, O. (1955) The Høstentorp Silver Hoard and its Period. A Study of a Danish Find of Scrap Silver from About 500 A.D. *Acta Archaeologica* 25, 1954, 171–219. København.

Walton, P. (1989) Dyes of the Viking Age: a Summary of Recent Work. *Dyes in History and Archaeology 7*, 14–20. York.

Walton, P. (1990a) Textile Production at Coppergate, York: Anglo-Saxon or Viking? in P. Walton and J. P. Wild (eds), *Textiles in Northern Archaeology. NESAT III: Textile Symposium in York*, 61–72. London.

Walton, P. (1990b) Dyes and wools in textiles from Norway and Denmark. *Journal of Danish Archaeology* 7, 1988, 144–158. Odense.

Walton, P. and Wild, J. P., eds, 1990, *Textiles in Northern Archaeology. NESAT III: Textile Symposium in York*. London.

Walton Rogers, P. (1998) Textiles, in G. Drinkall and M. Foreman, *The Anglo-Saxon Cemetery at Castledyke South, Barton-on-Humber*, 273–279. Sheffield Excavation Reports 6.

Walton Rogers, P. (1999) The Textiles, in C. Haughton and D. Powlesland, *West Heslerton. The Anglian Cemetery. Vol. I. The Excavation and Discussion of the Evidence*, 143–171. Nottingham.

Watt, M. (1984/85) Det antikvariske arbejde. *Fra Bornholms Museum 1984–1985*, 64–66. Rønne.

Watt, M. (1986) Guldmændenes parade. *Fra Bornholms Museum 1986*, 67–78. Rønne.

Watt, M. (1987) Guldageren. *Skalk* 1987/2, 3–9. Højbjerg.

Watt, M. (1989) Guldgubber fra Sorte Muld og Ravlunda. *Skepparpsåns Museiförening Bygdehistorik* Nr. 1, 129–137.

Watt, M. (1991a) Tilblivelse. *Skalk* 1991/1, 8–11. Højbjerg.

Watt, M. (1991b) Sorte Muld. Høvdingesæde og kultcentrum fra Bornholms yngre jernalder, in P. Mortensen and B. M. Rasmussen (eds), *Fra Stamme til Stat i Danmark 2. Høvdingesamfund og Kongemagt.* Jysk Arkæologisk Selskabs Skrifter XXII:2, 89–108. Højbjerg.

Watt, M. (1991c) Guldgubberne fra Sorte Muld, Bornholm. Tanker omkring et muligt hedensk kultcentrum fra yngre jernalder, in G. Steinsland et al., (eds), *Nordisk Hedendom. Et symposium*, 373–386. Odense.

Watt, M. (1992) Die Goldblechfiguren ("guldgubber") aus Sorte Muld, Bornholm, in K. Hauck (ed.), *Der historische Horizont der Götterbild-Amulette aus der Übergangsepoche von der Spätantike zum Frühmittelalter*. Berichte über das Colloquium vom 28.11.–1.12. 1988 in der Werner-Reimers-Stiftung, 195–227. Bad Homburg.

Watt, M. (1994) Det antikvariske arbejde 1991–1993. *Fra Bornholms Museum 1991–1993*, 65–88. Rønne.

Watt, M. (1999a) Gubber. *Reallexikon der Germanischen Altertumskunde* 13, 132–142.

Watt, M. (1999b) Guldgubber og patricer til guldgubber fra Uppåkra, in B. Hårdh (ed.), *Fynden i centrum. Keramik, Glas och Metall från Uppåkra*, 177–190. Uppåkrastudier 2. Acta Archaeologica Lundensia, Series 8^{0}, No. 30. Lund.

Watt, M. (1999c) Kings or gods? Iconographic evidence from Scandinavian gold foil figures. *Anglo-Saxon Studies in Archaeology and History* 10, 173–183. Oxford.

Watt, M. (2001) "Gummor" og "grodor". Om kønsbestemmelse af guldgubber, in B. Magnus, C. Orrling, M. Rasch and G. Tegnér (eds), *Vi får tacka Lamm*, 219–228. Stockholm.

Watt, M. (2002a) Images of women on "Guldgubber" from the Merovingian Age, in R. Simek and W. Heizmann (eds), *Mythological Woman. Studies in Memory of Lotte Motz*, 81–91. Wien.

Watt, M. (2002b) En glad gris – magtens dyr? in J. Pind, A. Nørgård Jørgensen, L. Jørgensen, B. Storgård, P. O. Rindel and J. Ilkjær (eds), *Drik – og du vil leve skønt. Festskrift til Ulla Lund Hansen på 60–årsdagen 18. august 2002*, 59–68. PNM. Studies in Archaeology & History vol. 7. København.

Watt, M. (2003) Guldgubbernes dragter – mode eller herskersymbolik?, in A. Hedeager Krag (ed.), *Dragt og magt*, 29–45. København.

Watt, M. (2004) The Gold-Figure Foils ("Guldgubbar") from Uppåkra, in L. Larsson (ed.), *Continuity for Centuries. A Ceremonial Building and its Context at Uppåkra, Southern Sweden*, 167–221. Uppåkrastudier 10. Acta Archaeologica Lundensia, Series 8, No. 30. Lund.

Watt, M. (forthcoming) *Sorte Muld-guldgubberne.* Nordiske Fortidsminder, Serie B. København

Weiner, A. B. and Schneider, J., eds, (1991) *Cloth and Human Experience*. London.

Werner, J. (1966) *Das Aufkommen von Bild and Schrift in Nordeuropa*. München.

Wiking-Faria, P. (2008) *Bockstensmannen och hans tid.* Värnamo.

Wild, J. P. (1968) Clothing in the North-West Provinces of the Roman Empire, *Bonner Jahrbücher 168*, 166–240.

Wild, J. P. (1985) The Clothing of Britannia, Gallia Belgica and Germania Inferior, in H. Temporini and W. Haase (eds), *Aufstieg und Niedergang der Römischen Welt*, Teilband II 12, 3, 362–422. Berlin.

Wild, J. P. and Bender Jørgensen, L. (1988) Clothes from the Roman Empire. Barbarians & Romans, in L. Bender Jørgensen and K. Tidow (eds), *Archaeological Textiles in Northern Europe. Report from the 2nd NESAT Symposium*, 65–98. Arkæologiske Skrifter 2. København.

Wilson, L. M. (1924) *The Roman Toga*. Baltimore.

Wilson, L. M. (1938) *The Clothing of the Ancient Romans.* Baltimore.

Windler, R., Rast-Eicher, A. and Mannering, U. (1995) Nessel und Flachs – Textilfunde aus einem frühmittelalterlichen Mädchengrab in Flürlingen (Kanton Zürich), *Archäologie der Schweiz* 18, 155–161.

Zachrisson, S. (1963) Från vendeltid till Vasatid. Redogörelse för de byggnadsarkeologiska undersökningarna av Eskilstuna kloster och slott 1961–62. *Eskilstuna Stads Museer. Årsbok*, 17–58.

Zagal-Mach Wolfe, U. I. (2013) *Grasping Technology, Assessing Craft. Developing a Research method to the Study of Craft-Tradition.* Acta Archaeological Ludensia Series 8^{0}, No 63. Lund.